广西防城季节性雨林物种及其分布格局

向悟生　王　斌　李先琨　丁　涛　李健星　郭屹立　李冬兴　等◎著

中国林业出版社
China Forestry Publishing House

图书在版编目（CIP）数据

广西防城季节性雨林物种及其分布格局 / 向悟生等著. -- 北京：中国林业出版社, 2021.12
ISBN 978-7-5219-1434-4

Ⅰ. ①广… Ⅱ. ①向… Ⅲ. ①季雨林-研究-防城港 Ⅳ. ① S718.54

中国版本图书馆 CIP 数据核字 (2021) 第 249559 号

责任编辑 于界芬	**电话** （010）83143542

出版发行	中国林业出版社有限公司
	（100009 北京西城区德内大街刘海胡同 7 号）
网　　址	http://www.forestry.gov.cn/lycb.html
印　　刷	北京博海升彩色印刷有限公司
版　　次	2021 年 12 月第 1 版
印　　次	2021 年 12 月第 1 次印刷
开　　本	889mm×1194mm　1/16
印　　张	19.25
字　　数	470 千字
定　　价	186.00 元

未经许可，不得以任何方式复制或抄袭本书之部分或全部内容。
版权所有　侵权必究

广西防城季节性雨林物种及其分布格局
著者名单

向悟生　王　斌　李先琨　丁　涛　李健星　郭屹立
李冬兴　夏　欣　杨泉光　陈　婷　黄甫昭　吴儒华
文淑均　潘子平　陈　皓　陆树华　曹　晓　张昊楠
郭　辰　高　军

GUANGXI FANGCHENG SEASONAL RAIN FORESTS:
SPECIES AND THEIR DISTRIBUTION PATTERNS

LIST OF AUTHORS

Wusheng Xiang, Bin Wang, Xiankun Li, Tao Ding, Jianxing Li, Yili Guo, Dongxing Li, Xin Xia, Quanguang Yang, Ting Chen, Fuzhao Huang, Ruhua Wu, Shujun Wen, Ziping Pan, Hao Chen, Shuhua Lu, Xiao Cao, Haonan Zhang, Chen Guo, Jun Gao

前　言

　　目前全世界对森林的关注达到了前所未有的高度，尤其是森林在支持地球生命中所起到的至关重要的角色越来越深入地被理解。季节性雨林是分布于热带区域的一种雨林类型，其分布区域的雨量和温度存在季节性差异，植物种类组成以常绿为主，在多雨季节其植物生长繁茂，具有较明显的雨林特征，在少雨季节群落的外貌仍维持常绿，因而与季雨林明显不同。季节性雨林在我国主要分布于海南、滇南、粤西南、桂南、藏东南、台南等区域。季节性雨林中的物种多样性十分丰富，在生物多样性保存与碳汇等方面具有重要作用，因此，我国研究工作者对其森林生态学方面开展了大量研究，积累了丰富的研究资料。

　　广西防城金花茶国家级自然保护区所处的十万大山山脉是国际生物多样性保护的热点区域，地带性植被为季节性雨林，具有丰富的物种多样性和独特的类群。保护区内动植物种类丰富，而较多古老孑遗和珍稀濒危特有种类，如国家重点保护野生植物狭叶坡垒（*Hopea chinensis*）和十万大山苏铁（*Cycas shiwandashanica*）等，如国家重点保护动物蟒蛇（*Python bivittatus*）和虎纹蛙（*Hoplobatrachus rugulosus*）等。特别是保护区内保存了大量被誉为"植物界的大熊猫"和"茶族皇后"的金花茶组植物资源，分布有金花茶（*Camellia petelotii*）、显脉金花茶（*Camellia euphlebia*）和东兴金花茶（*Camellia indochinensis* var. *tunghinensis*）等3种，聚集了大量野生金花茶资源。但之前对其金花茶种群资源及其赖以生存的季节性雨林生境缺乏长期监测平台，对其栖息地的上层树木组成和结构特征了解大多基于小面积、短期的植物群落样方调查多获得，难以真实反映季节性雨林上层树木与金花茶等下层植物的相互作用机制。而且，基于短期小面积的样方抽样调查，也无法清晰了解林下金花茶组植物的种群结构、动态与金花茶植物资源的保有量及其动态过程。有鉴于此，广西壮族自治区中国科学院广西植物研究所、生态环境部南京环境科学研究所和广西壮族自治区防城金花茶国家级自然保护区管理中心合作，在保护区建立了两个 1hm^2 永久监测样地，对金花茶组植物及其赖以生存的季节性雨林生境进行长期监测研究，以期更加全面准确地揭示自然状态下，金花茶组植物种群的结构和动态过程及其种群的维持机制。样地的建设参照《生物多样性观测技术导则-陆生维管植物》(HJ 710.1—2014)标准进行，并于2016年完成了第一次调查。样地的建成为野生金花茶种群资源及季节性雨林森林生态学研究提供了良好的野外科学平台，后续的长期监测和深入研究将为野生金花茶资源有效保护和持续利用提供必不可少的基础资料和科学理论指导。

本书介绍了防城金花茶保护区季节性雨林的地貌、土壤和植被等特征，基于2016年的植被调查和地形测量数据，展示了防城金花茶（JH）和防城十万山（SW）两个 $1hm^2$ 样地内的树种组成、径级结构、空间分布和典型形态特征等；以及样地内常见的草本、灌木和藤本植物的形态特征。本书共收录 DBH ≥ 1cm 的树木182种，草本和藤本植物72种。本书的植物生物学特征描述主要参考了《中国植物志》和《Flora of China》，植物名称依据《中国生物物种名录》（2021年版）核定。植物照片加上生物学特征的描述，可为读者认知金花茶栖息地的相关植物提供参考，也为该区域季节性雨林的深入研究提供了重要基础资料。

由于时间仓促，水平有限，疏漏和错误难以避免，请诸位专家和广大读者们予以批评指正为盼。

著者

2021年10月

PREFACE

Forests have attracted unprecedented attention in the world nowadays, especially the vital role that forests play in supporting life on the earth is deeply understood. Seasonal rain forest is a vegetation distributed in tropical regions, where there are seasonal differences in rainfall and temperature. Its plant species composition dominated by evergreen trees and grow thriving in rainy season, and which characteristics like the rainforest; and in less rain season, the physiognomy of the community is still evergreen which is different from monsoon forest obviously. In China, seasonal rain forests are mainly distributed in Hainan, southern Yunnan, southwest Guangdong, southern Guangxi, southeast Tibet, Tainan. The species diversity in seasonal rain forest is very high, which plays an important role in biodiversity conservation and carbon sink. Therefore, Chinese researchers focused the study of its forest ecology and accumulated a massive data.

Shiwandashan mountain in Guangxi, where Fangcheng Camellia National Nature Reserve is located, is a hot spot for international biodiversity conservation. The zonal vegetation is seasonal rain forest with high diversity and endemic species. The reserve preserves a large number of animals and plants, and many of them are ancient relics and rare endemic species, such as the Chinese key protected wild plants *Hopea chinensis* and *Cycas shiwandashanica*, and Chinese key protected wild animals *Python bivittatus* and *Hoplobatrachus rugulosus*. In particular, there are a large number of Camellia family plant resources known as "Giant Panda in Plant Kingdom" and "Queen of Tea Family" in the reserve. The reserve preserves *Camellia petelotii*, *Camellia euphlebia*, *Camellia indochinensis* var. *tunghinensis* with most of the wild Camellia resources in the world. However, the long-term monitoring platform for its Camellia population resources and the dependent seasonal rain habitat is still lacked. Most of the understanding on the species composition and structural characteristics of trees is based on small-scale and short-term plant community sample surveys, which cannot truly reveal the interaction mechanism of the trees of the seasonal rain forest and the understory such as Camellia plants. Moreover, the population structure and dynamics of the Camellia plants as well as the resources could not be clearly understood based on the short-term small-scale quadrat sampling survey. Therefore, Guangxi Institute of Botany, Chinese Academy of Sciences, Nanjing Institute of Environmental Sciences, MEE, cooperated with the management center of Fangcheng Camellia National Nature Reserve established two permanent monitoring plots of $1hm^2$ in the reserve to carry out long-term monitoring research on Camellia plants and their habitats. The construction of the plots was initiated with reference to the Technical Guidelines for Biodiversity Observation-Terrestrial Vascular Plants (HJ 710.1-2014), and the first survey was completed in 2016. The established plots provide a well field scientific platform for the study of wild Camellia population resources and forest ecology of seasonal rain forest. The following long-term monitoring and further study will provide essential basic data and scientific theoretical guidance for the effective protection and sustainable utilization of wild Camellia resources.

This book introduces the topography, soil and vegetation characteristics of seasonal rain forest in Fangcheng Camellia Nature Reserve. Based on the vegetation survey and topographic survey data in 2016, the tree species composition, diameter structure, spatial distribution and typical morphological characteristics in the two $1hm^2$ plots of Fangcheng Jinhuacha (JH) and Fangcheng Shiwanshan (SW) were described. The morphological characteristics of common herbs, shrubs and lianas in the sample plots were also described. In this book, we documented 182 species of trees and 72 species of herbs and lianas. We described the biological characteristics of plants mainly refer to Flora of China (Chinese and English version), the plant scientific names were approved according to the Catalogue of Life China 2021 Annual Checklist. These exquisite photos of plants with descriptions of their biological characteristics can provide the most direct perceptual information and basic information for readers to cognitive plants in the habitats of Camellia Sect. Chrysantha, and also key ecological information for the further study of seasonal rain forests in this area in the future.

Due to the limitation of knowledge and time, there might be some mistakes in this book. The authors kindly ask for the grant instruction from all readers.

Aditor
2021.10

目 录

1 广西防城金花茶国家级自然保护区简介 ……………………………………… 1
　1.1 地理位置和自然环境概况 …………………………………………………… 2
　1.2 自然保护区的建立与发展 …………………………………………………… 3
　1.3 生物多样性与主要植被类型 ………………………………………………… 3

2 广西防城季节性雨林2个1 hm² 监测样地 …………………………………… 5
　2.1 样地建设和群落调查 ………………………………………………………… 6
　2.2 地形和土壤 …………………………………………………………………… 7
　2.3 树种组成和群落结构 ………………………………………………………… 7
　2.4 草本层植物和藤本植物的多样性 …………………………………………… 12

3 广西防城金花茶1 hm² 样地（JH）的树种及其分布格局 …………………… 13

4 广西防城十万山1 hm² 样地（SW）的树种及其分布格局 ………………… 153

5 广西防城样地的草本层植物和藤本植物 …………………………………… 247

附录Ⅰ　植物名录 …………………………………………………………… 284
附录Ⅱ　植物中文名索引 …………………………………………………… 288
附录Ⅲ　植物学名索引 ……………………………………………………… 290

致　谢 ………………………………………………………………………… 293

CONTENTS

1 Brief Introduction to Guangxi Fangcheng Golden Camellia National Nature Reserve ········· 1
    1.1 Geographical location and general situation of natural environment ········· 2
    1.2 The establishment and development ········· 3
    1.3 Biodiversity and main vegetation types ········· 3

2 Two 1 hm^2 Seasonal Rainforest Monitoring Plot in Fangcheng, Guangxi ········· 5
    2.1 Plot establishment and plant investigation ········· 6
    2.2 Topography and soil ········· 7
    2.3 Tree species composition and community structure ········· 10
    2.4 Diversity of plants in the herb layer and lianas ········· 12

3 Tree Species and their Distribution Patterns in the Guangxi Fangcheng JH 1 hm^2 Plot ········· 13

4 Tree Species and their Distribution Patterns in the Guangxi Fangcheng SW 1 hm^2 Plot ········· 153

5 Plants in the Herb Layer and Lianas in the two Guangxi Fangcheng Plots ········· 247

Appendix Ⅰ   Checklist of Plant Names ········· 284
Appendix Ⅱ   Index of Plant Chinese Names ········· 288
Appendix Ⅲ   Index of Plant Scientific Names ········· 290

Acknowledgements ········· 294

广西防城季节性雨林物种及其分布格局
GUANGXI FANGCHENG SEASONAL RAIN FORESTS: SPECIES AND THEIR DISTRIBUTION PATTERNS

1 广西防城金花茶国家级自然保护区简介
Brief Introduction to Guangxi Fangcheng Golden Camellia National Nature Reserve

1.1 地理位置和自然环境概况

广西防城金花茶国家级自然保护区，位于广西防城港市防城区境内，地理坐标为108°2′2″~108°12′52″ E，21°43′52″~21°49′39″ N。保护区地处十万大山南麓蓝山支脉，东起深圳岭下的防城江，西至小峰水库边，北起松柏屯旁的防城江，南至那梭村的那他屯，东西长18.6 km，南北宽10.8 km。保护区范围横跨那梭、华石、大录和扶隆4个乡镇，涉及15个行政村65个村民小组，以及国营那梭农场和华石林场。保护区总面积为9098.6 hm²，其中核心区1479.1 hm²，缓冲区3459.2 hm²，实验区4160.3 hm²。

保护区地质构造主要是印支期褶皱，地貌总体格局为南陡北缓的单斜地形，以山地、丘陵为主，地势起伏明显，沟壑纵横交错。雨水汇集冲刷形成了众多中小河流，主要有西南面的东山江和东北面的防城江。

保护区属十万大山南坡的山前丘陵地带，北热带季风气候类型，温暖湿润，光照充足，热量丰富，雨量充沛。年平均气温21.8°C，1月平均气温12.6°C，7月平均气温28.2°C。≥0°C年积温为8100°C，最高达8163.5°C。年平均降水量为2900 mm，3~10月为多雨季节，总降雨量达2700 mm以上，时有山洪，7~8月是全年雨量高峰月，平均降雨量都在400~500 mm，每年11月至翌年2月平均降雨量都在100 mm以下，有时间歇出现秋旱或春旱。平均相对湿度为80%左右。

保护区内海拔跨幅大，土壤垂直分布明显。地带性土壤有砖红壤、红壤和黄壤，主要成土母岩有花岗岩、砂岩和页岩，其中海拔300 m以下为砖红壤，300~800 m为山地红壤，800 m以上为山地黄壤。

1.1 Location and Natural Environment

Guangxi Fangcheng Golden Camellia National Nature Reserve is located in Fangcheng District, Fangchenggang City, Guangxi, with the geographic coordinates of 108°2′2″~108°12′52″ E, 21°43′52″~21°49′39″ N. The reserve is located in the Lanshan Mountain branch at the southern foot of the Shiwandashan Mountain. It starts from Fangcheng River under Shenliling Mountain in the east, ends at Xiaofeng Reservoir in the west, starts from Fangcheng River beside Songbai Group in the north, and ends at Nata Group in Nasuo village in the south. The reserve is 18.6 km long from east to west and 10.8 km wide from north to south. The scope of the reserve spans four towns of Nasuo, Huashi, Dalu and Fulong, involving 65 villagers' groups in 15 administrative villages, as well as state-owned Nasuo Farm and Huashi Forest Farm. The total area of the reserve is 9098.6 ha, including 1479.1 ha in the core area, 3459.2 ha in the buffer area and 4160.3 ha in the experimental area.

The geological structure of the reserve is mainly Indosinian fold. The overall landscape pattern is a monoclinic terrain with steep south and gentle north. It is dominated by mountains and hills, with obvious relief and criss-crossing gullies. Rainwater has collected and scoured many small and medium-sized rivers, mainly Dongshan River in the southwest and Fangcheng River in the northeast.

The reserve belongs to the piedmont hilly area on the southern slope of Shiwandashan Mountain. It has a northern tropical monsoon climate with warm and humid climate, sufficient light, abundant heat and abundant rainfall. The annual average temperature is 21.8°C, the January average temperature is 12.6°C, and the July average temperature is 28.2°C. ≥ 0°C annual accumulated temperature is 8100°C, up to 8163.5°C. The annual average rainfall is 2900 mm, March to October is rainy season, with a total rainfall of more than 2700 mm and occasional flash floods. July to August is the peak rainfall month of the year, with an average rainfall of 400~500 mm. The annual average rainfall from November to February of the following year is below l00 mm, with occasional autumn drought or spring drought. The average relative humidity is about 80%.

The elevation span in the protected area is large and the vertical distribution of soil is obvious. Zonal soils include latosol, red soil, and yellow soil, and the main parent rocks are granite, sandstone, and shale. Among them, latosol below 300 m above sea level, red soil at 300~800 m above sea level, and yellow soil at 800 m above sea level.

1.2 自然保护区的建立与发展

广西防城金花茶国家级自然保护区的前身，为原防城各族自治县城乡建设环境保护局于 1984 年在防城区那梭镇那梭村上岳建立的金花茶保护点。1986 年广西壮族自治区人民政府批准建立自治区级自然保护区，隶属环保部门管理。1994 年晋升为国家级自然保护区，2010 年启动保护区范围调整工作，2016 年国务院批准保护区范围和功能区调整。2019 年保护区转隶至自治区林业局管理。

1.2 Establishment and Development

Guangxi Fangcheng Golden Camellia National Nature Reserve was formerly Golden Camellia Nature Protection Site established by the urban and rural construction environmental protection bureau of the former Fangcheng ethnic autonomous county in 1984 in Shangyue, Nasuo Village, Nasuo Town, Fangcheng District. In 1986, the People's Government of Guangxi Zhuang Autonomous Region approved the establishment of an autonomous region-level nature reserve, which is under the management of the environmental protection department. In 1994, it was promoted to a national nature reserve. In 2010, the scope adjustment of protected areas was started. In 2016, the State Council approved the adjustment of protected areas and functional areas. In 2019, the reserve was transferred to the Forestry Bureau of the autonomous region for management.

1.3 生物多样性与主要植被类型

保护区的地带性植被主要为北热带季节性雨林，在海拔 500 m 以下有广泛分布，其受到人为干扰后可退化形成次生森林类型，类似季雨林。此外，还有沟谷雨林、常绿阔叶林等植被。这些是金花茶和其他珍稀濒危动植物赖以生存的森林生态系统。

保护区分布有被誉为"植物界的大熊猫"和"茶族皇后"的金花茶组植物 3 种，金花茶（*Camellia petelotii*）、显脉金花茶（*Camellia euphlebia*）、东兴金花茶（*Camellia indochinensis* var. *tunghinensis*），在保护区内均有较为集中分布。保护区内植物群落保存完整，金花茶组植物分布区总面积达 764.0 hm^2，其中面积最大的是显脉金花茶（319.4 hm^2），其次是金花茶（297.9 hm^2），面积最少的是东兴金花茶（146.7 hm^2）。

据已发表论文和报告统计，保护区有国家重点保护野生植物 28 种，如十万大山苏铁（*Cycas balansae*）、狭叶坡垒（*Hopea chinensis*）、金毛狗（*Cibotium barometz*）、黑桫椤（*Alsophila podophylla*）、六角莲（*Dysosma pleiantha*）以及金花茶组（*Camellia* sect. *Archecamellia*）植物、兰科（Magnoliaceae）植物等，有国家重点保护野生动物 27 种，如蟒（*Python bivittatus*）、大灵猫（*Viverra zibetha*）、虎纹蛙（*Hoplobatrachus chinensis*）、黑鸢（*Milvus migrans*）、凤头鹰（*Accipiter trivirgatus*）等，具有重要的保护价值。

1.3 Biodiversity and Main Vegetation Types

The zonal vegetation in the nature reserve is mainly northern tropical seasonal rain forest, which is widely distributed below 500 m above sea level. After being disturbed by human activities, it can degenerate to form form secondary forest type, similar to monsoon forest. In addition, there are also ravine rainforest and evergreen broad-leaved forest. These are forest ecosystems on which Camellia chrysantha and other rare and endangered animals and plants depend.

There are three kinds of Golden Camellia plants known as "Giant Panda in Plant Kingdom" and "Queen of Tea Family", including *Camellia petelotii*, *Camellia euphlebia* and *Camellia indochinensis* var. *tunghinensis*, which are concentrated in the nature reserve. The plant community in the reserve is well preserved, and the total area of the golden camellia group plants is 764.0 hm^2, of which the largest area is *Camellia euphlebia*(319.4 hm^2), the next is *Camellia petelotii*(297.9 hm^2), and the smallest area is *Camellia indochinensis var. tunghinensis*(146.7 hm^2).

According to published papers and reports, there are 28 national key protected wild plants, such as *Cycas balansae*, *Hopea chinensis*, *Cibotium barometz*, *Alsophila podophylla*, *Dysosma pleiantha*, Camellia sect. Archecamellia plants, and Magnoliaceae plants in the reserve. And there are 27 kinds of national key protected animals, such as *Python bivittatus*, *Viverra zibetha*, *Hoplobatrachus chinensis*, *Milvus migrans*, and *Accipiter trivirgatus*. They have important protection value.

参考文献

孙鸿烈. 中国生态系统 [M]. 北京：科学出版社，2005.

苏宗明，李先琨，丁涛，宁世江，陈伟烈，莫新礼. 广西植被 [M]. 北京：中国林业出版社，2014.

王献溥，郭柯，温远光. 广西植被志要 [M]. 北京：高等教育出版社，2015.

谭伟福. 广西自然保护区 [M]. 北京：中国环境出版社，2014.

Reference

Sun H L. Ecosystem of China[M]. Beijing: Science Press, 2005. (in Chinese)

Su Z M, Li X K, Ding T, Ning S J, Chen W L, Mo X L. The vegetation of Guangxi [M]. Beijing: China Forestry Publishing House, 2014. (in Chinese)

Wang X P, Guo K, Wen Y G. Vegetation records of Guangxi [M]. Beijing: Higher Education Press, 2015. (in Chinese)

Tan W F. Guangxi nature reserve [M]. Beijing: China Environmental Press, 2014. (in Chinese)

广西防城季节性雨林物种及其分布格局
GUANGXI FANGCHENG SEASONAL RAIN FORESTS: SPECIES AND THEIR DISTRIBUTION PATTERNS

2 广西防城季节性雨林2个1 hm² 监测样地
Two 1 hm² seasonal rainforest monitoring plot in Fangcheng, Guangxi

2.1 样地建设和群落调查

2015 年 7 月,根据广西防城金花茶国家级自然保护区典型植被类型的分布情况,选定 2 个 1 hm² 森林样地研究区域。经踏查确定该区域植被的优势树种为鹅掌柴(*Schefflera heptaphylla*)、银柴(*Aporusa dioica*)、红鳞蒲桃(*Syzygium hancei*)、山蒲桃(*Syzygium levinei*)、锈毛梭子果(*Eberhardtia aurata*)等。该植被类型是原始热带季节性雨林和沟谷雨林退化后形成的次生森林,是目前保护区内金花茶种群分布的主要生境类型,具备很好的典型性和代表性。两个样地中心点的地理坐标分别为21°45′28″N、108°5′52″E;21°45′5″N、108°5′53″E。

选定样地后于 2015 年 9 月开始进行样地测量。先使用全站仪测量确定两个 100 m×100 m 样地的边框,然后在样地内标定小样方的节点,使用红线连接每个节点后形成小样方,作为植被调查的基本单元。由于南方山地的地形复杂,以往 20 m×20 m 节点的样方标定方法会影响后续树木坐标等参数的估测精度。因此,为了方便后续植被调查,采用全站仪精确标定所有 10 m×10 m 的节点。两个样地的闭合导线误差均小于 15 cm,10 m×10 m节点误差小于 10 cm。

样地植物群落调查参考全球森林监测网络 ForestGEO(the Forest Global Earth Observatory)固定样地建设的技术规范,对样地内所有胸径(DBH)≥1 cm 的木本植物个体进行定位、挂牌、测量和鉴定,记录植株编号、名称、胸径、树高、冠幅、坐标和生长状况等。

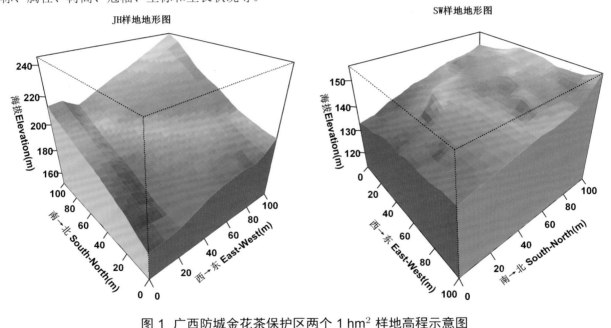

图 1 广西防城金花茶保护区两个 1 hm² 样地高程示意图

Figure 1 Schematic diagram of elevation of two 1 hm² plots in Fangcheng golden camellia reserve

2.1 Plot Establishment and Plant Census

In July 2015, according to the vegetation distribution area in Guangxi Fangchenggang Golden Camellia National Nature Reserve, the distribution areas of two 1 hm² forest plots were preliminarily selected. The forest community in this area are dominated by *Schefflera heptaphylla*, *Aporusa dioica*, *Syzygium hancei*, *Syzygium levinei*, *Eberhardtia aurata*, etc.. The vegetation type belongs to the secondary monsoon forest, which is the secondary forest formed by the degradation of tropical seasonal rain forest and ravine rain forest after human disturbance. It is the main habitat type of Golden Camellia plants in the reserve at present, which is very typical and representative. The geographic coordinates of the center points of the two plots are 21°45′28″N, 108°5′52″E; 21°45′5″N, 108°5′53″E.

In September 2015, two plots were surveyed and constructed. First, the total station was used to measure and determine the borders of two 100 m × 100 m plots, then the nodes of each small sample plot in the plot were calibrated, and different small sample plots were formed after connecting each node with red lines, which were the basic units of vegetation investigation. Because of the complex terrain in the mountainous areas of South China, the previous calibration methods of 20 m × 20 m nodes often lead to a large deviation in the estimation of coordinate parameters in subsequent vegetation survey. Therefore, in order to facilitate the later investigations, all 10 m × 10 m nodes were accurately calibrated by total station. The error of closed conductor was less than 15 cm, and the error of 10 m × 10 m node was less than 10 cm in both two plots.

The investigation of plant community in the plots refers to the technical specifications of the global forest monitoring network ForestGEO (The Forest Global Earth Observatory) for the construction of dynamic monitoring forest plots. All woody plant individuals with DBH ≥ 1 cm in the forest plot were located, listed, measured, and identified, and the plant number, name, DBH, tree height, crown width, coordinates and growth status were recorded.

2.2 地形和土壤

两个 1 hm² 样地包含了研究区域一些典型微地形类型。JH 样地的最大海拔高差为 73.9 m，样地整体靠近山脊的顶部，包含较陡的坡地地形和一条沟谷。SW 样地的最大海拔高差为 22.6 m，样地整体靠近山脊的底部，地形相对平缓，包含了一个小的山脊，样地的两条边线邻接有流水的溪谷。

两个样地的土壤类型均为砖红壤。样地土壤的有机碳含量在 4%~6% 之间，土壤含水量一般在 31%~33%；土壤总氮为 16~19 g/kg；总磷为 0.8~3.4 g/kg；水解性氮含量为 270~309 mg/kg，有效磷为 0.6~1.6 mg/kg；pH 值为 5.7~5.8；全钙含量为 15~18 g/kg；全镁含量为 6~8 g/kg。

土壤微生物多样性 OTUs 在 2800~3750 之间，Shannon 多样性指数为 8.1~9.2 之间；在细菌中，变形菌门（Proteobacteria）和酸杆菌门（Acidobacteria）是相对丰富门类。

2.2 Topography and Soil

Two 1 hm² plots contain some typical microtopography types in the study area.

In JH plot, the maximum elevation difference is 73.9 m, and the whole plot is close to the top of the ridge, including steep slope terrain and a valley. In SW plot, the maximum elevation difference is 22.6 m, and the whole plot is close to the bottom of the ridge, and the terrain is relatively gentle, including a small ridge, and the two side lines of the plot are adjacent to flowing valleys.

The soil types of the two plots are latosol. The organic carbon content of the sample soil is between 4%~6%, and the soil water content is generally 31%~33%; Soil total nitrogen is 16~19 g/kg, total phosphorus is 0.8~3.4 g/kg; Hydrolytic nitrogen content is 270~309 mg/kg, available phosphorus is 0.6~1.6 mg/kg; The pH value is 5.7~5.8; Total calcium content is 15~18 g/kg; Total magnesium content is 6~8 g/kg. OTUs of soil microbial diversity is between 2800~3750, Shannon diversity index is between 8.1~9.2; Among the bacteria, Proteobacteria, and Acidobacteria are relatively rich categories.

2.3 树种组成和群落结构

两个 1 hm² 样地中，总共记录到胸径 ≥1 cm 的植株个体 12618 株，分属于 182 种 119 属 55 科。其中，JH 样地有个体 6530 株，分属于 139 种 97 属 49 科；SW 样地有个体 6088 株，分属于 142 种 100 属 50 科。样地内有国家重点保护野生植物名录中 II 级保护种类 4 种：显脉金花茶（*Camellia euphlebia*）、东兴金花茶（*Camellia indochinensis* var. *tunghinensis*）、韶子（*Nephelium chryseum*）、紫荆木（*Madhuca pasquieri*）。有广西特有植物 4 种：赛短花润楠（*Machilus parabreviflora*）、硬叶糙果茶（*Camellia gaudichaudii*）、东兴金花茶（*Camellia indochinensis* var. *tunghinensis*）、少花山小橘（*Glycosmis oligantha*）。

表 1 两个 1hm² 样地不同胸径等级植物的统计数据
Table 1 Statistical data of plants with different DBH classes in two 1 hm² plots

径级 DBH (cm)	科数 No. of family	属数 No. of genera	物种数 No. of species	个体数 No. of individual	茎干数 No. of stem	平均胸径 Mean DBH (cm)	总基面积 Basal area (m²)
≥1	55	119	182	12618	15209	4.33	49.29
≥2	54	115	167	8144	9597	6.04	48.35
≥3	50	108	151	5441	6329	7.92	46.86
≥5	46	99	136	3274	3771	10.75	43.91
≥10	43	78	106	1466	1629	15.76	35.35
≥20	22	34	45	289	301	25.33	14.74
≥30	14	15	18	66	68	36.73	4.76
≥40	5	5	5	34	34	49.29	1.76
≥50	2	2	2	29	29	56.95	1.02

表 2 两个 1 hm² 样地重要值前 10 位的优势科
Table 2 Top 10 families with the highest important values in two 1 hm² plots

科名 Family	区系类型 Areal types	属数 No. of genera	物种数 No. of species	个体数 No. of individual	平均胸径 Mean DBH (cm)	基面积 Basal area (m²)	重要值 IV
茜草科 Rubiaceae	世界广布 Cosmopolitan	14	19	3933	2.47	3.85	16.40
樟科 Lauraceae	泛热带 Pantropic	5	16	1057	6.88	8.56	11.48
五加科 Araliaceae	热带亚洲和热带美洲间断 Trop. Asia & trop. Amer. disjuncted	2	2	1261	6.50	10.84	11.01
叶下珠科 Phyllanthaceae	泛热带 Pantropic	7	8	1295	4.24	3.77	7.41
桑科 Moraceae	世界广布 Cosmopolitan	4	16	579	4.52	2.25	5.96
锦葵科 Malvaceae	泛热带 Pantropic	4	4	654	4.75	2.40	4.07
山榄科 Sapotaceae	泛热带 Pantropic	3	3	439	6.71	2.85	3.63
桃金娘科 Myrtaceae	泛热带 Pantropic	3	6	458	4.44	1.47	3.29
蕈树科 Altingiaceae	北温带 North temperate	1	1	73	20.44	3.63	2.83
报春花科 Primulaceae	世界广布 Cosmopolitan	3	5	412	3.74	1.16	2.77

两个样地中主要优势科为茜草科（Rubiaceae）、樟科（Lauraceae）、五加科（Araliaceae）、叶下珠科（Phyllanthaceae）、桑科（Moraceae）、锦葵科（Malvaceae）、山榄科（Sapotaceae）、桃金娘科（Myrtaceae）、蕈树科（Altingiaceae）、报春花科（Primulaceae）等。含物种数相对就多的科有茜草科（Rubiaceae）、樟科（Lauraceae）、桑科（Moraceae）、桃金娘科（Myrtaceae）等，而属数较多的科有茜草科（Rubiaceae）、叶下珠科（Phyllanthaceae）、樟科（Lauraceae）。从优势科的组成看，优势科除世界广布类型外，其余大多为泛热带和热带亚洲分布，只有蕈树科（Altingiaceae）为北温带分布。

表 3　JH 1 hm² 样地重要值前 10 位的优势树种
Table 3　Top 10 species with the highest important values in JH 1 hm² plot

物种 Species	个体数 No. of individual	茎干数 No. of stem	平均胸径 Mean DBH (cm)	基面积 Basal area (m²)	最大树高 Max height (m)	重要值 IV
鹅掌柴 Schefflera heptaphylla	332	486	9.12	4.36	12.00	7.93
九节 Psychotria asiatica	1050	1380	2.30	0.82	5.00	6.57
肉实树 Sarcosperma laurinum	328	370	6.74	2.08	12.00	4.78
银柴 Aporosa dioica	301	334	4.98	0.97	10.00	3.11
黄樟 Cinnamomum parthenoxylon	71	87	14.15	1.74	12.48	3.10
黄椿木姜子 Litsea variabilis	392	430	3.33	0.64	8.00	3.07
水锦树 Wendlandia uvariifolia	218	286	5.13	0.88	8.00	2.53
罗伞 Brassaiopsis glomerulata	376	542	2.21	0.29	6.19	2.46
密花树 Myrsine seguinii	207	282	4.52	0.82	10.48	2.41
臀果木 Pygeum topengii	143	144	4.97	0.96	13.58	2.25

表 4　SW 1 hm² 样地重要值前 10 位的优势树种
Table 4　Top 10 species with the highest important values in SW 1 hm² plot

物种 Species	个体数 No. of individual	茎干数 No. of stem	平均胸径 Mean DBH (cm)	基面积 Basal area (m²)	最大树高 Max height (m)	重要值 IV
鹅掌柴 Schefflera heptaphylla	543	672	8.15	6.18	8.1	11.59
九节 Psychotria asiatica	1495	1867	2.08	0.81	2.5	9.35
华润楠 Machilus chinensis	194	205	12.02	3.21	9.0	5.74
银柴 Aporosa dioica	414	447	5.00	1.39	7.9	4.41
枫香树 Liquidambar formosana	64	73	19.16	2.60	12.0	4.26
假苹婆 Sterculia lanceolata	435	487	3.73	1.00	7.0	3.96
南山花 Prismatomeris tetrandra	377	415	1.94	0.16	3.0	2.41
杂色榕 Ficus variegata	254	284	3.61	0.42	4.4	2.15
黄樟 Cinnamomum parthenoxylon	44	48	14.57	1.04	8.6	2.04
红鳞蒲桃 Syzygium hancei	123	147	5.23	0.55	7.9	1.65

两个样地中，中上层的优势树种主要包括鹅掌柴（*Schefflera heptaphylla*）、黄樟（*Cinnamomum parthenoxylon*）、枫香树（*Liquidambar formosana*）、华润楠（*Machilus chinensis*）、肉实树（*Sarcosperma laurinum*）、银柴（*Aporusa dioica*）、水锦树（*Wendlandia uvariifolia*）、臀果木（*Pygeum topengii*）等种类。乔木中下层的优势种类主要包括九节（*Psychotria rubra*）、南山花（*Prismatomeris tetrandra*）、罗伞（*Brassaiopsis glomerulata*）、密花树（*Myrsine seguinii*）、黄椿木姜子（*Litsea variabilis*）、杂色榕（*Ficus variegata*）、假苹婆（*Sterculia lanceolata*）种类。鹅掌柴（*Schefflera heptaphylla*）、九节（*Psychotria rubra*）是两个样地重要值排第一、二位的优势种。在 JH 和 SW 样地中，前 10 个树种分别占重要值总和的 38.21% 和 47.56%。

按照随机抽样的方式，对 2 个 1 hm² 样地的种面积关系进行分析。JH 样地表现出更快的物种累积速率，当随机抽样面积达到 900 m² 时，已可随机抽取到约 60%(67 种) 左右的种类，达到 3600 m² 时能抽取到约 75%（110 种）的种类。而相比 JH 样地而言，SW 样地在随机抽样面积为 900 m² 时只能包含 40%(60 种) 左右的种类，而在 3600 m² 时只能抽取到 68%(99 种) 的树种种类。

两个样地共记录到 DBH≥1 cm 植株茎干数 15209 个，其中独立个体数 12618，分枝数 2591。个体数最多的均为九节（*Psychotria rubra*），占样地总个体数的 20.2%。两个样地中，排位前 10 的物种个体数约占样地总个体数的 70% 以上。

两个样地内所有个体的总径级分布呈现指数递减格局。JH 样地 1～5 cm 的个体 4729 棵，约占总个体数的 72.5%；SW 样地 1～5 cm 的个体 4615 棵，约占总个体数的 76.0%。整体来看，小径级个体数最多，随着胸径等级的增加个体数量逐渐下降，整个表现出群落径级结构比较稳定。

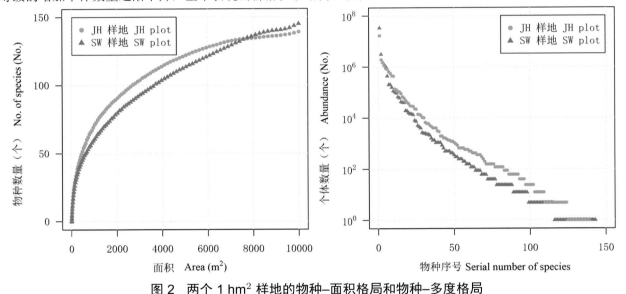

图 2　两个 1 hm² 样地的物种–面积格局和物种–多度格局

Figure 2　Species-area patterns and species-abundance patterns of two 1hm² plots

2.3 Tree Species Composition and Community Structure

In two 1 hm² plots, a total of 12618 individuals with DBH of ≥ 1 cm were recorded, belonging to 183 species, 119 genera, and 55 families. In JH plot, there are 6530 individuals, belonging to 139 species, 97 genera and 49 families. In SW plot, there are 6088 individuals, belonging to 143 species, 100 genera and 50 families. Among them, there are 4 species in National Key Protected Wild Plants, *Camellia euphlebia*, *Camellia indochinensis* var. *tunghinensis*, *Nephelium chryseum*, *Madhuca pasquieri*. And there are 4 Endemic plants in Guangxi, *Machilus parabreviflora*, *Camellia gaudichaudii*, *Camellia indochinensis* var. *tunghinensis*, *Glycosmis oligantha*.

In the two plots, the main dominant families are Rubiaceae, Lauraceae, Araliaceae, Phyllanthaceae, Moraceae, Malvaceae, Sapotaceae, Myrtaceae, Altingiaceae, Primulaceae, etc.. Families with relatively large number of species include Rubiaceae, Lauraceae, Moraceae, Myrtaceae, etc., while families with large number of genera include Rubiaceae, Phyllanthaceae, Lauraceae. From the flora composition of dominant families, the dominant families in the two plots are mostly distributed in Pantropical zone and Tropical Asia zone besides Cosmopolitan zone in the world, and only Altingiaceae is distributed in the North Temperate zone.

In the two plots, the dominant tree species in the upper and middle layers mainly include *Schefflera heptaphylla*, *Cinnamomum parthenoxylon*, *Liquidambar formosana*, *Machilus ichangensis* var. *leiophylla*, *Machilus chinensis*, *Sarcosperma laurinum*, *Aporusa dioica*, *Wendlandia uvariifolia*, *Pygeum topengii*, etc.. The dominant species in the middle and lower layers of trees mainly include *Psychotria rubra*, *Prismatomeris tetrandra*, *Ardisia quinquegona*, *Myrsine seguinii*, *Litsea variabilis*, *Ficus variegata*, *Sterculia lanceolata*, etc.. *Schefflera heptaphylla*, *Psychotria rubra* are the dominant species with the first and second important values in the two plots. The top 15 tree species account for 38.21% of the total important values in JH plot, and 47.56% in SW plot.

According to the method of random sampling, the relationship between species richness and plot area of 2 plots was analyzed. JH plot showed faster species accumulation rate. About 50% species could be randomly extracted when the random sampling area reached 900 m^2, and about 75% species could be extracted when the random sampling area reached 3600 m^2. SW plot can only contain about 40% species when the random sampling area is 900 m^2, while only 68% species can be extracted when 3600 m^2.

A total of 15209 stems of DBH≥1 cm plants were recorded in the two plots, including 12618 individuals and 2591 branches. The largest number of individuals in SW plot is *Psycotria rubra*, accounting for 20.2% of the total number of individuals in the plot. The number of species individuals in the top 10 of the two plots accounted for more than 70% of the total number of individuals in the plots.

The total diameter distribution of all individuals in the two plots showed exponential decreasing patterns. In JH plot, there are 4729 individuals of 1~5 cm, accounting for 72.5% of the total number of individuals. In SW plot, there are 4615 individuals of 1~5 cm, accounting for 76.0% of the total number of individuals. On the whole, the number of individuals in small diameter class is the largest. With the increase of DBH class, the number of individuals gradually decreases, and the whole community shows a relatively stable diameter class structure.

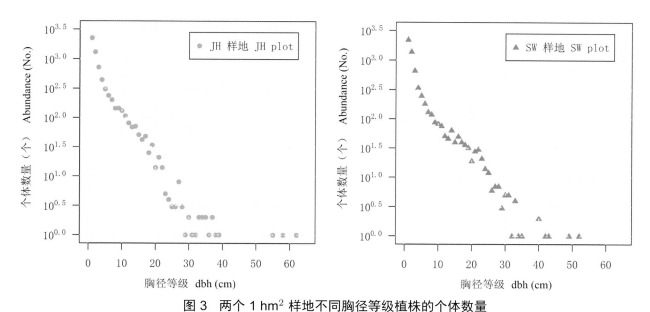

图 3 两个 1 hm² 样地不同胸径等级植株的个体数量

Figure 3 Number of plant individuals with different DBH classes in two 1 hm² plots

2.4 草本层植物和藤本植物的多样性

两个样地内位于森林群落草本层的植物有 49 种，隶属于 33 科 43 属，常见的有：铁芒萁（*Dicranopteris linearis*）、华南毛蕨（*Cyclosorus parasiticus*）、长花枝杜若（*Pollia secundiflora*）、曲枝假蓝（*Strobilanthes dalzielii*）等。其中，国家重点保护野生植物名录中 II 级保护种类有 5 种：宽叶线柱兰（*Zeuxine affinis*）、橙黄玉凤花（*Habenaria rhodocheila*）、长茎羊耳蒜（*Liparis viridiflora*）、竹叶兰（*Arundina graminifolia*、宽叶羊耳蒜（*Liparis latifolia*）。

两个样地内藤本植物有 23 种，隶属于 19 科 23 属，常见的有：苍白秤钩风（*Diploclisia glaucescens*）、天仙藤（*Fibraurea recisa*）、藤黄檀（*Dalbergia hancei*）、赤苍藤（*Erythropalum scandens*）等。另外，在挂牌监测的 DBH≥1 cm 植物中，有 4 种为攀援灌木状植物，分别为：假鹰爪（*Desmos chinensis*）、紫玉盘（*Uvaria macrophylla*）、老虎刺（*Pterolobium punctatum*）和思茅山橙（*Melodinus cochinchinensis*）。

2.4 Diversity of Plants in The Herb Layer and Lianas

In the two plot, there are 49 plants in the herb layer of the forest community, belonging to 33 families and 43 genera, the common species are *Dicranopteris linearis*, *Cyclosorus parasiticus*, *Pollia secundiflora*, and *Strobilanthes dalzielii* etc..

In the two plot, there are 23 species of lianas , belonging to 19 families and 23 genera. The common liana species are: *Diploclisia glaucescens*, *Fibraurea recisa*, *Dalbergia hancei*, and *Erythropalum scandens* etc.. In addition, among the plants with DBH≥1 cm included in dynamic monitoring, there are 4 species of climbing shrubs, which are *Desmos chinensis*, *Uvaria macrophylla*, *Pterolobium punctatum*, and *Melodinus cochinchinensis*.

3 广西防城金花茶 1 hm² 样地（JH）的树种及其分布格局

Tree Species and their Distribution Patterns in the Guangxi Fangcheng JH 1 hm² Plot

1 马尾松 | mǎ wěi sōng 松属

***Pinus massoniana* Lamb.**

松科 Pinaceae

样地名称（Plot name）＝ JH
个体数（Individual number/1 hm²）＝ 1
最大胸径（Max DBH）＝ 12.5 cm
重要值排序（Importance value rank）＝ 121

乔木，高达 45 m，胸径 1.5 m。树皮红褐色，下部灰褐色，裂成不规则的鳞状块片。针叶 2 针一束，稀 3 针一束，长 12~20 cm，细柔，微扭曲，两面有气孔线；叶鞘宿存。球果下垂，绿色，成熟时变为栗色。花期 4~5 月，球果翌年 10~12 月成熟。

Trees, to 45 m tall, trunk to 1.5 m d.b.h.. Bark red-brown toward apex of trunk, gray- or red-brown toward base, irregularly scaly and flaking. Needles 2 (or 3) per bundle, slightly twisted, 12–20 cm, stomatal lines present on all surfaces, base with persistent sheath. Seed cones pendulous, green, turning chestnut brown at maturity. Pollination Apr.–May, seed maturity Oct.–Dec. of next year.

树干　Trunk
摄影：王斌　Photo by: Wang Bin

枝叶　Branch and leaves
摄影：王斌　Photo by: Wang Bin

球果　Strobil
摄影：王斌　Photo by: Wang Bin

径级分布表　DBH class

胸径区间 Diameter class (cm)	个体数 No. of individuals in the plot	比例 Proportion (%)
1~2	0	0.00
2~5	0	0.00
5~10	0	0.00
10~20	1	100.00
20~35	0	0.00
35~50	0	0.00
≥50	0	0.00

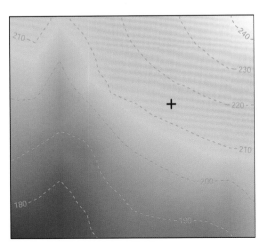

● 1~5 cm DBH　＋ 5~20 cm DBH　○ ≥20 cm DBH
个体分布图　Distribution of individuals

2 八角 | bā jiǎo （大茴角）

Illicium verum **Hook. f.**

五味子科 Schisandraceae

样地名称（Plot name）= JH
个体数（Individual number/1 hm^2）= 10
最大胸径（Max DBH）= 20.3 cm
重要值排序（Importance value rank）= 72

乔木，高达15 m。树皮深灰色。叶在顶端3~6片簇生，革质、厚革质，倒卵状椭圆形、倒披针形或椭圆形，长5~15 cm，宽2~5 cm，在阳光下可见密布透明油点。花粉红至深红色，心皮通常8。聚合果，呈八角形。正糙果花期3~5月，果熟期9~10月，春糙果花期8~10月，果熟期翌年3~4月。

Trees, to 15 m tall. Bark gray. Leaves in clusters of 3–6 at distal nodes. Leaf blade obovate-elliptic, oblanceolate, or elliptic, 5–15 × 2–5 cm, leathery to thickly leathery, transparent oil spots can be seen in the sun. Carpels usually 8. Aggregate fruit with ca. 8 follicle. Fl. Mar.–May and Aug.–Oct., fr. Sep.–Oct. and Mar.–Apr. of next year.

树干　Trunk
摄影：王斌　Photo by: Wang Bin

叶　Leaves
摄影：王斌　Photo by: Wang Bin

果　Fruit
摄影：王斌　Photo by: Wang Bin

径级分布表　DBH class

胸径区间 Diameter class (cm)	个体数 No. of individuals in the plot	比例 Proportion (%)
1~2	2	20.00
2~5	3	30.00
5~10	2	20.00
10~20	2	20.00
20~35	1	10.00
35~50	0	0.00
≥50	0	0.00

● 1~5 cm DBH　　+ 5~20 cm DBH　　○ ≥20 cm DBH
个体分布图　Distribution of individuals

3 香港木兰 | xiāng gǎng mù lán （香港玉兰）

长喙木兰属

Lirianthe championii **(Bentham) N. H. Xia & C. Y. Wu**

木兰科 Magnoliaceae

样地名称（Plot name）＝ JH
个体数（Individual number/1 hm²）＝ 3
最大胸径（Max DBH）＝ 2.6 cm
重要值排序（Importance value rank）＝ 101

灌木或小乔木。小枝绿色。叶薄革质，狭椭圆形，狭倒卵形或倒卵状披针形，长7~14（~20）cm，宽2.5~4.5（~6）cm，先端长渐尖，基部楔形，边缘起伏成波状；侧脉每边8~12条。托叶痕几达叶柄顶端。花直立，芳香，花被片9。果长3~4.5cm。花期4~6月，果期9~10月。

Shrubs or small trees. Young twigs green. Stipular scar nearly reaching apex of petiole. Leaf blade elliptic, narrowly oblong-elliptic, or narrowly obovate-elliptic, 7–14(–20) ×2–4.5(–6) cm, leathery, secondary veins 8–12 on each side of midvein, base cuneate to narrowly cuneate and slightly decurrent on petiole, apex acuminate. Flowers erect, very fragrant, tepals 9. Fruit 3–4.5 cm. Fl. Apr.–Jun., fr. Sep.–Oct..

树干　　Trunk
摄影：李健星　　Photo by：Li Jianxing

叶　　Leaves
摄影：丁涛　　Photo by：Ding Tao

花枝　　Flowering branch
摄影：丁涛　　Photo by：Ding Tao

径级分布表　DBH class

胸径区间 Diameter class (cm)	个体数 No. of individuals in the plot	比例 Proportion (%)
1~2	1	33.33
2~5	2	66.67
5~10	0	0.00
10~20	0	0.00
20~35	0	0.00
35~50	0	0.00
≥50	0	0.00

● 1~5 cm DBH　＋ 5~20 cm DBH　○ ≥20 cm DBH
个体分布图　Distribution of individuals

4 假鹰爪 | jiǎ yīng zhuǎ （鸡爪枫）

Desmos chinensis Lour.

番荔枝科 Annonaceae

样地名称（Plot name）= JH
个体数（Individual number/1 hm²）= 16
最大胸径（Max DBH）= 2.8 cm
重要值排序（Importance value rank）= 55

木质攀援植物，高达 4 m。枝皮粗糙，有纵条纹，有灰白色凸起的皮孔。叶薄纸质或膜质，长圆形或椭圆形，长 6~14 cm，宽 2~6.5 cm，顶端钝或急尖，基部圆形或稍偏斜，上面有光泽。花黄白色，单朵与叶对生或互生。果有柄，念珠状，内有种子 2~6 颗。花期 4~10 月，果期 6~12 月。

Woody climbers, to 4 m tall. Branch is rough with longitudinal stripes; raised grayish white lenticels. Leaf blade oblong to elliptic, 6–14×2–6.5 cm, membranous to thinly papery, adaxially glossy, base rounded to slightly oblique, apex acute to acuminate. Inflorescences superaxillary or leaf-opposed, 1-flowered. Fruiting moniliform, with 2–6 joints. Fl. Apr.–Oct., fr. Jun.–Dec..

叶背　　Leaf backs
摄影：丁涛　　Photo by：Ding Tao

花托　　Receptacle
摄影：丁涛　　Photo by：Ding Tao

果序　　Infructescence
摄影：李健星　　Photo by：Li Jianxing

径级分布表　DBH class

胸径区间 Diameter class (cm)	个体数 No. of individuals in the plot	比例 Proportion (%)
1~2	15	93.75
2~5	1	6.25
5~10	0	0.00
10~20	0	0.00
20~35	0	0.00
35~50	0	0.00
≥50	0	0.00

● 1~5 cm DBH　　+ 5~20 cm DBH　　○ ≥20 cm DBH
个体分布图　Distribution of individuals

5 紫玉盘 | zǐ yù pán （那大柒玉盘）

***Uvaria macrophylla* Roxb.**

番荔枝科 Annonaceae

样地名称（Plot name）= JH
个体数（Individual number/1 hm^2）= 5
最大胸径（Max DBH）= 2.0 cm
重要值排序（Importance value rank）= 99

攀援灌木，高达 18 m，枝条蔓延性。叶革质，长倒卵形或长椭圆形，长 9~30 cm，宽 3~15 cm，顶端急尖或钝，基部近心形或圆形。花 1~2 朵，与叶对生，暗紫红色或淡红褐色。果卵圆形或短圆柱形，暗紫褐色。花期 3~9 月，果期 7 月至翌年 3 月。

Climbing shrubs, to 18 m tall. Leaf blade obovate, oblong-obovate, elliptic, or broadly oblong, 9–30 × 3–15 cm, leathery, base shallowly cordate, apex acute, obtuse, or rounded and mostly apiculate. Inflorescences Leaf-opposed or rarely extra-axillary, petals dark red, purple, or purplish. Monocarps orange, ovoid to subterete. Fl. Mar.–Sep., fr. Jul.–Mar. of next year.

枝叶　　Branch and leaves
摄影：丁涛　　Photo by: Ding Tao

花　　Flower
摄影：丁涛　　Photo by: Ding Tao

果枝　　Fruiting branch
摄影：丁涛　　Photo by: Ding Tao

径级分布表　DBH class

胸径区间 Diameter class (cm)	个体数 No. of individuals in the plot	比例 Proportion (%)
1~2	4	80.00
2~5	1	20.00
5~10	0	0.00
10~20	0	0.00
20~35	0	0.00
35~50	0	0.00
≥50	0	0.00

● 1~5 cm DBH　　+ 5~20 cm DBH　　○ ≥20 cm DBH
个体分布图　Distribution of individuals

6 黄樟 | huáng zhāng

Cinnamomum parthenoxylon (Jack) Meisner
樟科 Lauraceae

樟属

样地名称（Plot name）＝ JH
个体数（Individual number/1 hm^2）＝ 71
最大胸径（Max DBH）＝ 30.7 cm
重要值排序（Importance value rank）＝ 6

常绿乔木，树干通直，高 10～20 m，胸径达 40 cm 以上。树皮暗灰褐色，上部为灰黄色，深纵裂，小片剥落，内皮带红色，具有樟脑气味。叶互生，椭圆状卵形或长椭圆状卵形，长 6～12 cm，宽 3～6 cm，先端通常急尖或短渐尖，基部楔形或阔楔形，革质。圆锥花序于枝条上部腋生或近顶生。果球形，黑色。花期 3～5 月，果期 4～10 月。

Evergreen trees, trunk straight, 10–20 m tall, up to 40 cm d.b.h.. Bark dark green-brown, gray-yellow on upper part, longitudinally deeply fissured, peeling off in lamellae, reddish inside, camphor-scented. Leaves alternate. Leaf blade usually elliptic-ovate or narrowly elliptic-ovate, 6–12 × 3–6 cm, leathery. Panicle axillary on upper part of branchlet or subterminal. Fruit black, globose. Fl. Mar.–May, fr. Apr.–Oct..

枝叶　　Branch and leaves
摄影：丁涛　　Photo by: Ding Tao

花　　Flowers
摄影：唐忠炳　　Photo by: Tang Zhongbing

果　　Fruits
摄影：唐忠炳　　Photo by: Tang Zhongbing

径级分布表　DBH class

胸径区间 Diameter class (cm)	个体数 No. of individuals in the plot	比例 Proportion (%)
1～2	2	2.82
2～5	4	5.63
5～10	7	9.86
10～20	38	53.52
20～35	20	28.17
35～50	0	0.00
≥50	0	0.00

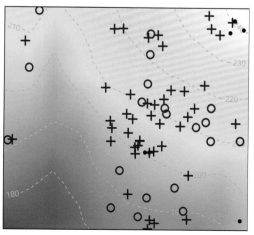

● 1～5 cm DBH　　＋ 5～20 cm DBH　　○ ≥20 cm DBH
个体分布图　Distribution of individuals

7 鼎湖钓樟 | dǐng hú diào zhāng （白胶木） 山胡椒属

Lindera chunii Merr.

樟科 Lauraceae

样地名称（Plot name）= JH
个体数（Individual number/1 hm²）= 3
最大胸径（Max DBH）= 7.1 cm
重要值排序（Importance value rank）= 110

常绿乔木，高 4.5~12 m。叶互生，椭圆形至长椭圆形；长 8~14 cm，宽 2.5~5 cm；先端尾状渐尖，基部楔形或急尖；纸质，三出脉，侧脉直达先端。伞形花序 1 个或 2~5 个生于叶腋短枝上。雄花序总梗长 1 mm；雌花序总梗长 2~2.5 mm。果椭圆形，无毛。花期 10 月到翌年 3 月，果期 5~8 月。

Evergreen trees, 4.5–12 m tall. Leaf blade elliptic to narrowly elliptic, caudate at apex, 8–14 × 2.5–5 cm, thinly papery, trinerved, base cuneate or subrounded, apex acuminate. Umbels (1 or) 2–5, inserted in leaf axil of short branchlets. Male flowers yellow-green; pedicels obconic, ca. 1 mm; female flowers yellowish; pedicels 2–2.5 mm. Fruit ellipsoid, glabrous. Fl. Oct.–Mar. of next year, fr. May–Aug..

果枝　　Fruiting branches
摄影：蒋裕良　Photo by: Jiang Yuliang

叶　　Leaves
摄影：蒋裕良　Photo by: Jiang Yuliang

叶背　　Leaf backs
摄影：蒋裕良　Photo by: Jiang Yuliang

径级分布表 DBH class

胸径区间 Diameter class (cm)	个体数 No. of individuals in the plot	比例 Proportion (%)
1~2	1	33.34
2~5	1	33.33
5~10	1	33.33
10~20	0	0.00
20~35	0	0.00
35~50	0	0.00
≥50	0	0.00

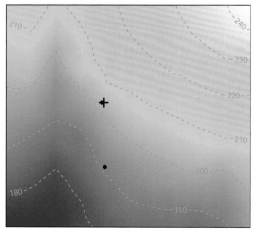

● 1~5 cm DBH　　＋ 5~20 cm DBH　　○ ≥20 cm DBH
个体分布图　Distribution of individuals

8 香叶树 | xiāng yè shù （香果树）

山胡椒属

***Lindera communis* Hemsl**

樟科 Lauraceae

样地名称（Plot name）= JH
个体数（Individual number/1 hm²）= 1
最大胸径（Max DBH）= 1.8 cm
重要值排序（Importance value rank）= 135

常绿灌木或乔木，高（1～）3～4（～5）m。树皮淡褐色。叶互生，通常披针形、卵形或椭圆形，长（3～）4～9（～12.5），宽（1～）1.5～3（～4.5）cm，先端渐尖、急尖、骤尖或有时近尾尖，基部宽楔形或近圆形；薄革质至厚革质。伞形花序。果卵形，无毛，成熟时红色。花期3～4月，果期9～10月。

Evergreen shrubs or trees, (1–) 3–4 (–5) m tall. Bark brownish. Leaves alternate. Leaf blade lanceolate, ovate, or elliptic, (3–) 4–9 (–12.5) × (1–) 1.5–3 (–4.5) cm, thinly leathery or thickly leathery, base broadly cuneate or subrounded, apex acuminate, acute, or sometimes nearly caudate-acuminate. Umbels solitary. Fruit ovate, glabrous, red at maturity. Fl. Mar.–Apr., fr. Sep.–Oct..

枝叶　Branch and leaves
摄影：丁涛　Photo by: Ding Tao

叶背　Leaf backs
摄影：丁涛　Photo by: Ding Tao

果　Fruits
摄影：李均　Photo by: Li Jun

径级分布表　DBH class

胸径区间 Diameter class (cm)	个体数 No. of individuals in the plot	比例 Proportion (%)
1～2	1	100.00
2～5	0	0.00
5～10	0	0.00
10～20	0	0.00
20～35	0	0.00
35～50	0	0.00
≥50	0	0.00

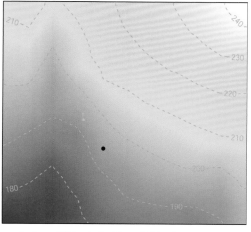

● 1～5 cm DBH　＋ 5～20 cm DBH　○ ≥20 cm DBH
个体分布图　Distribution of individuals

9 滇粤山胡椒 | diān yuè shān hú jiāo （山钓樟）

山胡椒属

***Lindera metcalfiana* C. K. Allen**

樟科 Lauraceae

样地名称（Plot name）= JH
个体数（Individual number/1 hm²）= 2
最大胸径（Max DBH）= 2.9 cm
重要值排序（Importance value rank）= 116

常绿乔木，高（2~5）3~12 m。叶互生，椭圆形或长椭圆形，长 8~16 cm，宽 2.5~2.6 cm，先端渐尖或尾尖，常呈镰刀状，基部宽楔形，革质，上面黄绿色，下面灰绿色，干时上面灰褐色。羽状脉，上面中脉突出，侧脉每边 6~10 条。雄伞形花序，有雄花 6 朵。雌花黄色。果球形。花期 3~5 月，果期 6~10 月。

Evergreen trees, (2–5) 3–12 m tall. Leaf blade papery or thinly leathery, alternate, 8–16 × 2.5–2.6 cm, elliptic to lanceolate, grayish brown or purple-brown when dry. Lateral veins 6–10 pairs, base cuneate, revolute on margin. Male flowers, 6-flowered; female flowers yellow. Fruits globose. Fl. Mar.–May, fr. Jun.–Oct..

枝叶　　Branch and leaves
摄影：蒋裕良　　Photo by: Jiang Yuliang

叶背　　Leaf backs
摄影：蒋裕良　　Photo by: Jiang Yuliang

花序　　Inflorescences
摄影：徐晔春　　Photo by: Xu Yechun

径级分布表　DBH class

胸径区间 Diameter class (cm)	个体数 No. of individuals in the plot	比例 Proportion (%)
1~2	1	50.00
2~5	1	50.00
5~10	0	0.00
10~20	0	0.00
20~35	0	0.00
35~50	0	0.00
≥50	0	0.00

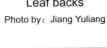

● 1~5 cm DBH　＋ 5~20 cm DBH　○ ≥20 cm DBH
个体分布图　Distribution of individuals

10 尖脉木姜子 | jiān mài mù jiāng zǐ

Litsea acutivena Hayata

樟科 Lauraceae

样地名称（Plot name）= JH
个体数（Individual number/1 hm²）= 2
最大胸径（Max DBH）= 3.2 cm
重要值排序（Importance value rank）= 115

常绿乔木，高达7 m。树皮褐色。叶互生或聚生枝顶，披针形、倒披针形或长圆状披针形，长4~11 cm，宽2~4 cm，先端急尖或短渐尖，基部楔形，革质。伞形花序生于当年生枝上端，簇生。果椭圆形，成熟时黑色。花期7~8月，果期12月至翌年2月。

Evergreen trees, up to 7 m tall. Bark brown. Leaf alternate or clustered toward apex of branchlet. Leaf blade lanceolate, oblanceolate, or oblong-lanceolate, 4–11 × 2–4 cm, base cuneate, apex acute or shortly acuminate. Umbels clustered toward apex of branchlet. Fruit ellipsoid, black at maturity. Fl. Jul.–Aug., fr. Dec.–Feb. of next year.

枝叶　　Branch and leaves
摄影：丁涛　　Photo by: Ding Tao

叶背　　Leaf backs
摄影：丁涛　　Photo by: Ding Tao

花序　　Inflorescences
摄影：唐忠炳　　Photo by: Tang Zhongbing

径级分布表　DBH class

胸径区间 Diameter class (cm)	个体数 No. of individuals in the plot	比例 Proportion (%)
1~2	1	50.00
2~5	1	50.00
5~10	0	0.00
10~20	0	0.00
20~35	0	0.00
35~50	0	0.00
≥50	0	0.00

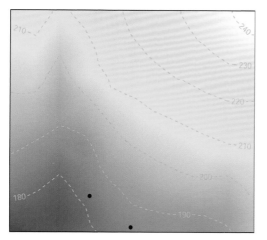

● 1~5 cm DBH　＋ 5~20 cm DBH　○ ≥20 cm DBH
个体分布图　Distribution of individuals

11 假柿木姜子 | jiǎ shì mù jiāng zǐ

木姜子属

***Litsea monopetala* (Roxb.) Pers.**

樟科 Lauraceae

样地名称（Plot name）= JH
个体数（Individual number/1 hm^2）= 1
最大胸径（Max DBH）= 21.9 cm
重要值排序（Importance value rank）= 104

常绿乔木，高达 18 m，胸径约 15 cm。树皮灰色或灰褐色。叶互生，宽卵形、倒卵形至卵状长圆形，长 8~20 cm，宽 4~12 cm，先端钝或圆，偶有急尖，基部圆或急尖，薄革质。伞形花序簇生叶腋，总梗极短。果长卵形，长约 7 mm，直径 5 mm。花期 11 月至翌年 5~6 月，果期 6~7 月。

Evergreen trees, up to 18 m tall, ca. 15 cm d.b.h.. Bark gray or gray-brown. Leaves alternate; densely hairy like branchlets. Leaf blade broadly ovate or obovate to ovate-oblong, 8–20 × 4–12 cm, base rounded or acute, apex obtuse or rounded, rarely acute. Umbels clustered on shortest branchlets. Fruit long ovoid, ca. 7 × 5 mm. Fl. Nov.–May or Jun. of next year, fr. Jun.–Jul..

树干　　　　Trunk
摄影：李健星　　Photo by：Li Jianxing

叶　　　　Leaves
摄影：丁涛　　Photo by：Ding Tao

叶背　　　　Leaf backs
摄影：丁涛　　Photo by：Ding Tao

径级分布表 DBH class

胸径区间 Diameter class (cm)	个体数 No. of individuals in the plot	比例 Proportion (%)
1~2	0	0.00
2~5	0	0.00
5~10	0	0.00
10~20	0	0.00
20~35	1	100.00
35~50	0	0.00
≥50	0	0.00

● 1~5 cm DBH　　+ 5~20 cm DBH　　○ ≥20 cm DBH
个体分布图 Distribution of individuals

12 豹皮樟 | chái pí zhāng （圆叶豹皮樟）

***Litsea rotundifolia* var. *oblongifolia* (Nees) C. K. Allen**

樟科 Lauraceae

样地名称（Plot name）＝ JH
个体数（Individual number/1 hm²）＝ 3
最大胸径（Max DBH）＝ 14.5 cm
重要值排序（Importance value rank）＝ 97

常绿灌木或小乔木，高可达 3 m。树皮灰色或灰褐色，常有褐色斑块。叶片宽卵形到圆形，2.2~4.5×1.5~4 cm，基部圆形，先端钝圆形或短渐尖，两面具不明显的网纹。伞形花序常 3 个簇生叶腋，几无总梗。果球形，直径约 6 mm，几无果梗，成熟时灰蓝黑色。花期 8~9 月，果期 9~11 月。

Evergreen shrubs or small trees, up to 3 m tall. Bark gray or gray-brown, often brownness spots. Leaf blade broadly ovaterounded to rotund, 2.2–4.5 × 1.5–4 cm, base rotund, apex obtuse-rounded or shortly acuminate, inconspicuously reticulateveined on both surfaces. Umbels often in cluster of 3, axillary. Fruit globose, ca. 6 mm in diam., subsessile, gray-blue-black at maturity. Fl. Aug.–Sep., fr. Sep.–Nov..

花　Flowers
摄影：李健星　Photo by: Li Jianxing

叶背　Leaf backs
摄影：丁涛　Photo by: Ding Tao

果　Fruits
摄影：扈文芳　Photo by: Hu Wenfang

径级分布表 DBH class

胸径区间 Diameter class (cm)	个体数 No. of individuals in the plot	比例 Proportion (%)
1~2	0	0.00
2~5	1	33.33
5~10	0	0.00
10~20	2	66.67
20~35	0	0.00
35~50	0	0.00
≥50	0	0.00

● 1~5 cm DBH　＋ 5~20 cm DBH　○ ≥20 cm DBH
个体分布图　Distribution of individuals

13 黄椿木姜子 | huáng chūn mù jiāng zǐ （雄鸡树） 木姜子属

Litsea variabilis Hemsl.

樟科 Lauraceae

样地名称（Plot name）= JH
个体数（Individual number/1 hm^2）= 392
最大胸径（Max DBH）= 19.0 cm
重要值排序（Importance value rank）= 5

常绿灌木或乔木，高达 15 m。叶对生或近对生，一般为椭圆形或倒卵形，长 5~14 cm，宽 2~4.5 cm，先端渐尖、钝或略圆，基部楔形或宽楔形，革质。伞形花序常 3~8 个集生叶腋。果球形，熟时黑色。花期 5~11 月，果期 9 月至翌年 5 月。

Evergreen shrubs or trees, up to 15 m tall. Leaves opposite or subopposite. Leaf blade usually elliptic, oblong, or obovate, 5–14 × 2–4.5 cm, base cuneate or broadly cuneate, apex acuminate, obtuse, or slightly rounded. Umbels often in cluster of 3–8, axillary. Fruit globose, black at maturity. Fl. May–Nov., fr. Sep.–May of next year.

果枝　Fruiting branches
摄影：丁涛　Photo by: Ding Tao

叶背　Leaf backs
摄影：丁涛　Photo by: Ding Tao

果　Fruits
摄影：丁涛　Photo by: Ding Tao

径级分布表 DBH class

胸径区间 Diameter class (cm)	个体数 No. of individuals in the plot	比例 Proportion (%)
1~2	163	41.58
2~5	147	37.50
5~10	64	16.33
10~20	18	4.59
20~35	0	0.00
35~50	0	0.00
≥50	0	0.00

● 1~5 cm DBH　　+ 5~20 cm DBH　　○ ≥20 cm DBH
个体分布图　Distribution of individuals

14 轮叶木姜子 | lún yè mù jiāng zǐ （槁树）

Litsea verticillata Hance
樟科 Lauraceae

样地名称（Plot name）＝ JH
个体数（Individual number/1 hm^2）＝ 29
最大胸径（Max DBH）＝ 17.5 cm
重要值排序（Importance value rank）＝ 38

常绿灌木或小乔木，高 2～5 m。叶 4～6 片轮生，披针形或倒披针状长椭圆形，长 7～25 cm，宽 2～6 cm，先端渐尖，基部急尖、钝或近圆，薄革质。伞形花序 2～10 个集生于小枝顶部。果卵形或椭圆形，顶端有小尖头。花期 4～11 月，果期 11 月至翌年 1 月。

Evergreen shrubs or small trees, 2–5 m tall. Leaves 4–6-verticillate. Leaf blade long lanceolate or long elliptic-oblanceolate, 7–25 × 2–6 cm, base acute, obtuse, or rotund, apex acuminate. Umbels in cluster of 2–10 at apex of branchlet. Fruit ovoid or ellipsoid. Fl. Apr.–Nov., fr. Nov.–Jan. of next year.

枝叶　Branch and leaves
摄影：丁涛　Photo by: Ding Tao

叶背　Leaf back
摄影：丁涛　Photo by: Ding Tao

花枝　Flowering branch
摄影：丁涛　Photo by: Ding Tao

径级分布表　DBH class

胸径区间 Diameter class (cm)	个体数 No. of individuals in the plot	比例 Proportion (%)
1～2	10	34.48
2～5	12	41.38
5～10	6	20.69
10～20	1	3.45
20～35	0	0.00
35～50	0	0.00
≥50	0	0.00

● 1～5 cm DBH　＋ 5～20 cm DBH　○ ≥20 cm DBH
个体分布图　Distribution of individuals

15 华润楠 | huá rùn nán （桢楠）

润楠属

***Machilus chinensis* (Champ. ex Benth.) Hemsl.**

樟科 Lauraceae

样地名称（Plot name）= JH
个体数（Individual number/1 hm^2）= 58
最大胸径（Max DBH）= 37.2 cm
重要值排序（Importance value rank）= 14

乔木，高约 8～11 m。嫩枝无毛。叶倒卵状长椭圆形至长椭圆状倒披针形，长 5～8（～10）cm，宽 2～3（～4）cm，先端钝或短渐尖，基部狭，革质，中脉在上面凹下，下面凸起，侧脉不明显，网状小脉在两面上形成蜂巢状浅窝穴。圆锥花序顶生；花白色。子房球形。果球形。花期 9 月，果期翌年 2 月。

Trees, 8–11 m tall. Branchlets glabrous. Leaf blade obovate-oblong to oblong-oblanceolate, 5–8 (–10) × 2–3 (–4) cm, leathery, midrib raised abaxially, concave adaxially, reticulate veinlets foveolate on both surfaces, apex obtuse or shortly acuminate. Panicles usually terminal, branched at upper part of peduncle. Flowers white. Fruit globose. Fl. Sep., fr. Feb. of next year.

树干　　Trunk
摄影：李健星　　Photo by：Li Jianxing

叶背　　Leaf backs
摄影：蒋裕良　　Photo by：Jiang Yuliang

果枝　　Fruiting branches
摄影：杨平　　Photo by：Yang Ping

径级分布表　DBH class

胸径区间 Diameter class (cm)	个体数 No. of individuals in the plot	比例 Proportion (%)
1～2	4	6.90
2～5	3	5.17
5～10	9	15.52
10～20	35	60.34
20～35	6	10.35
35～50	1	1.72
≥50	0	0.00

● 1～5 cm DBH　　+ 5～20 cm DBH　　○ ≥20 cm DBH
个体分布图　Distribution of individuals

16 赛短花润楠 | sài duǎn huā rùn nán

润楠属

***Machilus parabreviflora* Hung T. Chang**

樟科 Lauraceae

样地名称（Plot name）= JH
个体数（Individual number/1 hm²）= 15
最大胸径（Max DBH）= 21.9 cm
重要值排序（Importance value rank）= 34

灌木，高约 2 m。嫩枝无毛。叶聚生于枝梢，线状倒披针形，长 6～11（～12）cm，宽 1～2（～2.7）cm，先端尾状渐尖，基部窄楔形下延，革质，上面榄绿色，有光泽，下面灰白色或黄褐色。圆锥花序近顶生，长 2～4 cm，无毛或有微柔毛；总梗长 1～3 cm。子房无毛，果球形，直径 8 mm。花期 7～9 月，果期 10～12 月。

Shrubs, ca. 2 m tall. Branchlets glabrous. Leaf blade abaxially gray-white or yellowish brown, adaxially shiny, narrowly oblanceolate, 6–11 (–12) × 1–2 (–2.7) cm, leathery. Panicles subterminal, 2–4 cm, glabrous or puberulent; peduncle 1–3 cm. Ovary glabrous. Fruit globose, ca. 8 mm in diam.. Fl. Jul.–Sep., fr. Oct.–Dec..

枝叶　Branches and leaves
摄影：丁涛　Photo by: Ding Tao

叶背　Leaf backs
摄影：丁涛　Photo by: Ding Tao

果序　Infructescence
摄影：丁涛　Photo by: Ding Tao

径级分布表　DBH class

胸径区间 Diameter class (cm)	个体数 No. of individuals in the plot	比例 Proportion (%)
1～2	1	6.67
2～5	1	6.67
5～10	1	6.67
10～20	10	66.66
20～35	2	13.33
35～50	0	0.00
≥50	0	0.00

● 1～5 cm DBH　＋ 5～20 cm DBH　○ ≥20 cm DBH

个体分布图　Distribution of individuals

17 绒毛润楠 | róng máo rùn nán （绒楠）

润楠属

***Machilus velutina* Champ. ex Benth.**

樟科 Lauraceae

样地名称（Plot name）= JH
个体数（Individual number/1 hm²）= 1
最大胸径（Max DBH）= 13.3 cm
重要值排序（Importance value rank）= 119

乔木，高可达18 m。枝、芽、叶下面和花序均密被锈色绒毛。叶狭倒卵形、椭圆形或狭卵形，长5～11（～18）cm，宽2～5（～5.5）cm，先端渐狭或短渐尖，基部楔形，革质，中脉上面稍凹下，下面突起。花序在小枝顶端。果球形，紫红色。花期10～12月，果期翌年2～3月。

Trees, up to 18 m tall. All parts densely ferruginous tomentose. Leaf blade lustrous adaxially, narrowly obovate, elliptic, or narrowly ovate, 5–11 (–18) × 2–5 (–5.5) cm, leathery, midrib abaxially raised, slightly concave adaxially, base cuneate, apex attenuate or shortly acuminate. Inflorescences terminal. Fruit purplish red, globose. Fl. Oct.–Dec., fr. Feb.–Mar. of next year.

花枝　Flowering branches
摄影：陆昭岑　Photo by: Lu Zhaochen

叶背　Leaf backs
摄影：陆昭岑　Photo by: Lu Zhaochen

花序　Inflorescence
摄影：陆昭岑　Photo by: Lu Zhaochen

径级分布表　DBH class

胸径区间 Diameter class (cm)	个体数 No. of individuals in the plot	比例 Proportion (%)
1～2	0	0.00
2～5	0	0.00
5～10	0	0.00
10～20	1	100.00
20～35	0	0.00
35～50	0	0.00
≥50	0	0.00

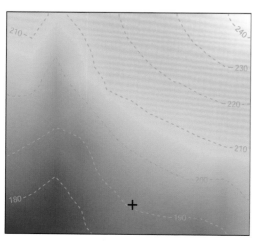

● 1～5 cm DBH　＋ 5～20 cm DBH　○ ≥20 cm DBH
个体分布图　Distribution of individuals

18 细枝龙血树 | xì zhī lóng xuè shù

龙血树属

***Dracaena elliptica* Thunb.**

天门冬科 Asparagaceae

样地名称（Plot name）＝ JH
个体数（Individual number/1 hm²）＝ 1
最大胸径（Max DBH）＝ 1.3 cm
重要值排序（Importance value rank）＝ 137

灌木状，高 1～5 m。茎常具许多分枝，具疏的环状叶痕。叶生于分枝上部或近顶端，叶柄长 1 cm。狭椭圆状披针形或条状披针形，长 10～15 cm，宽 2～3 cm 或更宽，中脉明显。圆锥花序生于分枝顶端，较短，长 7～10 cm，花通常单生；花梗长达 1 cm。

Plants shrubby, 1–5 m tall. Stems branched; internodes longer than wide. Leaves spaced along distal part of branches, distinctly petiolate; petiole ca. 1 cm. Leaf blade linear-lanceolate or narrowly elliptic-lanceolate, 10–15 × 2–3 cm, midvein distinct. Inflorescence terminal, branched, 7–10 cm; rachis glabrous. Flowers solitary; pedicel ca. 10 mm.

枝叶　Branch and leaves
摄影：丁涛　Photo by：Ding Tao

叶　Leaves
摄影：丁涛　Photo by：Ding Tao

叶背　Leaf back
摄影：丁涛　Photo by：Ding Tao

径级分布表　DBH class

胸径区间 Diameter class (cm)	个体数 No. of individuals in the plot	比例 Proportion (%)
1～2	1	100.00
2～5	0	0.00
5～10	0	0.00
10～20	0	0.00
20～35	0	0.00
35～50	0	0.00
≥50	0	0.00

● 1～5 cm DBH　＋ 5～20 cm DBH　○ ≥20 cm DBH
个体分布图　Distribution of individuals

19 狭叶泡花树 | xiá yè pāo huā shù

泡花树属

***Meliosma angustifolia* Merr.**

清风藤科 Sabiaceae

样地名称（Plot name）= JH
个体数（Individual number/1 hm²）= 26
最大胸径（Max DBH）= 17.0 cm
重要值排序（Importance value rank）= 33

常绿小乔木或大乔木，高可达 20 m。树皮暗灰褐色。叶为奇数羽状复叶，连柄长 20~30 cm，有小叶 13~23 片；小叶革质，狭椭圆形、椭圆状披针形或披针形，长 5~12 cm，宽 1.5~3 cm，先端钝渐尖，基部稍偏斜，狭楔形，或很少圆钝，全缘或少有 1~2 个小齿。圆锥花序顶生或腋生。核果倒卵形。花期 3~5 月，果期 6~10 月。

Evergreen large or small trees, up to 20 m tall. Bark gray. Leaves odd pinnate, 20–30 cm. Leaflets 13–23, narrowly elliptic, elliptic-lanceolate, or lanceolate, 5–12 × 1.5–3 cm, leathery, base somewhat narrowly cuneate or rarely rounded, oblique, margin entire or sometimes 1- or 2-toothed, apex obtuse-acuminate. Panicle terminal or axillary. Drupe obovoid. Fl. Mar.–May, fr. Jun.–Oct..

树干　Trunk
摄影：李健星　Photo by: Li Jianxing

复叶背部　Compound leaf backs
摄影：李健星　Photo by: Li Jianxing

嫩复叶　New compound leaves
摄影：丁涛　Photo by: Ding Tao

径级分布表　DBH class

胸径区间 Diameter class (cm)	个体数 No. of individuals in the plot	比例 Proportion (%)
1~2	6	23.08
2~5	4	15.38
5~10	8	30.77
10~20	8	30.77
20~35	0	0.00
35~50	0	0.00
≥50	0	0.00

● 1~5 cm DBH　+ 5~20 cm DBH　○ ≥20 cm DBH
个体分布图　Distribution of individuals

20 小果山龙眼 | xiǎo guǒ shān lóng yǎn （红叶树）

山龙眼属

***Helicia cochinchinensis* Lour.**

山龙眼科 Proteaceae

样地名称（Plot name）= JH
个体数（Individual number/1 hm²）= 5
最大胸径（Max DBH）= 15.1 cm
重要值排序（Importance value rank）= 85

乔木或灌木，高 3~20 m。枝和叶均无毛。叶薄革质或纸质，长圆形、倒卵状椭圆形、长椭圆形或披针形，长 5~12 （~15）cm，宽 2.5~5.5 cm，顶端短渐尖、尖头或钝，基部楔形，稍下延；总状花序，腋生；果椭圆状，果皮革质，蓝黑色或黑色。花期 6~10 月，果期 11 月至翌年 3 月。

Shrubs or trees, 3–20 m tall. Branchlets and leaves glabrous. Leaf blade elliptic, obovate-oblong, narrowly oblong, or lanceolate, 5–12 (–15) × 2.5–5.5 cm, papery to ± leathery, base cuneate and somewhat decurrent into petiole, apex shortly acuminate, ± acute, or obtuse; Inflorescences axillary. Fruit bluish black to black, ellipsoid, pericarp, leathery. Fl. Jun.–Oct., fr. Nov.–Mar. of next year.

树干　　　Trunk
摄影：李健星　　Photo by：Li Jianxing

枝叶　　　Branch and leaves
摄影：丁涛　　Photo by：Ding Tao

果　　　Fruits
摄影：丁涛　　Photo by：Ding Tao

径级分布表 DBH class

胸径区间 Diameter class (cm)	个体数 No. of individuals in the plot	比例 Proportion (%)
1~2	3	60.00
2~5	0	0.00
5~10	0	0.00
10~20	2	40.00
20~35	0	0.00
35~50	0	0.00
≥50	0	0.00

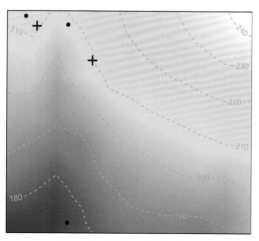

● 1~5 cm DBH　　+ 5~20 cm DBH　　○ ≥20 cm DBH
个体分布图　Distribution of individuals

21 假山龙眼 | jiǎ shān lóng yǎn

Heliciopsis henryi (Diels) W. T. Wang
山龙眼科 Proteaceae

样地名称（Plot name）= JH
个体数（Individual number/1 hm²）= 23
最大胸径（Max DBH）= 9.3 cm
重要值排序（Importance value rank）= 52

乔木，高 5~15 m。叶薄革质，倒披针形或长圆形，长 15~24 cm，宽 4~8 cm，顶端短钝尖或圆钝，有时微凹，基部楔形。花序分枝。果椭圆，褐色，长 3.5~4.5 cm，直径 2.5~3 cm；外果皮厚约 1 mm，中果皮厚约 2 mm，干后残留海绵状纤维，内果皮厚 1~2 mm。花期 4~5 月，果期 10 月至翌年 5 月。

Trees, 5–15 m tall. Leaf blade oblanceolate to oblong, 15–24 × 4–8 cm, ± leathery, base cuneate, margin entire, apex obtusely acute, obtuse, or sometimes retuse. Inflorescences ramiflorous. Fruit brownish, ellipsoid, 3.5–4.5× 2.5–3 cm; exocarp ca. 1 mm thick; mesocarp ca. 2 mm thick, spongy when dry; endocarp 1–2 mm thick. Fl. Apr.–May, fr. Oct.–May of next year.

树干 Trunk
摄影：李健星 Photo by: Li Jianxing

叶 Leaves
摄影：李健星 Photo by: Li Jianxing

枝叶 Branch and leaves
摄影：李健星 Photo by: Li Jianxing

径级分布表 DBH class

胸径区间 Diameter class (cm)	个体数 No. of individuals in the plot	比例 Proportion (%)
1~2	6	26.09
2~5	13	56.52
5~10	4	17.39
10~20	0	0.00
20~35	0	0.00
35~50	0	0.00
≥50	0	0.00

● 1~5 cm DBH   + 5~20 cm DBH   ○ ≥20 cm DBH
个体分布图 Distribution of individuals

22 枫香树 | fēng xiāng shù （山枫香树）

***Liquidambar formosana* Hance**

蕈树科 Altingiaceae

样地名称（Plot name）= JH
个体数（Individual number/1 hm²）= 9
最大胸径（Max DBH）= 62.0 cm
重要值排序（Importance value rank）= 16

乔木，高达 30 m，胸径最大可达 1 m。树皮灰褐色。叶薄革质，阔卵形，掌状 3 裂，基部心形；叶脉两面均显著。雄性短穗状花序，常多个排成总状；雌性头状花序，有花 24~43 朵。头状果序圆球形，直径 3~4 cm。花期 3~6 月，果期 7~9 月。

Trees, to 30 m tall, trunk sometimes 1 m in diam.. Bark gray-brown. Leaf blade broadly ovate, base rounded, apex caudate-acuminate, cordate, subcordate or truncate, veins prominent on both surfaces. Male inflorescence a short spike, several arranged in a raceme; female inflorescence 24–43-flowered. Infructescence globose, 3–4 cm wide. Fl. Mar.–Jun., fr. Jul.–Sep..

树干　　Trunk
摄影：李健星　　Photo by: Li Jianxing

叶　　Leaves
摄影：王斌　　Photo by: Wang Bin

枝叶　　Branches and leaves
摄影：李健星　　Photo by: Li Jianxing

径级分布表　DBH class

胸径区间 Diameter class (cm)	个体数 No. of individuals in the plot	比例 Proportion (%)
1~2	0	0.00
2~5	0	0.00
5~10	1	11.11
10~20	3	33.33
20~35	1	11.11
35~50	1	11.11
≥50	3	33.34

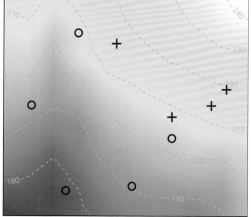

● 1~5 cm DBH　+ 5~20 cm DBH　○ ≥20 cm DBH
个体分布图　Distribution of individuals

23 鼠刺 | shǔ cì （牛皮桐）

***Itea chinensis* Hook. Arn.**

鼠刺科 Iteaceae

样地名称（Plot name）= JH
个体数（Individual number/1 hm²）= 13
最大胸径（Max DBH）= 9.9 cm
重要值排序（Importance value rank）= 65

灌木或小乔木，高 4~10 m。幼枝黄绿色，无毛；老枝棕褐色，具纵棱条。叶薄革质，倒卵形或卵状椭圆形，长 5~12（~15）cm，宽 3~6 cm，先端锐尖，基部楔形。腋生总状花序，通常短于叶。蒴果长圆状披针形，具纵条纹。花期 3~5 月，果期 5~12 月。

Shrubs or small trees, 4–10 m tall. Young branchlets yellow-green, glabrous; old branchlets brown, striate. Leaf blade, obovate or ovate-elliptic, 5–12 (–15) × 3–6 cm, thinly leathery, base cuneate, apex acute. Racemes axillary, usually shorter than leaves. Capsule striate, oblong-lanceolate. Fl. Mar.–May, fr. May–Dec..

花枝　Flowering branch
摄影：丁涛　Photo by：Ding Tao

叶背　Leaf backs
摄影：丁涛　Photo by：Li Jianxing

花序　Inflorescence
摄影：丁涛　Photo by：Ding Tao

径级分布表　DBH class

胸径区间 Diameter class (cm)	个体数 No. of individuals in the plot	比例 Proportion (%)
1~2	5	38.46
2~5	4	30.77
5~10	4	30.77
10~20	0	0.00
20~35	0	0.00
35~50	0	0.00
≥50	0	0.00

● 1~5 cm DBH　+ 5~20 cm DBH　○ ≥20 cm DBH
个体分布图　Distribution of individuals

24 碟腺棋子豆 | dié xiàn qí zǐ dòu （白花合欢）

猴耳环属

***Archidendron kerrii* (Gagnep.) I. C. Nielsen**

豆科 Fabaceae

样地名称（Plot name）= JH
个体数（Individual number/1 hm²）= 33
最大胸径（Max DBH）= 9.8 cm
重要值排序（Importance value rank）= 41

灌木或小乔木，高 3~8 m。二回羽状复叶；总叶柄长 2~5 cm，顶部羽片着生处有 1 个圆形腺体；羽片 1 对；小叶 1~3 对，对生，纸质，椭圆形，长 6~14 cm，宽 3~6 cm，先端短尖或渐尖，基部渐狭或急尖，两面无毛。头状花序排成疏散的圆锥花序，腋生或顶生。荚果圆柱形。花期 5 月，果期 8 月。

Shrubs or small trees, to 3–8 m tall. Leaf petiole 2–5 cm; petiolar gland plate-form, at inser-tion of pinna and first leaflet pair; pinnae 1 pair; leaflets 1–3 pairs, opposite, elliptic, 6–14 × 3–6 cm, papery, both surfaces glabrous, base cuneate or acute, apex acu-minate or acute. Heads arranged in axillary or terminal loose panicles. Legume cylindric. Fl. May, fr. Aug..

树干　　Trunk
摄影：李健星　Photo by: Li Jianxing

复叶　　Compound leaf
摄影：李健星　Photo by: Li Jianxing

荚果　　Legume
摄影：李健星　Photo by: Li Jianxing

径级分布表　DBH class

胸径区间 Diameter class (cm)	个体数 No. of individuals in the plot	比例 Proportion (%)
1~2	8	24.24
2~5	11	33.33
5~10	14	42.43
10~20	0	0.00
20~35	0	0.00
35~50	0	0.00
≥50	0	0.00

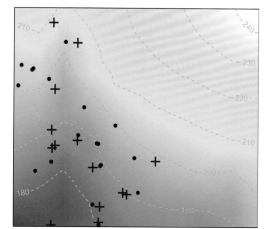

● 1~5 cm DBH　　+ 5~20 cm DBH　　○ ≥20 cm DBH
个体分布图　Distribution of individuals

25 亮叶猴耳环 | liàng yè hóu ěr huán

Archidendron lucidum (Benth.) I. C. Nielsen
豆科 Fabaceae

样地名称（Plot name）= JH
个体数（Individual number/1 hm^2）= 15
最大胸径（Max DBH）= 17.2 cm
重要值排序（Importance value rank）= 51

乔木，高 2~10 m。羽片 1~2 对；总叶柄近基部、每对羽片下和小叶片下的叶轴上均有圆形而凹陷的腺体；小叶斜卵形或长圆形，长 5~9（~11）cm，宽 2~4.5 cm，顶生的一对最大，对生，余互生且较小，先端渐尖而具钝小尖头，基部略偏斜。头状花序球形，排成圆锥花序。荚果旋卷成环状，边缘在种子间缢缩。花期 4~6 月，果期 7~12 月。

Trees, 2–10 m tall. Pinnae 1 or 2 pairs. Leaf rachis and base of petiole with round, sunken glands, obliquely ovate or oblong, 5–9 (–11) × 2–4.5 cm, apical ones larger, opposite, proximal ones alternate and smaller, base oblique, apex acuminate, mucronate. Heads globose, arranged in panicles. Legume twisted into a circle, margin between seeds constricted. Fl. Apr.–Jun., fr. Jul.–Dec..

枝叶　Branch and leaves
摄影：杨平　Photo by: Yang Ping

复叶　Compound leaf
摄影：丁涛　Photo by: Ding Tao

荚果　Legume
摄影：杨平　Photo by: Yang Ping

径级分布表　DBH class

胸径区间 Diameter class (cm)	个体数 No. of individuals in the plot	比例 Proportion (%)
1~2	6	40.00
2~5	5	33.33
5~10	3	20.00
10~20	1	6.67
20~35	0	0.00
35~50	0	0.00
≥50	0	0.00

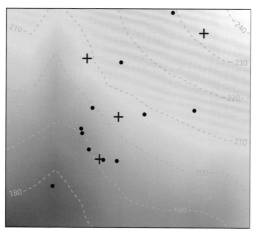

● 1~5 cm DBH　＋ 5~20 cm DBH　○ ≥20 cm DBH
个体分布图　Distribution of individuals

26 香花枇杷 | xiāng huā pí pá （山枇杷）

***Eriobotrya fragrans* Champ. ex Benth.**

蔷薇科 Rosaceae

样地名称（Plot name）= JH
个体数（Individual number/1 hm^2）= 3
最大胸径（Max DBH）= 11.4 cm
重要值排序（Importance value rank）= 96

小乔木或灌木，高可达10 m。叶片革质，长圆椭圆形，长7～15 cm，宽2.5～5 cm，先端急尖或短渐尖，基部楔形或渐狭，边缘在中部以上具不明显疏锯齿，中部以下全缘，中脉在两面皆隆起，侧脉9～11对。圆锥花序顶生，长7～9 cm，总花梗和花梗均密生棕色绒毛。果实球形，表面具颗粒状突起，并有绒毛。花期4～5月，果期8～9月。

Small trees or shrubs, up to 10 m tall. Leaf blade oblong-elliptic, 7–15 × 2.5–5 cm, leathery, midvein prominent on both surfaces, lateral veins 9–11 pairs, base cuneate or attenuate, margin entire basally, remotely inconspicuously serrate apically, apex acute or shortly acuminate. Panicle 7–9 cm in diam.. Pome brown, globose, tomentose and granular-punctate. Fl. Apr.–May, fr. Aug.–Sep..

枝叶　　Branch and leaves
摄影：唐忠炳　　Photo by：Tang Zhongbing

叶背　　Leaf back
摄影：李健星　　Photo by：Li Jianxing

花序　　Inflorescence
摄影：唐忠炳　　Photo by：Tang Zhongbing

径级分布表　DBH class

胸径区间 Diameter class (cm)	个体数 No. of individuals in the plot	比例 Proportion (%)
1～2	2	66.67
2～5	0	0.00
5～10	0	0.00
10～20	1	33.33
20～35	0	0.00
35～50	0	0.00
≥50	0	0.00

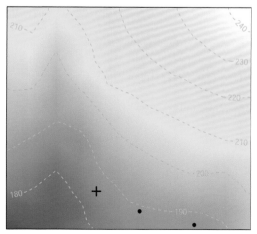

● 1～5 cm DBH　　+ 5～20 cm DBH　　○ ≥20 cm DBH
个体分布图　Distribution of individuals

27 臀果木 | tún guǒ mù （臀形果）

Pygeum topengii Merr.

蔷薇科 Rosaceae

样地名称（Plot name）= JH
个体数（Individual number/1 hm²）= 143
最大胸径（Max DBH）= 36.6 cm
重要值排序（Importance value rank）= 10

乔木，高可达 25 m。树皮深灰色至灰褐色。叶片革质，卵状椭圆形或椭圆形，先端短渐尖而钝，基部宽楔形，两边略不相等，全缘，近基部有 2 枚黑色腺体。总状花序单生或 2 至数个簇生。果实肾形，顶端凹陷，无毛，深褐色。花期 6~9 月，果期冬季。

Trees, to 25 m tall. Bark dark gray to grayish brown. Leaf blade ovate-elliptic to elliptic, leathery with 2 black nectaries near base, base broadly cuneate and asymmetric, margin entire, apex shortly acuminate and with an apical obtuse tip. Racemes solitary or to several in a fascicle. Drupe dark brown, reniform, glabrous, apically depressed. Fl. Jun.–Sep., fr. winter.

树干　Trunk
摄影：李健星　Photo by: Li Jianxing

叶背　Leaf backs
摄影：李健星　Photo by: Li Jianxing

枝叶　Branch and leaves
摄影：李健星　Photo by: Li Jianxing

径级分布表　DBH class

胸径区间 Diameter class (cm)	个体数 No. of individuals in the plot	比例 Proportion (%)
1~2	77	53.85
2~5	37	25.87
5~10	9	6.29
10~20	11	7.69
20~35	8	5.60
35~50	1	0.70
≥50	0	0.00

● 1~5 cm DBH　+ 5~20 cm DBH　○ ≥20 cm DBH
个体分布图　Distribution of individuals

28 石斑木 | shí bān mù （春花木）

Rhaphiolepis indica (L.) Lindl.
蔷薇科 Rosaceae

样地名称（Plot name）= JH
个体数（Individual number/1 hm²）= 6
最大胸径（Max DBH）= 15.5 cm
重要值排序（Importance value rank）= 69

灌木，稀小乔木，高可达 4 m。叶片，卵形、长圆形，稀倒卵形或长圆披针形，先端圆钝，急尖、渐尖或长尾尖，基部渐狭连于叶柄，边缘具细钝锯齿。顶生圆锥花序或总状花序。果实球形，紫黑色。花期 4 月，果期 7~8 月。

Shrubs, rarely small trees, up to 4 m tall. Leaf blade ovate, oblong, rarely obovate, oblong-lanceolate, narrowly elliptic or lanceolate-elliptic, leather. base attenuate, margin crenulate, serrate, or obtusely serrate, apex obtuse, acute, acuminate, or long caudate. Panicle or racemes terminal. Pome purplish black, globose. Fl. Apr., fr. Jul.–Aug..

石斑木属

树干　　Trunk
摄影：李健星　　Photo by: Li Jianxing

叶　　Leaves
摄影：李健星　　Photo by: Li Jianxing

花　　Flower
摄影：杨平　　Photo by: Yang Ping

径级分布表　DBH class

胸径区间 Diameter class (cm)	个体数 No. of individuals in the plot	比例 Proportion (%)
1~2	0	0.00
2~5	1	16.67
5~10	1	16.67
10~20	4	66.66
20~35	0	0.00
35~50	0	0.00
≥50	0	0.00

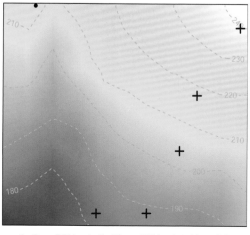

● 1~5 cm DBH　　+ 5~20 cm DBH　　○ ≥20 cm DBH
个体分布图　Distribution of individuals

29 假玉桂 | jiǎ yù guì （米吃）　　朴属

Celtis timorensis Span.

大麻科 Cannabaceae

样地名称（Plot name）= JH
个体数（Individual number/1 hm²）= 32
最大胸径（Max DBH）= 22.1 cm
重要值排序（Importance value rank）= 36

常绿乔木，高达 20 m。树皮灰白至灰褐色，木材有恶臭。叶革质，卵状椭圆形或卵状长圆形，长 5～15 cm，宽 2.5～7.5 cm，基部宽楔形至近圆形，稍不对称，基部一对侧脉延伸达 3/4 以上，但不达先端，先端渐尖至尾尖。小聚伞花序。果宽卵状，先端残留花柱基部而成一短喙状，成熟时黄色、橙红色至红色。花期 4～5 月，果期 10～11 月。

Evergreen trees, to 20 m tall. Bark grayish white, or grayish brown. Leaf blade ovate-elliptic to ovate-oblong, 5–15 × 2.5–7.5 cm, ± coriaceous, base broadly cuneate to ± rounded and distinctly asymmetric, apex acuminate to caudate-acuminate. Inflorescence a branched cyme. Drupe yellow, becoming red to orange-red when mature, broadly ovoid, base rounded, apex conic-acute. Fl. Apr.–May, fr. Oct.–Nov..

树干　　Trunk
摄影：李健星　　Photo by: Li Jianxing

叶背　　Leaf backs
摄影：李健星　　Photo by: Li Jianxing

枝叶　　Branch and leaves
摄影：李健星　　Photo by: Li Jianxing

径级分布表　DBH class

胸径区间 Diameter class (cm)	个体数 No. of individuals in the plot	比例 Proportion (%)
1～2	6	18.75
2～5	15	46.88
5～10	5	15.63
10～20	5	15.62
20～35	1	3.12
35～50	0	0.00
≥50	0	0.00

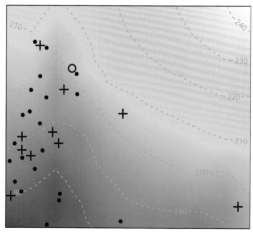

• 1～5 cm DBH　　+ 5～20 cm DBH　　○ ≥20 cm DBH
个体分布图 Distribution of individuals

30 白颜树 | bái yán shù （大叶颜树）

白颜树属

***Gironniera subaequalis* Planch.**

大麻科 Cannabaceae

样地名称（Plot name）= JH
个体数（Individual number/1 hm²）= 7
最大胸径（Max DBH）= 22.1 cm
重要值排序（Importance value rank）= 73

落叶乔木，高 10~20 m，稀达 30 m，胸径 25~50 cm，稀达 100 cm。树皮灰或深灰色，较平滑；小枝黄绿色，疏生黄褐色长粗毛。叶革质，长 10~25 cm，宽 4.5~10 cm，先端短尾状渐尖，托叶成对，脱落后在枝上留有一环托叶痕。雌雄异株，聚伞花序成对腋生，核果具宿存的花柱及花被。花期 2~4 月，果期 7~11 月。

Trees, 10–20 (–30) m tall, d.b.h.. 25–50 (–100) cm, dioecious. Bark gray to dark gray, smooth. Branchlets yellowish green or brown, covered with hirsute hairs. Stipules opposite, a ring of stipe mark on the branch after falling off. Leaf blade 10–25 × 4.5–10 cm, leathery, apex shortly caudate-acuminate. Drupes perianth and style persistent. Fl. Feb.–Apr., fr. Jul.-Nov..

果枝　　Fruiting branches
摄影：丁涛　　Photo by: Ding Tao

叶背　　Leaf backs
摄影：丁涛　　Photo by: Ding Tao

果　　Fruit
摄影：丁涛　　Photo by: Ding Tao

径级分布表　DBH class

胸径区间 Diameter class (cm)	个体数 No. of individuals in the plot	比例 Proportion (%)
1~2	1	14.29
2~5	3	42.86
5~10	2	28.57
10~20	0	0.00
20~35	1	14.28
35~50	0	0.00
≥50	0	0.00

● 1~5 cm DBH　＋ 5~20 cm DBH　○ ≥20 cm DBH
个体分布图　Distribution of individuals

31 大果榕 | dà guǒ róng

Ficus auriculata Lour.
桑科 Moraceae

样地名称（Plot name）＝ JH
个体数（Individual number/1 hm²）＝ 6
最大胸径（Max DBH）＝ 7.4 cm
重要值排序（Importance value rank）＝ 94

乔木，高 4～10 m。树皮灰褐色，粗糙。叶互生，厚纸质，广卵状心形，长 15～55 cm，宽（10～）15～27 cm，先端钝，具短尖，基部心形，边缘具整齐细锯齿，表面无毛，背面多被开展短柔毛。榕果簇生于树干基部或老茎短枝上，具明显的纵棱 8～12 条。花期 8 月至翌年 3 月，果期 5～8 月。

Trees, 4–10 m tall. Bark grayish brown, rough. Leaves alternate. Leaf blade broadly ovate-cordate, 15–55 × (10–) 15–27 cm, thickly papery, abaxially with short spreading pubescence, base cordate to occasionally rounded, margin regularly shallowly dentate, apex obtuse and mucronate. Figs on specialized leafless branchlets at base of trunk and main branches, with 8–12 conspicuous longitudinal ridges. Fl. Aug.–Mar., fr. May–Aug..

果　　　　　　　　　Fruits
摄影：杨平　　　　　Photo by：Yang Ping

枝叶　　　　　　　　Branch and leaves
摄影：杨平　　　　　Photo by：Yang Ping

果　　　　　　　　　Fruit
摄影：丁涛　　　　　Photo by：Ding Tao

径级分布表　DBH class

胸径区间 Diameter class (cm)	个体数 No. of individuals in the plot	比例 Proportion (%)
1～2	2	33.33
2～5	1	16.67
5～10	3	50.00
10～20	0	0.00
20～35	0	0.00
35～50	0	0.00
≥50	0	0.00

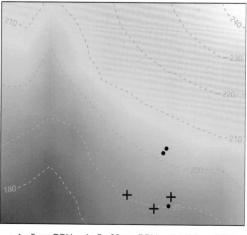

● 1～5 cm DBH　　＋ 5～20 cm DBH　　○ ≥20 cm DBH
个体分布图　Distribution of individuals

32 尖叶榕 | jiān yè róng

Ficus henryi Warb.
桑科 Moraceae

样地名称（Plot name）= JH
个体数（Individual number/1 hm²）= 6
最大胸径（Max DBH）= 19.4 cm
重要值排序（Importance value rank）= 75

乔木，高 3~10 m。幼枝黄褐色，无毛，具薄翅。叶倒卵状长圆形至长圆状披针形，长 7~16 cm，宽 2.5~5 cm，先端渐尖或尾尖，基部楔形，两面均被点状钟乳体。榕果单生叶腋，球形至椭圆形。榕果成熟呈橙红色。花期 5~6 月，果期 7~9 月。

Trees, 3–10 m tall. Branchlets yellowish brown, narrowly winged, glabrous. Leaf blade obovate-oblong to narrowly lanceolate, 7–16 × 2.5–5 cm, both surfaces with cystoliths, base cuneate, apex acuminate to caudate. Figs axillary, solitary, erect, reddish orange when mature, globose to ellipsoid, smooth. Fl. May–Jun., fr. Jul.–Sep..

树干　Trunk
摄影：李健星　Photo by: Li Jianxing

枝叶　Branch and leaves
摄影：李健星　Photo by: Li Jianxing

果　Fruits
摄影：朱鑫鑫　Photo by: Zhu Xinxin

径级分布表　DBH class

胸径区间 Diameter class (cm)	个体数 No. of individuals in the plot	比例 Proportion (%)
1~2	0	0.00
2~5	1	16.67
5~10	2	33.33
10~20	3	50.00
20~35	0	0.00
35~50	0	0.00
≥50	0	0.00

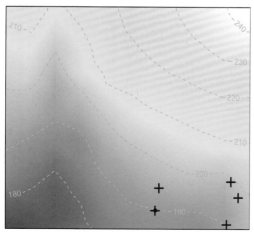

● 1~5 cm DBH　+ 5~20 cm DBH　○ ≥20 cm DBH
个体分布图　Distribution of individuals

33 粗叶榕 | cū yè róng

Ficus hirta Vahl

桑科 Moraceae

样地名称（Plot name）= JH
个体数（Individual number/1 hm^2）= 5
最大胸径（Max DBH）= 13.4 cm
重要值排序（Importance value rank）= 100

灌木或小乔木。嫩枝中空，小枝、叶和榕果均被金黄色开展的长硬毛。叶互生，纸质，多型，长椭圆状披针形或广卵形，长 10~25 cm，有时全缘或 3~5 深裂，先端急尖或渐尖，基部圆形，浅心形或宽楔形，表面疏生贴伏粗硬毛，背面密或疏生开展的白色或黄褐色绵毛和糙毛。榕果成对腋生或生于已落叶枝上，表面光滑。花果期 4~6 月。

Shrubs or small trees. Branchlets leafless in middle, golden yellow or brown hirsute. Leaves alternate. Leaf blade simple, 10–25 cm, glabrous or golden yellow hirsute, base cuneate, rounded, or shallowly cordate, margin entire or with small serrations, apex acute to acuminate. Figs axillary on normal Leafy shoots, paired, globose to ± globose, smooth. Fl. and fr. Apr.–Jun..

枝叶　　　Branch and leaves
摄影：杨平　　Photo by：Yang Ping

叶背　　　Leaf back
摄影：杨平　　Photo by：Yang Ping

果　　　Fruits
摄影：杨平　　Photo by：Yang Ping

径级分布表 DBH class

胸径区间 Diameter class (cm)	个体数 No. of individuals in the plot	比例 Proportion (%)
1~2	3	60.00
2~5	1	20.00
5~10	0	0.00
10~20	1	20.00
20~35	0	0.00
35~50	0	0.00
≥50	0	0.00

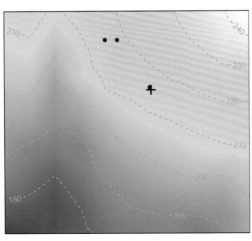

● 1~5 cm DBH　　+ 5~20 cm DBH　　○ ≥20 cm DBH
个体分布图　Distribution of individuals

34 青藤公 | qīng téng gōng （山榕）

榕属

***Ficus langkokensis* Drake**

桑科 Moraceae

样地名称（Plot name）＝ JH
个体数（Individual number/1 hm²）＝ 12
最大胸径（Max DBH）＝ 18.0 cm
重要值排序（Importance value rank）＝ 54

乔木，高 6~15 m。树皮红褐色或灰黄色，小枝细，黄褐色，被锈色糠屑状毛。叶互生，纸质，椭圆状披针形至椭圆形，长 7~19 cm，宽 2~6 cm，顶端尾状渐尖，基部阔楔形，全缘。榕果成对或单生于叶腋，球形，被锈色糠屑状毛，顶端具脐状凸起。花果期全年。

Trees, 6–15 m tall. Bark reddish brown to grayish yellow. Branchlets yellowish brown, slender, with reddish yellow scurfy hairs. Leaves alternate. Leaf blade elliptic-lanceolate to elliptic, 7–19 × 2–6 cm, papery, base broadly cuneate, margin entire, apex caudate to acuminate. Figs axillary on leafy shoots, paired or solitary, globose, with rust-colored scurfy hairs, apical pore navel-like, convex. Fl. and fr. throughout year.

树干　Trunk
摄影：李健星　Photo by: Li Jianxing

叶背　Leaf backs
摄影：李健星　Photo by: Li Jianxing

枝叶　Branch and leaves
摄影：李健星　Photo by: Li Jianxing

径级分布表　DBH class

胸径区间 Diameter class (cm)	个体数 No. of individuals in the plot	比例 Proportion (%)
1~2	2	16.67
2~5	6	50.00
5~10	1	8.33
10~20	3	25.00
20~35	0	0.00
35~50	0	0.00
≥50	0	0.00

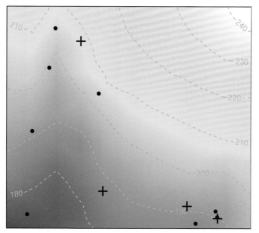

● 1~5 cm DBH　＋ 5~20 cm DBH　○ >20 cm DBH
个体分布图　Distribution of individuals

35 杂色榕 | zá sè róng

Ficus variegata Blume
桑科 Moraceae

样地名称（Plot name）= JH
个体数（Individual number/1 hm²）= 107
最大胸径（Max DBH）= 11.1 cm
重要值排序（Importance value rank）= 20

乔木，高 7~15 m，树皮灰色，光滑。叶互生，叶柄长 2.5~6.8 cm。叶片厚纸质，广卵形至卵状椭圆形，长 10~17 cm，基部圆形至浅心形，全缘，顶端渐尖或钝；基生叶脉 4 条。榕果簇生于老茎发出的瘤状短枝。花果期春季至秋季。

Trees, 7–15 m tall. Bark gray to grayish brown, smooth. Leaves alternate; petiole 2.5–6.8 cm. Leaf blade broadly ovate to ovate-elliptic, 10–17 cm, thickly papery, base rounded to shallowly cordate, margin entire, apex acute, acuminate, or obtuse. Basal lateral veins 4. Figs clustered on shortly tuberculate branchlets from old stem. Fl. and fr. spring–winter.

树干　　Trunk
摄影：李健星　　Photo by: Li Jianxing

叶　　Leaves
摄影：李健星　　Photo by: Li Jianxing

果序　　Infructescence
摄影：李健星　　Photo by: Li Jianxing

径级分布表　DBH class

胸径区间 Diameter class (cm)	个体数 No. of individuals in the plot	比例 Proportion (%)
1~2	34	31.78
2~5	44	41.12
5~10	26	24.30
10~20	3	2.80
20~35	0	0.00
35~50	0	0.00
≥50	0	0.00

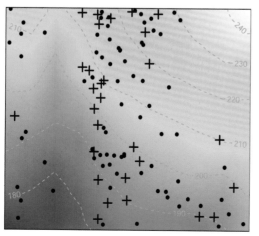

● 1~5 cm DBH　+ 5~20 cm DBH　○ ≥20 cm DBH
个体分布图　Distribution of individuals

36 变叶榕 | biàn yè róng （细叶牛乳树） 榕属

Ficus variolosa **Lindl. ex Benth.**

桑科 Moraceae

样地名称（Plot name）= JH
个体数（Individual number/1 hm²）= 35
最大胸径（Max DBH）= 18.5 cm
重要值排序（Importance value rank）= 31

灌木或乔木，高 3～10 m，树皮灰褐色，光滑；小枝节间短。叶薄革质，狭椭圆形至椭圆状披针形，长 5～12 cm，宽 1.5～4 cm，先端钝或钝尖，基部楔形，全缘。榕果成对或单生叶腋，球形，表面有瘤体。花期 12 月至翌年 6 月，果期全年。

Shrubs or trees, 3–10 m tall. Bark grayish brown, smooth. Branchlet internodes short. Leaf blade narrowly elliptic to elliptic-lanceolate, 5–12 × 1.5–4 cm, ± leathery, base cuneate, margin entire, apex obtuse to blunt. Figs axillary on normal Leafy shoots, paired or solitary, globose, tuberculate. Fl. Dec.–Jun. of next year, fr. throughout year.

果枝 Fruiting branch
摄影：丁涛 Photo by: Ding Tao

叶 Leaves
摄影：丁涛 Photo by: Ding Tao

果 Fruits
摄影：丁涛 Photo by: Ding Tao

径级分布表 DBH class

胸径区间 Diameter class (cm)	个体数 No. of individuals in the plot	比例 Proportion (%)
1～2	8	22.86
2～5	13	37.14
5～10	10	28.57
10～20	4	11.43
20～35	0	0.00
35～50	0	0.00
≥50	0	0.00

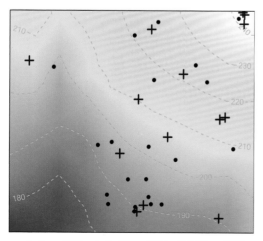

● 1～5 cm DBH　+ 5～20 cm DBH　O ≥20 cm DBH
个体分布图 Distribution of individuals

37 白肉榕 | bái ròu róng （突脉榕）

***Ficus vasculosa* Wall. ex Miq.**

桑科 Moraceae

样地名称（Plot name）= JH
个体数（Individual number/1 hm²）= 41
最大胸径（Max DBH）= 37.6 cm
重要值排序（Importance value rank）= 25

乔木，高 10~15 m，胸径 10~15 cm。树皮灰色，平滑。叶革质，长 4~11 cm，宽 2~4 cm，先端钝或渐尖，基部楔形，全缘或为不规则分裂；侧脉 10~12 对，两面突起，网脉在表面甚明显。榕果球形，基部缢缩为短柄，榕果成熟时黄色或黄红色。花果期 5~7 月。

Trees, 10–15 m tall, d.b.h. 10–15 cm. Bark gray, smooth. Leaf blade simple or irregularly lobed, 4–11 × 2–4 cm, leathery, base cuneate, apex obtuse to acuminate; secondary veins 10–12 on each side of midvein, reticulate veins prominent on both surfaces. Figs yellow to yellowish red when mature, globose, base attenuate into a short stalk. Fl. and fr. May–Jul..

树干　　Trunk
摄影：李健星　　Photo by: Li Jianxing

枝叶　　Branch and leaves
摄影：李健星　　Photo by: Li Jianxing

果序　　Infructescence
摄影：李健星　　Photo by: Li Jianxing

径级分布表　DBH class

胸径区间 Diameter class (cm)	个体数 No. of individuals in the plot	比例 Proportion (%)
1~2	10	24.39
2~5	16	39.02
5~10	7	17.07
10~20	5	12.20
20~35	2	4.88
35~50	1	2.44
≥50	0	0.00

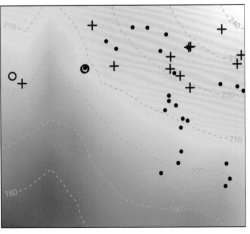

● 1~5 cm DBH　　+ 5~20 cm DBH　　○ ≥20 cm DBH
个体分布图　Distribution of individuals

38 黄葛树 | huáng gé shù （绿黄葛树） 榕属

***Ficus virens* Aiton**

桑科 Moraceae

样地名称（Plot name）= JH
个体数（Individual number/1 hm^2）= 1
最大胸径（Max DBH）= 6.8 cm
重要值排序（Importance value rank）= 127

乔木，有板根或支柱根，幼时附生。叶薄革质或皮纸质，卵状披针形至椭圆状卵形，长 10~15 cm，宽 4~7 cm，先端短渐尖，基部钝圆或楔形至浅心形，基生叶脉短，侧脉（5~）7~10（~11）对，背面突起，网脉稍明显。榕果球形，成熟时紫红色。花果期 4~7 月。

Trees, epiphytic when young, with buttress or prop roots. Leaf blade obovate, narrowly lanceolate, ovate-lanceolate, or elliptic-ovate, 10–15 × 4–7 cm, thinly leathery to thickly papery, base bluntly rounded, cuneate, or cordate, margin entire, apex acuminate to shortly acuminate; basal lateral veins short, secondary veins (5–) 7–10 (–11) on each side of midvein. Figs axillary on leafy branchlets, paired or solitary or in clusters on leafless older branchlets, purple red when mature. Fl. and fr. Apr.–Jul..

枝叶　Branch and leaves
摄影：蒋裕良　Photo by: Jiang Yuliang

叶背　Leaf backs
摄影：陆昭岑　Photo by: Lu Zhaochen

果　Fruits
摄影：陆昭岑　Photo by: Lu Zhaochen

径级分布表　DBH class

胸径区间 Diameter class (cm)	个体数 No. of individuals in the plot	比例 Proportion (%)
1~2	0	0.00
2~5	0	0.00
5~10	1	100.00
10~20	0	0.00
20~35	0	0.00
35~50	0	0.00
≥50	0	0.00

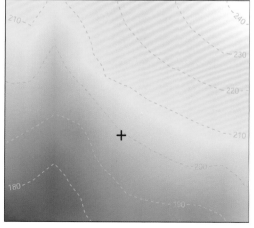

● 1~5 cm DBH　+ 5~20 cm DBH　○ ≥20 cm DBH
个体分布图　Distribution of individuals

39 上思青冈 | shàng sī qīng gāng

Quercus delicatula Chun Tsiang

壳斗科 Fagaceae

样地名称（Plot name）= JH
个体数（Individual number/1 hm²）= 2
最大胸径（Max DBH）= 35.7 cm
重要值排序（Importance value rank）= 82

乔木，高达 13 m。叶片纸质，卵状椭圆形、长椭圆形，有时倒卵状椭圆形，长 6~9 cm，宽 2~3.5 cm，顶端尾状尖，基部楔形，全缘或顶部有数对浅钝锯齿。果序长约 1 cm，着生 1~2 果。壳斗杯形，包着坚果 1/3；小苞片合生成 7~8 条同心环带。坚果椭圆形，两端均圆形，果脐平。花期 4~5 月，果期 10~11 月。

Trees, to 13 m tall. Leaf blade ovate, oblong-elliptic, or sometimes obovate-elliptic, 6–9 × 2–3.5 cm, papery, base cuneate, margin entire or shallowly crenate toward apex, apex caudate. Infructescence ca. 1 cm, with 1 or 2 fruits. Cupule cupular, enclosing ca. 1/3 of nut; bracts in 7 or 8 rings. Nut ellipsoid, base and apex rounded, flat; stylopodium persistent, umbonate. Fl. Apr.–May, fr. Oct.–Nov..

叶　Leaves
摄影：丁涛　Photo by: Ding Tao

叶背　Leaf backs
摄影：丁涛　Photo by: Ding Tao

果　Fruits
摄影：刘冰　Photo by: Liu Bing

径级分布表　DBH class

胸径区间 Diameter class (cm)	个体数 No. of individuals in the plot	比例 Proportion (%)
1~2	1	50.00
2~5	0	0.00
5~10	0	0.00
10~20	0	0.00
20~35	0	0.00
35~50	1	50.00
≥50	0	0.00

● 1~5 cm DBH　＋ 5~20 cm DBH　○ ≥20 cm DBH
个体分布图　Distribution of individuals

40 疏花卫矛 | shū huā wèi máo

卫矛属

***Euonymus laxiflorus* Champ. Benth.**

卫矛科 Celastraceae

样地名称（Plot name）= JH
个体数（Individual number/1 hm²）= 22
最大胸径（Max DBH）= 4.1 cm
重要值排序（Importance value rank）= 49

落叶灌木，高 3～12 m，胸径达 18 cm。叶薄革质，卵状椭圆形、长方椭圆形或窄椭圆形，长 6～10（～12）cm，宽 2.5～3.5 cm，先端钝渐尖，基部阔楔形或稍圆。聚伞花序分枝疏松。蒴果紫红色，倒圆锥状；种子长圆状，种皮枣红色，假种皮橙红色。花期 3～8 月，果期 5～11 月。

Deciduous shrubs, 3–12 m tall, up to 18 cm d.b.h. Leaf blade thinly leathery, elliptic-obovate or ovate, 6–10 (–12) × 2.5–3.5 cm, base attenuate, apex caudate or with a long tail; dichotomously branched with few flowers. Capsule obovoid, base attenuate, purple. Seeds ovoid, dark brown, partially covered by orange aril. Fl. Mar.–Aug., fr. May–Nov..

树干　　　Trunk
摄影：李健星　Photo by：Li Jianxing

枝叶　　　Branch and leaves
摄影：李健星　Photo by：Li Jianxing

花序　　　Inflorescence
摄影：孟德昌　Photo by：Meng Dechang

径级分布表　DBH class

胸径区间 Diameter class (cm)	个体数 No. of individuals in the plot	比例 Proportion (%)
1～2	16	72.73
2～5	6	27.27
5～10	0	0.00
10～20	0	0.00
20～35	0	0.00
35～50	0	0.00
≥50	0	0.00

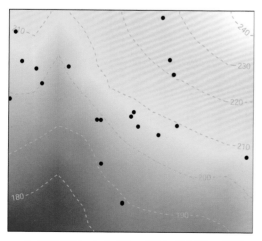

● 1～5 cm DBH　＋ 5～20 cm DBH　○ ≥20 cm DBH
个体分布图　Distribution of individuals

41 中华杜英 | zhōng huá dù yīng （华杜英）

***Elaeocarpus chinensis* (Gardner Champ.) Hook. f. ex Benth.**

杜英科 Elaeocarpaceae

样地名称（Plot name）= JH
个体数（Individual number/1 hm^2）= 19
最大胸径（Max DBH）= 14.5 cm
重要值排序（Importance value rank）= 47

常绿乔木，高 3~7 m。叶薄革质，卵状披针形或披针形，长 5~8 cm，宽 2~3 cm，先端渐尖，基部圆形，稀为阔楔形，上面绿色有光泽，下面有细小黑腺点，侧脉 4~6 对，在上面隐约可见，在下面稍突起。总状花序生于无叶的上一年枝条上。核果椭圆形，长不到 1 cm。花期 5~6 月，果期 6~9 月。

Evergreen trees, 3–7 m tall. Leaf blade ovate-lanceolate or lanceolate, 5–8 × 2–3 cm, papery, abaxially black glandular punctate, lateral veins 4–6 per side, slightly raised abaxially, base rounded, rarely broadly cuneate, apex acuminate. Racemes in axils of fallen leaves. Drupe ellipsoid, shorter than 1 cm. Fl. May–Jun., fr. Jun.–Sep..

树干　Trunk
摄影：丁涛　Photo by：Ding Tao

枝叶　Branch and leaves
摄影：丁涛　Photo by：Ding Tao

果序　Infructescences
摄影：唐忠炳　Photo by：Tang Zhongbing

径级分布表　DBH class

胸径区间 Diameter class (cm)	个体数 No. of individuals in the plot	比例 Proportion (%)
1~2	2	10.53
2~5	7	36.84
5~10	9	47.37
10~20	1	5.26
20~35	0	0.00
35~50	0	0.00
≥50	0	0.00

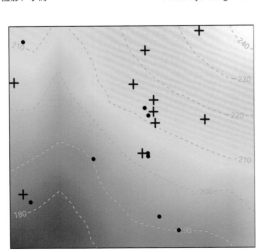

● 1~5 cm DBH　＋ 5~20 cm DBH　○ ≥20 cm DBH
个体分布图　Distribution of individuals

42 山杜英 | shān dù yīng （羊屎树）

杜英属

Elaeocarpus sylvestris **(Lour.) Poir. in Lamarck**

杜英科 Elaeocarpaceae

样地名称（Plot name）= JH
个体数（Individual number/1 hm²）= 1
最大胸径（Max DBH）= 1.0 cm
重要值排序（Importance value rank）= 139

乔木，高约 15 m。叶纸质，倒卵形或倒披针形，长 4～12 cm，宽 2～7 cm，上下两面均无毛，干后黑褐色，不发亮，先端钝，或略尖，基部窄楔形，下延，侧脉 4～5 对。总状花序生于枝顶叶腋内。核果细小，椭圆形，长 1～1.2 cm，内果皮薄骨质，有腹缝沟 3 条。花期 4～5 月，果期 5～8 月。

Trees, to 15 m tall. Leaf blade black-brown when dry, obovate or oblanceolate, 4–12 × 2–7 cm, papery, both surfaces glabrous, lateral veins usually 4 or 5 per side, base narrowly cuneate, decurrent, margin crenate or sinuately crenate, apex obtuse or shortly acuminate. Racemes in axils of fallen and current leaves. Drupe ellipsoid, 1–1.2 cm; endocarp thinly bony, with 3 ventral sutures. Fl. Apr.–May, fr. May–Aug..

果枝　　　　　　　　　　Fruiting branch
摄影：杨平　　　　　　　Photo by：Yang Ping

叶　　　　　　　　　　Leaves
摄影：杨平　　　　　　Photo by：Yang Ping

果　　　　　　　　　　Fruit
摄影：陆昭岑　　　　　Photo by：Lu Zhaochen

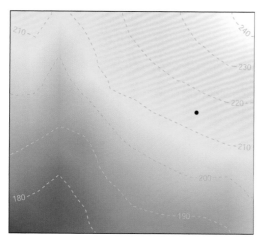

● 1～5 cm DBH　　＋ 5～20 cm DBH　　○ ≥20 cm DBH
个体分布图　Distribution of individuals

径级分布表　DBH class

胸径区间 Diameter class (cm)	个体数 No. of individuals in the plot	比例 Proportion (%)
1～2	1	100.00
2～5	0	0.00
5～10	0	0.00
10～20	0	0.00
20～35	0	0.00
35～50	0	0.00
≥50	0	0.00

43 旁杞木 | páng qǐ mù （百六齿）

***Carallia pectinifolia* W. C. Ko**

红树科 Rhizophoraceae

样地名称（Plot name）= JH
个体数（Individual number/1 hm^2）= 75
最大胸径（Max DBH）= 16.2 cm
重要值排序（Importance value rank）= 24

灌木或小乔木。小枝和枝干燥时紫褐色，有明显的纺锤形的木栓质皮孔。叶纸质，矩圆形，稀倒披针形，长 5～13 cm，宽 2.5～5.5 cm，顶端渐尖或尾状，基部阔楔形，边缘有篦状小齿。二歧聚伞花序。果实球形，成熟时红色，有宿存的红色花萼裂片。花果期春夏两季。

Shrubs or small trees. Branches and branchlets purplish brown when dried; lenticels fusiform, conspicuous. Leaf blade oblong to rarely oblanceolate, 5–13 × 2.5–5.5 cm, papery, base broadly cuneate, margin serrate, apex acuminate. Inflorescences dichasial cymes. Fruit red, globose; persistent calyx red. Fl. and fr. spring-summer.

果枝　　Fruiting branch
摄影：丁涛　　Photo by：Ding Tao

叶　　Leaves
摄影：李健星　　Photo by：Li Jianxing

花枝　　Flowering branch
摄影：丁涛　　Photo by：Ding Tao

径级分布表　DBH class

胸径区间 Diameter class (cm)	个体数 No. of individuals in the plot	比例 Proportion (%)
1～2	25	33.33
2～5	44	58.67
5～10	5	6.67
10～20	1	1.33
20～35	0	0.00
35～50	0	0.00
≥50	0	0.00

● 1～5 cm DBH　　+ 5～20 cm DBH　　○ ≥20 cm DBH
个体分布图　Distribution of individuals

44 巴豆 | bā dòu （小巴豆）

巴豆属

Croton tiglium L.

大戟科 Euphorbiaceae

样地名称（Plot name）= JH
个体数（Individual number/1 hm²）= 8
最大胸径（Max DBH）= 7.7 cm
重要值排序（Importance value rank）= 88

灌木或小乔木，高达 7 m。叶纸质，卵形，基部阔楔形至近圆形，边缘有细锯齿；基出脉 3 （~5）条，侧脉 3~4 对；基部两侧叶缘上各有 1 枚盘状腺体；叶柄长 2.5~5 cm。总状花序，顶生，苞片钻状，子房密被星状柔毛，花柱 2 深裂。蒴果卵状长圆形。花期 2~7 月，果期 5~9 月。

Shrubs or small trees, up to 7 m tall. Leaf blade ovate, papery, base cuneate or broadly so, rounded, with discoid glands, basal veins 3 (–5), lateral veins 3 or 4, petiole 2.5–5 cm. Racemes terminal, bracts subulate, ovary densely stellate-hairy; styles bipartite. Capsules ellipsoidal, oblong-ovoid. Fl. Jan.–Jul., fr. May–Sep..

花枝　Flowering branch
摄影：丁涛　Photo by: Ding Tao

叶　Leaf
摄影：丁涛　Photo by: Ding Tao

花序　Inflorescence
摄影：丁涛　Photo by: Ding Tao

径级分布表　DBH class

胸径区间 Diameter class (cm)	个体数 No. of individuals in the plot	比例 Proportion (%)
1~2	1	12.50
2~5	3	37.50
5~10	4	50.00
10~20	0	0.00
20~35	0	0.00
35~50	0	0.00
≥50	0	0.00

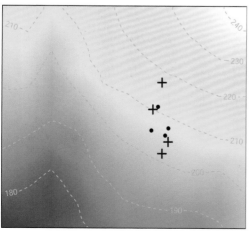

● 1~5 cm DBH　＋ 5~20 cm DBH　○ ≥20 cm DBH
个体分布图　Distribution of individuals

45 粗毛野桐 | cū máo yě tóng

Hancea hookeriana Seem.
大戟科 Euphorbiaceae

样地名称（Plot name）= JH
个体数（Individual number/1 hm²）= 2
最大胸径（Max DBH）= 31.6 cm
重要值排序（Importance value rank）= 78

灌木或小乔木，高 1.5～10 m。叶对生，同对的叶形状和大小极不相同，小型叶退化成托叶状，钻形，长 1～1.2 cm；大型叶近革质，长圆状披针形，顶端渐尖，基部钝或圆形，叶基部有时具褐色斑状腺体；羽状脉，侧脉 8～9 对。蒴果三棱状球形，密生稍硬而直的软刺，被灰黄色星状毛。花期 3～5 月，果期 8～10 月。

Shrubs or small trees, 1.5–10 m tall. Leaves opposite, each pair very unequal; smaller leaves subulate, 1–1.2 cm; larger leaves oblong-lanceolate, thinly leathery; apex acuminate, base obtuse, sometimes with 2 brown spotted glands; veins 8 or 9 pairs; Capsule 3-locular, gray-yellow pilosulose and densely softly spiny. Fl. Mar.–May, fr. Aug.–Oct..

果枝　Fruiting branch
摄影：杨平　Photo by: Yang Ping

花序　Inflorescence
摄影：杨平　Photo by: Yang Ping

果　Fruit
摄影：杨平　Photo by: Yang Ping

径级分布表 DBH class

胸径区间 Diameter class (cm)	个体数 No. of individuals in the plot	比例 Proportion (%)
1～2	0	0.00
2～5	0	0.00
5～10	0	0.00
10～20	1	50.00
20～35	1	50.00
35～50	0	0.00
≥50	0	0.00

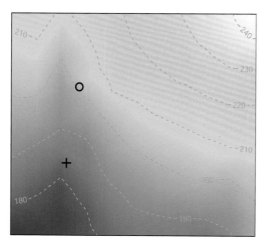

● 1～5 cm DBH　+ 5～20 cm DBH　○ ≥20 cm DBH
个体分布图 Distribution of individuals

46 轮苞血桐 | lún bāo xuè tóng （安达曼血桐） 血桐属

Macaranga andamanica Kurz

大戟科 Euphorbiaceae

样地名称（Plot name）= JH
个体数（Individual number/1 hm²）= 10
最大胸径（Max DBH）= 15.6 cm
重要值排序（Importance value rank）= 68

灌木，高 1～5 m。嫩枝无毛，具疏生颗粒状腺体。叶纸质，长圆形或椭圆状披针形，长 7～14 cm，宽 2.5～5.5 cm，顶端渐尖，基部微耳状心形，两侧各具斑状腺体 1 个，叶缘疏生细腺齿，两面无毛。雄花序纤细。雌花序通常一朵花。蒴果双球形，具颗粒状腺体。花果期全年。

Shrubs, 1–5 m tall. Branchlets pubescent, glabrescent. Leaf blade oblong-lanceolate or oblong, 7–14 × 2.5–5.5 cm, thickly papery, often drying dull brown, abaxially sparsely dark glandular-scaly, midrib pilose, base slightly auriculate-cordate, with 2 glands, apex acuminate. Male inflorescences very slender, female inflorescences often 1-flowered. Capsule 2-lobed, densely glandular-scaly. Fl. and fr. throughout year.

果　　Fruit
摄影：丁涛　　Photo by: Ding Tao

叶　　Leaves
摄影：丁涛　　Photo by: Ding Tao

花序　　Inflorescence
摄影：丁涛　　Photo by: Ding Tao

径级分布表　DBH class

胸径区间 Diameter class (cm)	个体数 No. of individuals in the plot	比例 Proportion (%)
1～2	3	30.00
2～5	2	20.00
5～10	3	30.00
10～20	2	20.00
20～35	0	0.00
35～50	0	0.00
≥50	0	0.00

● 1～5 cm DBH　　+ 5～20 cm DBH　　○ ≥20 cm DBH
个体分布图　Distribution of individuals

47 印度血桐 | yìn dù xuè tóng （盾叶木） 血桐属

Macaranga indica Wight

大戟科 Euphorbiaceae

样地名称（Plot name）= JH
个体数（Individual number/1 hm²）= 7
最大胸径（Max DBH）= 18.1 cm
重要值排序（Importance value rank）= 66

乔木，高 10~25 m。叶薄革质，卵圆形，长 14~25 cm，宽 13~23 cm，顶端短渐尖，基部钝圆，盾状着生，具斑状腺体 2 个，叶缘具疏生腺齿，上面无毛或沿叶脉具疏毛，下面被柔毛和具颗粒状腺体；掌状脉 9 条。雄花序圆锥状，长 10~15 cm；雌花序圆锥状，长 5~6 cm。萼果球形，具颗粒状腺体。花期 8~10 月，果期 10~11 月。

Trees, 10–25 m tall. Leaf blade ovate-orbicular, 14–25 × 13–23 cm, thinly leathery, abaxially pubescent and glandular-scaly, adaxially glabrous or pilose along veins, base rounded and broadly peltate, with glands, margin serrulate, apex acuminate; palmate veins 9. Male inflorescences branched, 10–15 cm; female inflorescences branched, 5–6 cm. Capsule globose, sparsely glandular-scaly. Fl. Aug.–Oct., fr. Oct.–Nov..

树干　Trunk
摄影：李健星　Photo by: Li Jianxing

叶　Leaves
摄影：丁涛　Photo by: Ding Tao

果序　Infructescence
摄影：蒋裕良　Photo by: Jiang Yuliang

径级分布表 DBH class

胸径区间 Diameter class (cm)	个体数 No. of individuals in the plot	比例 Proportion (%)
1~2	0	0.00
2~5	2	28.57
5~10	1	14.29
10~20	4	57.14
20~35	0	0.00
35~50	0	0.00
≥50	0	0.00

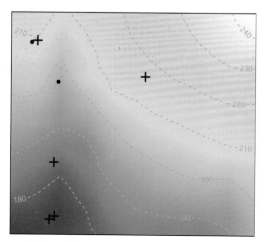

● 1~5 cm DBH　+ 5~20 cm DBH　○ ≥20 cm DBH
个体分布图　Distribution of individuals

48 勐仑三宝木 | měng lún sān bǎo mù （孟仑三宝木）

三宝木属

Trigonostemon bonianus Gagnep.

大戟科 Euphorbiaceae

样地名称（Plot name）= JH
个体数（Individual number/1 hm²）= 12
最大胸径（Max DBH）= 3.1 cm
重要值排序（Importance value rank）= 67

灌木或小乔木，高 2~4 m。叶纸质，卵形、椭圆形至长圆状披针形，长 10~17 cm，宽 2.5~5 cm，顶端短尖至渐尖，有时尖头骤狭呈尾状，基部阔楔形至近圆形，全缘或有不明显疏细锯齿，齿端有腺；侧脉每边 3~5 条。聚伞圆锥花序生于枝条近顶端。蒴果近球形，无毛，具 3 纵沟，果皮薄壳质；种子椭圆状。花期 5~8 月，果期 8~12 月。

Shrubs or small trees, 2–4 m tall. Leaf blade narrowly elliptic to oblong-lanceolate, 10–17 × 2.5–5 cm, papery, glabrous on both surfaces, base cuneate, margin entire or sparsely serrulate, apex acuminate to caudate-acuminate, acumen obtuse; veins from base 3, lateral veins 3–5. Racemes terminal, spreading; ovary glabrous; styles 3, short; stigmas capitate. Fl. May–Aug., fr. Aug.–Dec..

花枝　Flowering branch
摄影：丁涛　Photo by: Ding Tao

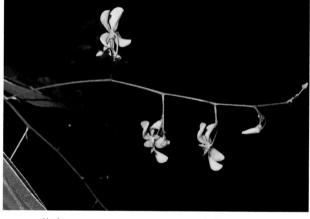

花序　Inflorescence
摄影：丁涛　Photo by: Ding Tao

果　Fruit
摄影：李健星　Photo by: Li Jianxing

径级分布表　DBH class

胸径区间 Diameter class (cm)	个体数 No. of individuals in the plot	比例 Proportion (%)
1~2	10	83.33
2~5	2	16.67
5~10	0	0.00
10~20	0	0.00
20~35	0	0.00
35~50	0	0.00
≥50	0	0.00

● 1~5 cm DBH　+ 5~20 cm DBH　○ ≥20 cm DBH
个体分布图　Distribution of individuals

49 木油桐 | mù yóu tóng （皱桐）

油桐属

Vernicia montana Lour.

大戟科 Euphorbiaceae

样地名称（Plot name）= JH
个体数（Individual number/1 hm²）= 1
最大胸径（Max DBH）= 4.2 cm
重要值排序（Importance value rank）= 130

常绿乔木，高达 20 m。枝条无毛，散生突起皮孔。叶柄长 7~17 cm，无毛，顶端有 2 枚具柄的杯状腺体。叶阔卵形，长 8~20 cm，宽 6~18 cm，顶端短尖至渐尖，基部心形至截平，全缘或 2~5 裂。花序生于当年生已发叶的枝条上；核果卵球状，具 3 条纵棱，棱间有粗疏网状皱纹。花期 4~6 月，果期 7~10 月。

Evergreen trees, up to 20 m tall. Branches glabrous, with sparsely elevated lenticels. Petiole 7–17 cm, glabrous, apex with 2 stalked and cupular glands. Leaf blade broadly ovate, 8–20 × 6–18 cm, apex acute to acuminate, base cordate to truncate, margin entire or 2–5-fid. Inflorescences produced with new leaves. Drupes ovoid, longitudinally 3-angular, between angles with sparsely reticulate wrinkles. Fl. Apr.–Jun., fr. Jul.–Oct..

果枝　　Fruiting branches
摄影：杨平　　Photo by: Yang Ping

枝叶　　Branch and leaves
摄影：丁涛　　Photo by: Ding Tao

花　　Flowers
摄影：陆昭岑　　Photo by: Lu Zhaochen

径级分布表 DBH class

胸径区间 Diameter class (cm)	个体数 No. of individuals in the plot	比例 Proportion (%)
1~2	0	0.00
2~5	1	100.00
5~10	0	0.00
10~20	0	0.00
20~35	0	0.00
35~50	0	0.00
≥50	0	0.00

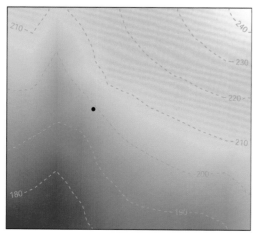

● 1~5 cm DBH　＋ 5~20 cm DBH　○ ≥20 cm DBH
个体分布图　Distribution of individuals

50 黄毛五月茶 | huáng máo wǔ yuè chá （旱禾子树）

***Antidesma fordii* Hemsl.**

叶下珠科 Phyllanthaceae

样地名称（Plot name）= JH
个体数（Individual number/1 hm²）= 262
最大胸径（Max DBH）= 10.1 cm
重要值排序（Importance value rank）= 13

小乔木，高达 7 m。小枝、叶柄、托叶、花序轴被黄色绒毛，其余均被长柔毛或柔毛。叶片长圆形、椭圆形或倒卵形，长 7～25 cm，宽 3～10.5 cm，顶端短渐尖或尾状渐尖，基部近圆或钝；叶柄长 1～3 mm。花序顶生或腋生。核果纺锤形。花期 3～7 月，果期 7 月至翌年 1 月。

Small trees, up to 7 m tall. Young twigs, petioles, and inflorescence axes densely yellow tomentose. Petiole 1–3 mm. Leaf blade oblong, sometimes elliptic, slightly ovate or obovate, 7–25× 3–10.5 cm, papery, base rounded to obtuse, sometimes truncate, apex acuminate to caudate. Inflorescences terminal and axillary. Drupes ellipsoid, laterally compressed. Fl. Mar.–Jul., fr. Jul.–Jan. of next year.

树干　Trunk
摄影：李健星　Photo by：Li Jianxing

花枝　Flowering branch
摄影：丁涛　Photo by：Ding Tao

果序　Infructescence
摄影：李健星　Photo by：Li Jianxing

径级分布表　DBH class

胸径区间 Diameter class (cm)	个体数 No. of individuals in the plot	比例 Proportion (%)
1～2	87	33.21
2～5	162	61.83
5～10	12	4.58
10～20	1	0.38
20～35	0	0.00
35～50	0	0.00
≥50	0	0.00

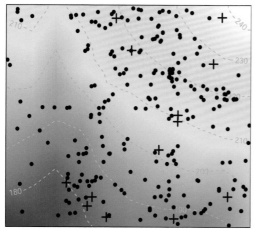

● 1～5 cm DBH　+ 5～20 cm DBH　○ ≥20 cm DBH
个体分布图　Distribution of individuals

51 银柴 | yín chái （大沙叶）

Aporosa dioica (Roxb.) Müll. Arg.
叶下珠科 Phyllanthaceae

样地名称（Plot name）= JH
个体数（Individual number/1 hm²）= 301
最大胸径（Max DBH）= 15.5 cm
重要值排序（Importance value rank）= 4

乔木，高达 9 m，在次森林中常呈灌木状，高约 2 m。叶片革质，椭圆形、长椭圆形、倒卵形或倒披针形，长 6~12 cm，宽 3.5~6 cm，顶端圆至急尖，基部圆或楔形，全缘或具有稀疏的浅锯齿；叶柄长 5~12 mm，顶端两侧各具 1 个小腺体。蒴果椭圆状。花果期几乎全年。

Trees, to 9 m tall, usually shrubby in secondary forest, ca. 2 m tall. Leaf blade elliptic, narrowly ovate, oblong-elliptic, obovate, or oblanceolate, 6–12 × 3.5–6 cm, leathery, base rounded or cuneate, margin entire or sparsely shallowly dentate, apex rounded to acute. Petiole 5–12 mm, apex bilateral with 2 glands. Capsules ellipsoid. Fl. and fr. almost throughout year.

树干 Trunk
摄影：李健星 Photo by: Li Jianxing

叶 Leaves
摄影：丁涛 Photo by: Ding Tao

果 Fruits
摄影：丁涛 Photo by: Ding Tao

径级分布表 DBH class

胸径区间 Diameter class (cm)	个体数 No. of individuals in the plot	比例 Proportion (%)
1~2	74	24.58
2~5	102	33.89
5~10	90	29.90
10~20	35	11.63
20~35	0	0.00
35~50	0	0.00
≥50	0	0.00

● 1~5 cm DBH   + 5~20 cm DBH   ○ ≥20 cm DBH
个体分布图 Distribution of individuals

52 重阳木 | chóng yáng mù （水蚬木）

秋枫属

Bischofia polycarpa (H. Lév.) Airy Shaw

叶下珠科 Phyllanthaceae

样地名称（Plot name）= JH
个体数（Individual number/1 hm²）= 10
最大胸径（Max DBH）= 39.2 cm
重要值排序（Importance value rank）= 35

落叶乔木，高达 15 m。三出复叶；叶柄长 9～13.5 cm；顶生小叶通常较两侧的大，小叶片纸质，卵形或椭圆状卵形，有时长圆状卵形，顶端突尖或短渐尖，基部圆或浅心形，边缘具钝细锯齿每 1 cm 长 4～5 个。花雌雄异株，组成总状花序。果实浆果状，圆球形，成熟时褐红色。花期 4～5 月，果期 10～11 月。

Deciduous trees, to 15 m tall. Leaves palmately 3-foliolate; petiole 9–13.5 cm; terminal Leaflets usually larger than bilateral ones. Leaflet blades ovate or elliptic-ovate, sometimes oblong-ovate, papery, base rounded or shallowly cordate, apex acute or shortly acuminate, margins with 4 or 5 teeth per cm. Plants dioecious. Inflorescences pendent racemes. Fruits globose, brown-red when mature. Fl. Apr.–May, fr. Oct.–Nov..

树干 Trunk
摄影：李健星 Photo by: Li Jianxing

叶背 Leaf back
摄影：王斌 Photo by: Wang Bin

果序 Infructescences
摄影：李健星 Photo by: Li Jianxing

径级分布表 DBH class

胸径区间 Diameter class (cm)	个体数 No. of individuals in the plot	比例 Proportion (%)
1～2	1	10.00
2～5	2	20.00
5～10	1	10.00
10～20	2	20.00
20～35	3	30.00
35～50	1	10.00
≥50	0	0.00

● 1～5 cm DBH   + 5～20 cm DBH   ○ ≥20 cm DBH
个体分布图 Distribution of individuals

53 黑面神 | hēi miàn shén （鬼划符）

Breynia fruticosa (L.) Hook. f.

叶下珠科 Phyllanthaceae

样地名称（Plot name）＝ JH
个体数（Individual number/1 hm²）＝ 85
最大胸径（Max DBH）＝ 7.6 cm
重要值排序（Importance value rank）＝ 29

灌木，高 1～3 m。茎皮灰褐色；枝条上部常呈扁压状，紫红色。叶片革质，卵形、阔卵形或菱状卵形，长 3～7 cm，宽 1.8～3.5 cm，两端钝或急尖。花小，单生或 2～4 朵簇生于叶腋内，雌花位于小枝上部，雄花则位于小枝的下部。蒴果圆球状。花期全年，果期 5～12 月。

Shrubs, 1–3 m tall. Stem gray-brown; branches compressed at upper part, purple. Leaf blade ovate, broadly ovate, or rhombic-ovate, 3–7 × 1.8–3.5 cm, leathery, base obtuse or acute, apex (obtuse or) acute to subacuminate. Flowers small, solitary or 2–4-flowered in axillary clusters, male in proximal axils, female in distal axils, inserted in different branchlets. Capsules globose, apex rounded. Fl. throughtout year, fr. May–Dec..

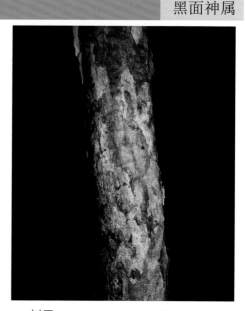

树干　Trunk
摄影：丁涛　Photo by: Ding Tao

花枝　Flowering branch
摄影：丁涛　Photo by: Ding Tao

果枝　Fruiting branches
摄影：丁涛　Photo by: Ding Tao

径级分布表　DBH class

胸径区间 Diameter class (cm)	个体数 No. of individuals in the plot	比例 Proportion (%)
1～2	23	27.06
2～5	55	64.71
5～10	7	8.23
10～20	0	0.00
20～35	0	0.00
35～50	0	0.00
≥50	0	0.00

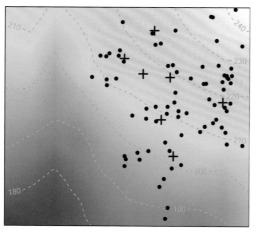

● 1～5 cm DBH　＋ 5～20 cm DBH　○ ≥20 cm DBH
个体分布图　Distribution of individuals

54 禾串树 | hé chuàn shù （禾串土蜜树）

土蜜树属

***Bridelia balansae* Tutcher**

叶下珠科 Phyllanthaceae

样地名称（Plot name）＝ JH
个体数（Individual number/1 hm²）＝ 16
最大胸径（Max DBH）＝ 17.7 cm
重要值排序（Importance value rank）＝ 56

乔木，高达 17 m，树干通直，胸径达 30 cm。树皮黄褐色，近平滑，内皮褐红色；小枝具有凸起的皮孔，无毛。叶片近革质，椭圆形或长椭圆形，长 5～15 cm，宽 1.5～5.5 cm，顶端渐尖或尾状渐尖，基部钝，边缘反卷。聚伞花序腋生。核果长卵形，成熟时紫黑色。花期 5～8 月，果期 9～11 月。

Trees, up to 17 m tall, straight trunk, ca. 30 cm in d.b.h.. Bark fulvous, nearly smooth; branchlets glabrous with elevated lenticels. Leaf blade elliptic or elliptic-lanceolate, 5–15 × 1.5–5.5 cm, leathery or nearly so, base cuneate, rarely obtuse, margin slightly revolute, apex acute, acuminate, or caudate-acuminate. Glomerules axillary, drupes oblong-ovoid, purple-black when mature. Fl. May–Aug., fr. Sep.–Nov..

树干　　Trunk
摄影：李健星　　Photo by: Li Jianxing

叶背　　Leaf backs
摄影：丁涛　　Photo by: Ding Tao

花　　Flowers
摄影：李健星　　Photo by: Li Jianxing

径级分布表　DBH class

胸径区间 Diameter class (cm)	个体数 No. of individuals in the plot	比例 Proportion (%)
1～2	8	50.00
2～5	5	31.25
5～10	2	12.50
10～20	1	6.25
20～35	0	0.00
35～50	0	0.00
≥50	0	0.00

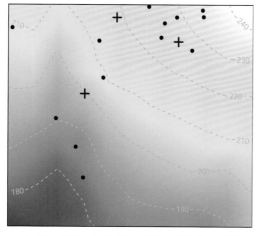

● 1～5 cm DBH　＋ 5～20 cm DBH　○ ≥20 cm DBH
个体分布图　Distribution of individuals

55 膜叶土蜜树 | mó yè tǔ mì shù

土蜜树属

***Bridelia glauca* Blume**

叶下珠科 Phyllanthaceae

样地名称（Plot name）= JH
个体数（Individual number/1 hm²）= 1
最大胸径（Max DBH）= 1.9 cm
重要值排序（Importance value rank）= 131

乔木，高达 15 m。叶片膜质或近膜质，倒卵形、长圆形或椭圆状披针形，长 5～15 cm，宽 2.5～7.5 cm，顶端急尖至渐尖，基部钝至圆；侧脉每边 7～12（～18）条，在叶背凸起。花白色，雌雄同株。核果椭圆状，长 6～11 mm，顶端具小尖头，基部有宿存萼片，1 室。花期 3～9 月，果期 9～12 月。

Trees, up to 15 m tall. Leaf blade elliptic-ovate, oblong, or obovate, 5–15 × 2.5–7.5 cm, membranous or thickly papery, base acute, obtuse, or often truncate, apex acute to acuminate; lateral veins 7–12 (–18) pairs. Flowers monoecious. Drupes ellipsoidal, 6–11 mm, base with persistent sepals, apex with mucro, 1-celled, stalk usually slender. Fl. Mar.–Sep., fr. Sep.–Dec..

枝叶　　Branch and leaves
摄影：朱鑫鑫　Photo by: Zhu Xinxin

叶背　　Leaf backs
摄影：朱鑫鑫　Photo by: Zhu Xinxin

果　　Fruits
摄影：朱鑫鑫　Photo by: Zhu Xinxin

径级分布表　DBH class

胸径区间 Diameter class (cm)	个体数 No. of individuals in the plot	比例 Proportion (%)
1～2	1	100.00
2～5	0	0.00
5～10	0	0.00
10～20	0	0.00
20～35	0	0.00
35～50	0	0.00
≥50	0	0.00

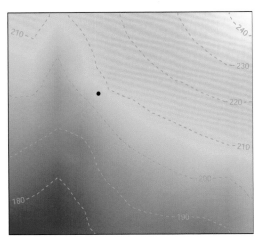

● 1～5 cm DBH　＋ 5～20 cm DBH　○ ≥20 cm DBH
个体分布图　Distribution of individuals

56 毛果算盘子 | máo guǒ suàn pán zǐ （漆大姑） 算盘子属

***Glochidion eriocarpum* Champ. ex Benth.**

叶下珠科 Phyllanthaceae

样地名称（Plot name）= JH
个体数（Individual number/1 hm^2）= 14
最大胸径（Max DBH）= 11.7 cm
重要值排序（Importance value rank）= 64

灌木或小乔木，高达5m，小枝密被淡黄色长柔毛。叶片纸质，卵形、狭卵形或宽卵形，长4～8cm，宽1.5～3.5cm，顶端渐尖或急尖，基部钝、截形或圆形，两面均被长柔毛，下面毛被较密。花单生或2～4朵簇生于叶腋内。蒴果扁球状，密被长柔毛。花果期几乎全年。

Shrubs or small trees, to 5 m tall. Branchlets densely spreading yellowish or gray-yellow villous. Leaf blade ovate, narrowly ovate, or broadly ovate, 4–8 × 1.5–3.5 cm, papery, densely yellowish or gray-villous, but denser abaxially, base obtuse, truncate, or rounded, apex acuminate or acute. Flowers axillary, solitary or in 2–4-flowered clusters. Capsules depressed globose, densely yellowish or gray-villous. Fl. and fr. almost throughout year.

树干　　Trunk
摄影：李健星　　Photo by: Li Jianxing

枝叶　　Branch and leaves
摄影：李健星　　Photo by: Li Jianxing

花序　　Inflorescences
摄影：李健星　　Photo by: Li Jianxing

径级分布表　DBH class

胸径区间 Diameter class (cm)	个体数 No. of individuals in the plot	比例 Proportion (%)
1～2	3	21.43
2～5	9	64.29
5～10	1	7.14
10～20	1	7.14
20～35	0	0.00
35～50	0	0.00
≥50	0	0.00

● 1～5 cm DBH　＋ 5～20 cm DBH　○ ≥20 cm DBH
个体分布图　Distribution of individuals

57 山桂花 | shān guì huā （大叶山桂花）

山桂花属

Bennettiodendron leprosipes (Clos) Merr.
杨柳科 Salicaceae

样地名称（Plot name）= JH
个体数（Individual number/1 hm^2）= 8
最大胸径（Max DBH）= 15.8 cm
重要值排序（Importance value rank）= 92

常绿灌木或小乔木，高 2~6 m，有时达 15 m。树皮灰褐色，不裂，有臭味。叶纸质或薄纸质，倒卵状长圆形或长圆状椭圆形，长（5~）10~23 cm，宽 4~7.5 cm，先端短渐尖，基部渐狭，边缘有粗齿和带不整齐的腺齿。圆锥花序顶生。浆果成熟时红色，球形。花期 3~4 月，果期 5~11 月。

Evergreen shrubs or small trees, 2–6 (–15) m tall. Bark gray-brown, fetid, not flaking. Leaf blade mostly narrowly to broadly elliptic, elliptic-oblong, or obovate, usually (5–) 10–23 × 4–7.5 cm, papery or thinly papery, base usually acute-cuneate, less often obtuse-cuneate, rarely rounded, margin sparsely obtusely serrate, apex obtuse. Inflorescence terminal, paniculate. Berry red when mature, globose. Fl. Mar.–Apr., fr. May–Nov..

树干 Trunk
摄影：李健星 Photo by: Li Jianxing

叶背 Leaf backs
摄影：李健星 Photo by: Li Jianxing

花序 Inflorescences
摄影：椰子 Photo by: Ye Zi

径级分布表 DBH class

胸径区间 Diameter class (cm)	个体数 No. of individuals in the plot	比例 Proportion (%)
1~2	4	50.00
2~5	1	12.50
5~10	2	25.00
10~20	1	12.50
20~35	0	0.00
35~50	0	0.00
≥50	0	0.00

● 1~5 cm DBH   ＋ 5~20 cm DBH   ○ ≥20 cm DBH
个体分布图 Distribution of individuals

58 膜叶脚骨脆 | mó yè jiǎo gǔ cuì （红花木） 脚骨脆属

Casearia membranacea Hance

杨柳科 Salicaceae

样地名称（Plot name）= JH
个体数（Individual number/1 hm²）= 1
最大胸径（Max DBH）= 8.1 cm
重要值排序（Importance value rank）= 124

常绿乔木或灌木，高 4~18 m。叶纸质，长椭圆形或卵状长椭圆形，长 5~12（~14）cm，宽 2.5~5（~6）cm，先端短尖而钝，有的锐尖，基部宽楔形至钝，通常不对称，边缘浅波状或有不明显的钝齿。蒴果卵状或卵状椭圆形。花期 4 至翌年 1 月，果期 5 月至翌年 1 月。

Evergreen trees or shrubs, 4–18 m tall. Leaf blade papery, often elliptic or oblong, 5–12 (–14) × 2.5–5 (–6) cm, apex broadly acute to rounded, base mostly acute to obtuse, nearly rounded or asymmetric, margin serrulate, crenulate or nearly entire. Capsule ellipsoid to oblong. Fl. Apr.–Jan. of next year, fr. May–Jan. of next year.

枝叶　Branch and leaves
摄影：丁涛　Photo by：Ding Tao

叶　Leaves
摄影：丁涛　Photo by：Ding Tao

叶背　Leaf backs
摄影：丁涛　Photo by：Ding Tao

径级分布表　DBH class

胸径区间 Diameter class (cm)	个体数 No. of individuals in the plot	比例 Proportion (%)
1~2	0	0.00
2~5	0	0.00
5~10	1	100.00
10~20	0	0.00
20~35	0	0.00
35~50	0	0.00
≥50	0	0.00

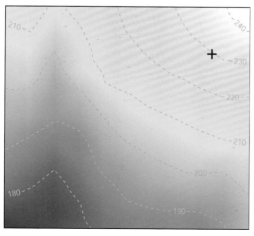

● 1~5 cm DBH　　＋ 5~20 cm DBH　　○ ≥20 cm DBH
个体分布图　Distribution of individuals

59 爪哇脚骨脆 | zhuǎ wājiǎo gǔ cuì （毛叶脚骨脆）

脚骨脆属

Casearia velutina Blume

杨柳科 Salicaceae

样地名称（Plot name）= JH
个体数（Individual number/1 hm^2）= 25
最大胸径（Max DBH）= 11.9 cm
重要值排序（Importance value rank）= 50

乔木或灌木，高达 10 m。小枝棕黄色，密生短柔毛，有棱脊，性脆。叶纸质，卵状长圆形，稀卵形，长 7~20 cm，宽 4~8 cm，先端渐尖，或急尖，基部圆形，边缘有锐齿，侧脉 8~12 对。花小，两性，淡紫色，数朵簇生于叶腋。花期 2~12 月，果期 4~6 月。

Trees or shrubs, to 10 m tall. Terminal bud densely pubescent, twig tips and branchlets densely to sparsely pubescent, hairs spreading, yellowish brown. Leaf blade variable in shape and size, elliptic to oblong, rarely ovate, 7–20 × 4–8 cm, thickly papery. Leaf lateral veins 8–12 pairs, base acute to rounded, sides convex to concave. Flowers (1 to) few to many in axillary sessile or subsessile glomerules. Fl. Feb.–Dec., fr. Apr.–Jun..

枝叶　Branch and leaves
摄影：丁涛　Photo by：Ding Tao

花序　Inflorescences
摄影：李均　Photo by：Li Jun

花枝　Flowering branch
摄影：李均　Photo by：Li Jun

径级分布表　DBH class

胸径区间 Diameter class (cm)	个体数 No. of individuals in the plot	比例 Proportion (%)
1~2	3	12.00
2~5	7	28.00
5~10	10	40.00
10~20	5	20.00
20~35	0	0.00
35~50	0	0.00
≥50	0	0.00

● 1~5 cm DBH　+ 5~20 cm DBH　○ ≥20 cm DBH
个体分布图　Distribution of individuals

60 箣柊 | cè zhōng （土乌药）

Scolopia chinensis **(Lour.) Clos**

杨柳科 Salicaceae

样地名称（Plot name）= JH
个体数（Individual number/1 hm^2）= 1
最大胸径（Max DBH）= 1.8 cm
重要值排序（Importance value rank）= 134

常绿小乔木或灌木，高 2~6 m。叶革质，椭圆形至长圆状椭圆形，长 4~7 cm，宽 2~4 cm，先端圆或钝，基部近圆形至宽楔形，两侧各有腺体 1 个，全缘或有细锯齿，两面光滑无毛，三出脉，侧脉纤细与网脉两面均明显。总状花序腋生或顶生。浆果圆球形。花期 6~9 月，果期 10 月到翌年 4 月。

Evergreen shrubs or small trees, 2–6 m tall. Leaf blade elliptic to oblong-elliptic, 4–7 × 2–4 cm, leathery, both surfaces glabrous, base broadly acute to subrounded, margin entire to serrulate, with a pair of glands at junction of blade and petiole, apex broadly acute to rounded. Racemes axillary or terminal. Berry orbicular-globose. Fl. Jun.–Sep., fr. Oct.–Apr. of next year.

果枝　　Fruiting branch
摄影：杨平　　Photo by：Yang Ping

花序　　Inflorescence
摄影：杨平　　Photo by：Yang Ping

果　　Fruits
摄影：杨平　　Photo by：Yang Ping

径级分布表　DBH class

胸径区间 Diameter class (cm)	个体数 No. of individuals in the plot	比例 Proportion (%)
1~2	1	100.00
2~5	0	0.00
5~10	0	0.00
10~20	0	0.00
20~35	0	0.00
35~50	0	0.00
≥50	0	0.00

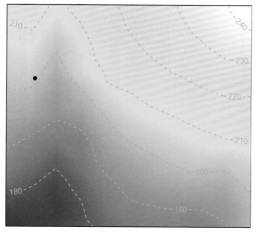

● 1~5 cm DBH　　+ 5~20 cm DBH　　○ ≥20 cm DBH
个体分布图　Distribution of individuals

61 南岭柞木 | nán lǐng zhà mù （光叶柞木） 柞木属

***Xylosma controversa* Clos**

杨柳科 Salicaceae

样地名称（Plot name）＝ JH
个体数（Individual number/1 hm²）＝ 8
最大胸径（Max DBH）＝ 12.6 cm
重要值排序（Importance value rank）＝ 71

常绿灌木或小乔木，高4～15 m。树皮棕灰色；幼时有枝刺，结果株无刺。叶薄革质，雌雄株稍有区别，通常雌株的叶有变化，菱状椭圆形至卵状椭圆形，先端渐尖，基部楔形或圆形，边缘有锯齿。花小，总状花序腋生。浆果黑色，球形，顶端有宿存花柱。花期7～11月，果期8～12月。

Evergreen shrubs or small trees, 4–15 m tall. Bark brown-gray; branches spiny when young, unarmed when old. Leaf blade broadly ovate to ovate-elliptic, leathery, base usually obtuse to rounded, less often acute, margin serrate, apex acute, tip usually acuminate. Inflorescence axillary, racemose, short. Berry black, globose, calyx and disk persistent at least while fruit attached to plant; styles persistent. Fl. Jul.–Nov., fr. Aug.–Dec..

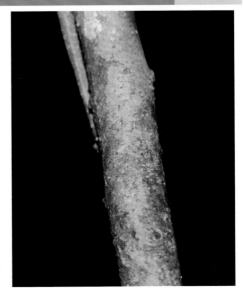

树干　　Trunk
摄影：李健星　　Photo by: Li Jianxing

叶　　Leaves
摄影：李健星　　Photo by: Li Jianxing

叶背　　Leaf backs
摄影：李健星　　Photo by: Li Jianxing

径级分布表　DBH class

胸径区间 Diameter class (cm)	个体数 No. of individuals in the plot	比例 Proportion (%)
1～2	2	25.00
2～5	4	50.00
5～10	1	12.50
10～20	1	12.50
20～35	0	0.00
35～50	0	0.00
≥50	0	0.00

● 1～5 cm DBH　　＋ 5～20 cm DBH　　○ ≥20 cm DBH
个体分布图　Distribution of individuals

62 岭南山竹子 | lǐng nán shān zhú zǐ 藤黄属

Garcinia oblongifolia Champ. ex Benth.

藤黄科 Clusiaceae

样地名称（Plot name）＝ JH
个体数（Individual number/1 hm²）＝ 33
最大胸径（Max DBH）＝ 11.1 cm
重要值排序（Importance value rank）＝ 40

乔木或灌木，高 5～15 m，胸径可达 30 cm。树皮深灰色。老枝通常具断环纹。叶片近革质，长圆形，倒卵状长圆形至倒披针形，长 5～10 cm，宽 2～3.5 cm，顶端急尖或钝，基部楔形。花单性，异株，单生或成伞状聚伞花序。浆果卵球形或圆球形，基部萼片宿存，顶端承以隆起的柱头。花期 4～5 月，果期 10～12 月。

Trees or shrubs, 5–15 m tall, to 30 cm in diam.. Bark dark gray. Branchlets usually with interrupted rings. Leaf blade oblong, obovate-oblong to oblanceolate, 5–10 × 2–3.5 cm, subleathery, base cuneate, apex acute or obtuse. Plant dioecious. Flowers solitary or in an umbel-like cyme. Fruit ovoid or globose, subtended by persistent sepals at base and crowned by convex stigma. Fl. Apr.–May, fr. Oct.–Dec..

果枝　Fruiting branch
摄影：丁涛　Photo by: Ding Tao

叶背　Leaf backs
摄影：丁涛　Photo by: Ding Tao

果　Fruit
摄影：丁涛　Photo by: Ding Tao

径级分布表　DBH class

胸径区间 Diameter class (cm)	个体数 No. of individuals in the plot	比例 Proportion (%)
1～2	7	21.21
2～5	17	51.52
5～10	8	24.24
10～20	1	3.03
20～35	0	0.00
35～50	0	0.00
≥50	0	0.00

● 1～5 cm DBH　＋ 5～20 cm DBH　○ ≥20 cm DBH
个体分布图　Distribution of individuals

63 黄牛木 | huáng niú mù （黄芽木）

Cratoxylum cochinchinense (Lour.) Blume
金丝桃科 Hypericaceae

样地名称（Plot name）= JH
个体数（Individual number/1 hm²）= 38
最大胸径（Max DBH）= 14.9 cm
重要值排序（Importance value rank）= 26

落叶灌木或乔木，高 1.5~18 （~25）m。树干下部有簇生的长枝刺；叶柄间线痕连续或间有中断。叶片椭圆形至长椭圆形或披针形，长 3~10.5 cm，宽 1~4 cm，先端骤然锐尖或渐尖，基部钝形至楔形，坚纸质，有透明腺点及黑点。聚伞花序腋生或腋外生及顶生，具梗。蒴果椭圆形，棕色。花期 4~5 月，果期 6 月以后。

Deciduous shrubs or trees, 1.5–18 (–25) m tall. Trunk with clusters of long thorns on lower part. Interpetiolar scars not always continuous. Leaf blade elliptic to oblong or lanceolate, 3–10.5 × 1–4 cm, papery, abaxially with pellucid or dark glands, base obtuse to cuneate, apex abruptly acute or acuminate. Cymes axillary or extra-axillary and terminal, pedunculate. Capsule brown, ellipsoid. Fl. Apr.–May, fr. after Jun..

花枝　Flowering branches
摄影：杨平　Photo by: Yang Ping

枝叶　Branch and leaves
摄影：李健星　Photo by: Li Jianxing

花　Flowers
摄影：杨平　Photo by: Yang Ping

径级分布表　DBH class

胸径区间 Diameter class (cm)	个体数 No. of individuals in the plot	比例 Proportion (%)
1~2	4	10.53
2~5	12	31.58
5~10	12	31.58
10~20	10	26.31
20~35	0	0.00
35~50	0	0.00
≥50	0	0.00

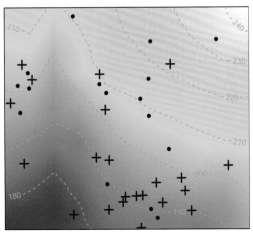

● 1~5 cm DBH　＋ 5~20 cm DBH　○ ≥20 cm DBH
个体分布图　Distribution of individuals

64 桃金娘 | táo jīn niáng （山稔）

***Rhodomyrtus tomentosa* (Aiton) Hassk.**

桃金娘科 Myrtaceae

样地名称（Plot name）＝ JH
个体数（Individual number/1 hm²）＝ 9
最大胸径（Max DBH）＝ 3.2 cm
重要值排序（Importance value rank）＝ 83

灌木，高 1～2 m。叶对生，革质，叶片椭圆形或倒卵形，长 3～8 cm，宽 1～4 cm，先端圆或钝，常微凹入，有时稍尖，基部阔楔形，离基三出脉，直达先端且相结合。花有长梗。浆果卵状壶形，熟时紫黑色。花期 4～5 月。果期 7～8 月。

Shrubs, 1–2 m tall. Leaves opposite. Leaf blade elliptic to obovate, 3–8 × 1–4 cm, leathery, originating near leaf blade base, and meeting at apex, base broadly cuneate, apex rounded to obtuse and often slightly emarginate or sometimes slightly apiculate. Flowers stipitate. Berry purplish black when mature, urceolate. Fl. Apr.–May, fr. Jul.–Aug..

树干　　　Trunk
摄影：李健星　Photo by：Li Jianxing

花枝　　　Flowering branch
摄影：李健星　Photo by：Li Jianxing

果　　　Fruit
摄影：李健星　Photo by：Li Jianxing

径级分布表　DBH class

胸径区间 Diameter class (cm)	个体数 No. of individuals in the plot	比例 Proportion (%)
1～2	5	55.56
2～5	4	44.44
5～10	0	0.00
10～20	0	0.00
20～35	0	0.00
35～50	0	0.00
≥50	0	0.00

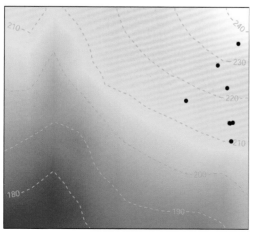

● 1～5 cm DBH　＋ 5～20 cm DBH　○ ≥20 cm DBH
个体分布图　Distribution of individuals

65 黑嘴蒲桃 | hēi zuǐ pú táo

蒲桃属

Syzygium bullockii **(Hance) Merr. L. M. Perry**

桃金娘科 Myrtaceae

样地名称（Plot name）= JH
个体数（Individual number/1 hm^2）= 33
最大胸径（Max DBH）= 16.5 cm
重要值排序（Importance value rank）= 32

灌木至小乔木，高达 5 m。叶片革质，椭圆形至卵状长圆形，长 4~12 cm，宽 2.5~5.5 cm，侧脉多数，以 70° 开角斜向上，离边缘 1~2 mm 处相结合成边脉，基部圆形或微心形，先端渐尖，尖头钝。圆锥花序顶生。果实椭圆形，花期 3~8 月，果期 8 月至翌年 2 月。

Shrubs to small trees, to 5 m tall. Leaf blade elliptic to ovate-oblong, 4–12 × 2.5–5.5 cm, leathery, secondary veins numerous and at an angle of ca. 70° from midvein, intramarginal veins 1–2 mm from margin, base rounded to slightly cordate, apex acuminate and with an obtuse acumen. Inflorescences terminal. Fruit ellipsoid. Fl. Mar.–Aug., fr. Aug.–Feb. of next year.

枝叶　Branch and leaves
摄影：丁涛　Photo by: Ding Tao

花序　Inflorescence
摄影：扈文芳　Photo by: Hu Wenfang

果　Fruits
摄影：扈文芳　Photo by: Hu Wenfang

径级分布表　DBH class

胸径区间 Diameter class (cm)	个体数 No. of individuals in the plot	比例 Proportion (%)
1~2	11	33.33
2~5	12	36.37
5~10	6	18.18
10~20	4	12.12
20~35	0	0.00
35~50	0	0.00
≥50	0	0.00

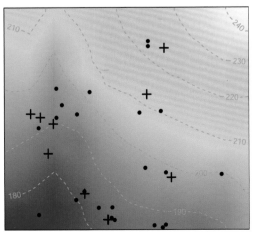

● 1~5 cm DBH　+ 5~20 cm DBH　○ ≥20 cm DBH
个体分布图　Distribution of individuals

66 红鳞蒲桃 | hóng lín pú táo

Syzygium hancei **Merr. L. M. Perry**

桃金娘科 Myrtaceae

样地名称（Plot name）= JH
个体数（Individual number/1 hm²）= 83
最大胸径（Max DBH）= 16.4 cm
重要值排序（Importance value rank）= 18

灌木或乔木，高达 20 m。叶片革质，狭椭圆形至长圆形或为倒卵形，长 3~7 cm，宽 1.5~4 cm，先端钝或略尖，基部阔楔形或较狭窄，上面干后暗褐色，不发亮，有多数细小而下陷的腺点，下面同色。圆锥花序腋生。果实球形，直径 5~6 mm。花期 7~9 月，果期 11 月至翌年 1 月。

Shrubs or trees, to 20 m tall. Leaf blade narrowly elliptic, oblong, or obovate, 3–7 × 1.5–4 cm, leathery, both surfaces dark brown when dry, adaxially not glossy and with numerous small impressed glands, base broadly cuneate to narrow, apex obtuse to slightly acute. Inflorescences axillary, paniculate cymes. Fruit globose, 5–6 mm in diam.. Fl. Jul.–Sep., fr. Nov.–Jan. of next year.

树干　Trunk
摄影：李健星　Photo by：Li Jianxing

枝叶　Branch and leaves
摄影：孙观灵　Photo by：Sun Guanling

果　Fruits
摄影：孙观灵　Photo by：Sun Guanling

径级分布表 DBH class

胸径区间 Diameter class (cm)	个体数 No. of individuals in the plot	比例 Proportion (%)
1~2	14	16.87
2~5	29	34.94
5~10	28	33.73
10~20	12	14.46
20~35	0	0.00
35~50	0	0.00
≥50	0	0.00

● 1~5 cm DBH　+ 5~20 cm DBH　○ ≥20 cm DBH
个体分布图 Distribution of individuals

67 狭叶蒲桃 | xiá yè shān pú táo

蒲桃属

***Syzygium levinei* (Merr.) Merr. L. M. Perry**
桃金娘科 Myrtaceae

样地名称（Plot name）= JH
个体数（Individual number/1 hm^2）= 57
最大胸径（Max DBH）= 14.9 cm
重要值排序（Importance value rank）= 28

常绿乔木，高 14~24 m。叶片革质，椭圆形或卵状椭圆形，长 4~8 cm，宽 1.5~3.5 cm，先端急锐尖，基部阔楔形，上面干后灰褐色，下面同色或稍淡，两面有细小腺点，侧脉以 45° 开角斜向上，脉间相隔 2~3 mm。圆锥花序顶生和上部腋生。果实近球形，长 7~8 mm。花期 6~9 月，果期翌年 2~5 月。

Evergreen trees, 14–24 m tall. Leaf blade elliptic to ovate-elliptic, 4–8 × 1.5–3.5 cm, leathery, both surfaces grayish brown when dry and with minute glands, secondary veins 2–3 mm apart and at an angle of ca. 45° from midvein, base broadly cuneate, apex acute. Inflorescences terminal or axillary on apical parts of branchlets, paniculate cymes. Fruit subglobose, 7–8 mm long. Fl. Jun.–Sep., fr. from Feb. to May of next year.

树干　　Trunk
摄影：李健星　　Photo by: Li Jianxing

叶背　　Leaf backs
摄影：李健星　　Photo by: Li Jianxing

枝叶　　Branch and leaves
摄影：李健星　　Photo by: Li Jianxing

径级分布表　DBH class

胸径区间 Diameter class (cm)	个体数 No. of individuals in the plot	比例 Proportion (%)
1~2	18	31.58
2~5	22	38.60
5~10	15	26.31
10~20	2	3.51
20~35	0	0.00
35~50	0	0.00
≥50	0	0.00

● 1~5 cm DBH　＋ 5~20 cm DBH　○ ≥20 cm DBH
个体分布图　Distribution of individuals

68 柏拉木 | bǎi lā mù （野锦香）

***Blastus cochinchinensis* Lour.**
野牡丹科 Melastomataceae

样地名称（Plot name）＝ JH
个体数（Individual number/1 hm²）＝ 17
最大胸径（Max DBH）＝ 7.1 cm
重要值排序（Importance value rank）＝ 61

灌木，高 0.6～3 m。茎圆柱形。叶片纸质或近坚纸质，长 6～12（～18）cm，宽 2～4（～5）cm，背面浓密腺，正面稀少腺但是后脱落的，次脉在中脉两边各 1（或 2）。伞状聚伞花序，腋生，总梗长约 2 mm 至几无。蒴果椭圆形，4 裂，为宿存萼所包。花期 6～8 月，果期 10～12 月。

Shrubs, 0.6–3 m tall. Stems terete. Leaf blade, 6–12 (–18) × 2–4 (–5) cm, papery to substiffly papery, abaxially densely glandular, adaxially sparsely glandular but glabrescent, secondary veins 1 (or 2) on each side of midvein. Inflorescences axillary, umbellate cymose; peduncle ca. 2 mm to almost absent. Ovary inferior, urceolate, 4-celled, slightly glandular. Capsule elliptic to ovoid, 4-sided. Fl. Jun.–Aug., fr. Oct.–Dec..

树干　　Trunk
摄影：李健星　　Photo by: Li Jianxing

枝叶　　Branch and leaves
摄影：李健星　　Photo by: Li Jianxing

果枝　　Fruiting branch
摄影：李健星　　Photo by: Li Jianxing

径级分布表　DBH class

胸径区间 Diameter class (cm)	个体数 No. of individuals in the plot	比例 Proportion (%)
1～2	11	64.71
2～5	5	29.41
5～10	1	5.88
10～20	0	0.00
20～35	0	0.00
35～50	0	0.00
≥50	0	0.00

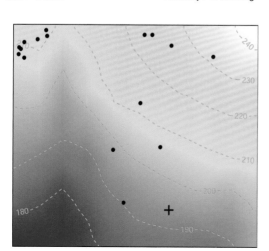

● 1～5 cm DBH　　＋ 5～20 cm DBH　　○ ≥20 cm DBH
个体分布图　Distribution of individuals

69 毛棯 | máo rěn （毛荵）

野牡丹属

Melastoma sanguineum Sims

野牡丹科 Melastomataceae

样地名称（Plot name）= JH
个体数（Individual number/1 hm²）= 9
最大胸径（Max DBH）= 11.9 cm
重要值排序（Importance value rank）= 80

灌木，高 1.5~3 m。茎、小枝、叶柄、花梗及花萼均被平展的长粗毛。叶片坚纸质，卵状披针形至披针形，顶端长渐尖或渐尖，基部钝或圆形，长（4.5~）8~15（~22）cm，宽（1.7~）2.5~5（~8）cm，两面具糙伏毛，次脉在中脉两边各 2。伞房花序。果杯状球形，胎座肉质。花期几乎全年，果期 8~10 月。

Shrubs, 1.5–3 m tall. Stems, branches, petioles, pedicels, and calyces densely hirsute, trichomes basally flattened. Leaf blade ovate-lanceolate to lanceolate, (4.5-) 8–15 (–22) × (1.7-) 2.5–5 (–8) cm, stiffly papery, both surfaces strigose, secondary veins 2 on each side of midvein, base obtuse to rounded, margin entire, apex long acuminate to acuminate. Inflorescences terminal. Fruit urceolate-turbinate, succulent. Fl. all year, fr. Aug.–Oct..

果枝　　　　　Fruiting branch
摄影：丁涛　　Photo by：Ding Tao

花　　　　　Flower
摄影：丁涛　　Photo by：Ding Tao

果　　　　　Fruits
摄影：向悟生　Photo by：Xiang Wusheng

径级分布表　DBH class

胸径区间 Diameter class (cm)	个体数 No. of individuals in the plot	比例 Proportion (%)
1~2	3	33.33
2~5	4	44.45
5~10	1	11.11
10~20	1	11.11
20~35	0	0.00
35~50	0	0.00
≥50	0	0.00

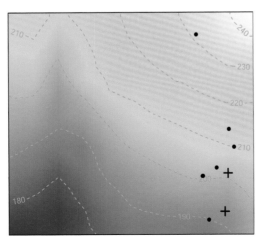

● 1~5 cm DBH　＋ 5~20 cm DBH　○ ≥20 cm DBH
个体分布图　Distribution of individuals

70 毛叶榄 | máo yè lǎn

Canarium subulatum Guillaumin

橄榄科 Burseraceae

样地名称（Plot name）= JH
个体数（Individual number/1 hm²）= 1
最大胸径（Max DBH）= 12.3 cm
重要值排序（Importance value rank）= 122

乔木，高 7～25（～35）m。小叶 3～6 对，纸质至革质，披针形或椭圆形（至卵形），长 6～14 cm，宽 2～5.5 cm，先端渐尖至骤狭渐尖，尖头长约 2 cm；基部楔形至圆形，全缘。花序腋生，雄花序为聚伞圆锥花序，雌花序为总状。果卵圆形至纺锤形，成熟时黄绿色；外果皮厚，干时有皱纹。花期 4～5 月，果期 10～12 月。

Trees, 7–25 (–35) m tall. Leaflets 3–6 pairs; blades lanceolate, elliptic, or ovate, 6–14 × 2–5.5 cm, base rounded or obliquely cuneate, margin entire, apex acuminate with acumen up to ca. 2 cm. Inflorescences axillary. Male flowers numerous in cymose panicles; female inflorescence racemose. Drupe ovoid or spindle-shaped, yellow-green; exocarp thick, wrinkled when dry. Fl. Apr.–May, fr. Oct.–Dec..

树干　　　Trunk
摄影：李健星　Photo by: Li Jianxing

花序　　　Inflorescence
摄影：李健星　Photo by: Li Jianxing

果　　　Fruits
摄影：李健星　Photo by: Li Jianxing

径级分布表 DBH class

胸径区间 Diameter class (cm)	个体数 No. of individuals in the plot	比例 Proportion (%)
1～2	0	0.00
2～5	0	0.00
5～10	0	0.00
10～20	1	100.00
20～35	0	0.00
35～50	0	0.00
≥50	0	0.00

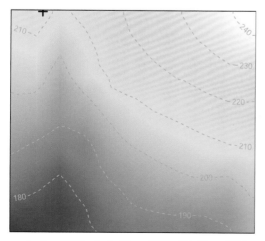

● 1～5 cm DBH　　＋ 5～20 cm DBH　　○ ≥20 cm DBH
个体分布图　Distribution of individuals

71 野漆 | yě qī （漆木）

Toxicodendron succedaneum (L.) Kuntze
漆树科 Anacardiaceae

样地名称（Plot name）= JH
个体数（Individual number/1 hm²）= 16
最大胸径（Max DBH）= 13.2 cm
重要值排序（Importance value rank）= 44

乔木或灌木，高达 1~2 （~10) m。叶柄长 6~9 cm。奇数羽状复叶互生，常集生小枝顶端，无毛，有小叶 5~15 对，叶轴和叶柄圆柱形；小叶对生或近对生，长 3~16 cm，宽 0.9~5.5 cm。圆锥花序长 7~15 cm，核果大，不对称，压扁，先端偏离中心，外果皮薄，淡黄色，无毛，中果皮厚，蜡质，白色，果核坚硬。花期 5 月，果期 7~10 月。

Trees or shrubs, to 1–2 (–10) m tall. Petiole 6–9 cm. Leaf blade imparipinnately compound; leaflets 5–15, opposite or subopposite; leaflet, 3–16 × 0.9–5.5 cm. Inflorescence paniculate, 7–15 cm. Drupe large, asymmetrical, compressed, apex eccentric; epicarp thin, yellow, glabrous; mesocarp thick, white, waxy. Fl. May, fr. Jul.-Oct..

枝叶　Branch and leaves
摄影：丁涛　Photo by：Ding Tao

复叶　Compound leaf
摄影：丁涛　Photo by：Ding Tao

小叶　leaflets
摄影：丁涛　Photo by：Ding Tao

径级分布表　DBH class

胸径区间 Diameter class (cm)	个体数 No. of individuals in the plot	比例 Proportion (%)
1~2	2	12.50
2~5	3	18.75
5~10	7	43.75
10~20	4	25.00
20~35	0	0.00
35~50	0	0.00
≥50	0	0.00

● 1~5 cm DBH　＋ 5~20 cm DBH　○ ≥20 cm DBH
个体分布图　Distribution of individuals

72 赤才 | chì cái

Lepisanthes rubiginosa (Roxb.) Leenh.
无患子科 Sapindaceae

样地名称（Plot name）= JH
个体数（Individual number/1 hm²）= 4
最大胸径（Max DBH）= 1.5 cm
重要值排序（Importance value rank）= 93

常绿灌木或小乔木，高 2~3 （~7）m。小叶 2~8 对，革质，第一对（近基）卵形，明显较小，向上渐大，椭圆状卵形至长椭圆形，长 3~20 cm，顶端钝或圆，很少短尖，全缘。花序通常为复总状。果爿长 12~14 mm，宽 5~7 mm，红色。花期春季，果期夏季。

Evergreen shrubs or small trees, usually 2–3 (–7)m tall. Leaflets 2–8 pairs; first pair (near base) ovate, evidently smaller, gradually larger toward leaf apex, elliptic-ovate to narrowly elliptic, 3–20 cm, leathery, base broadly cuneate to rounded, margin entire, apex obtuse or rounded, rarely acute. Inflorescences compound racemose. Fertile schizocarps red, 12–14 × 5–7 mm. Fl. spring, fr. summer.

鳞花木属

树干　　Trunk
摄影：李健星　Photo by: Li Jianxing

小叶背面　　Leaflet backs
摄影：李健星　Photo by: Li Jianxing

果序　　Infructescences
摄影：椰子　Photo by: Ye Zi

径级分布表　DBH class

胸径区间 Diameter class (cm)	个体数 No. of individuals in the plot	比例 Proportion (%)
1~2	4	100.00
2~5	0	0.00
5~10	0	0.00
10~20	0	0.00
20~35	0	0.00
35~50	0	0.00
≥50	0	0.00

● 1~5 cm DBH　＋ 5~20 cm DBH　○ ≥20 cm DBH
个体分布图　Distribution of individuals

73 韶子 | sháo zǐ

韶子属

Nephelium chryseum Blume
无患子科 Sapindaceae

样地名称（Plot name）= JH
个体数（Individual number/1 hm²）= 14
最大胸径（Max DBH）= 11.0 cm
重要值排序（Importance value rank）= 57

常绿乔木，高 10~20 m 或更高。叶连柄长 20~40 cm；小叶常 4 对，很少 2 或 3 对，薄革质，长圆形，长 6~18 cm，宽 2.5~7.5 cm，两端近短尖，全缘，背面粉绿色，被柔毛。花序多分枝，雄花序与叶近等长，雌花序较短。果椭圆形，红色。花期春季，果期夏季。

Evergreen trees, 10–20 m tall or more. Leaves with petiole 20–40 cm. Leaflets (2–) 4 pairs; blades 6–18 × 2.5–7.5 cm, thinly leathery, glaucous, base and apex nearly acute. Inflorescences many branched, male ones nearly as long as leaves, female ones shorter. Fruit red, ellipsoid. Fl. spring, fr. summer.

树干　　　Trunk
摄影：李健星　　Photo by: Li Jianxing

复叶　　　Compound leaves
摄影：丁涛　　Photo by: Ding Tao

小叶背面　　Leaflet backs
摄影：丁涛　　Photo by: Ding Tao

径级分布表　DBH class

胸径区间 Diameter class (cm)	个体数 No. of individuals in the plot	比例 Proportion (%)
1~2	11	78.57
2~5	1	7.15
5~10	1	7.14
10~20	1	7.14
20~35	0	0.00
35~50	0	0.00
≥50	0	0.00

● 1~5 cm DBH　　＋ 5~20 cm DBH　　○ ≥20 cm DBH
个体分布图　Distribution of individuals

74 假黄皮 | jiǎ huáng pí

Clausena excavata N. L. Burman

芸香科 Rutaceae

样地名称（Plot name）= JH
个体数（Individual number/1 hm²）= 9
最大胸径（Max DBH）= 2.5 cm
重要值排序（Importance value rank）= 74

灌木，高 1～2 m。小枝及叶轴均密被向上弯的短柔毛且散生微凸起的油点。叶有小叶 21～27 片，小叶柄长 2～5 mm。小叶片不对称，斜卵形，斜披针形或斜四边形，长 2～9 cm，宽 1～3 cm，边缘波浪状，两面被毛或仅叶脉有毛，老叶几无毛。花序顶生。果椭圆形，长 12～18 mm，宽 8～15 mm。花期 4～5 及 7～8 月，盛果期 8～10 月。

Shrubs, 1–2 m tall. Branchlets and leaf rachises pubescent, with oil glands. Leaves 21–27-foliolate; petiolules 2–5 mm. Leaflet blades ovate, lanceolate, or rhomboid, asymmetric, 2–9 × 1–3 cm, both surfaces pubescent or only pubescent along veins, base oblique, margin repand. Inflorescences terminal. Fruit ellipsoid, 12–18 × 8–15 mm. Fl. Apr.–May and Jul.–Aug., fr. Aug.–Oct..

花枝 Flowering branch
摄影：杨平 Photo by：Yang Ping

复叶 Compound leaves
摄影：杨平 Photo by：Yang Ping

果 Fruit
摄影：杨平 Photo by：Yang Ping

径级分布表 DBH class

胸径区间 Diameter class (cm)	个体数 No. of individuals in the plot	比例 Proportion (%)
1～2	6	66.67
2～5	3	33.33
5～10	0	0.00
10～20	0	0.00
20～35	0	0.00
35～50	0	0.00
≥50	0	0.00

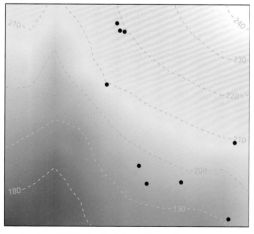

● 1～5 cm DBH ✚ 5～20 cm DBH ○ ≥20 cm DBH
个体分布图 Distribution of individuals

75 少花山小橘 | shǎo huā shān xiǎo jú

山小橘属

***Glycosmis oligantha* C. C. Huang**

芸香科 Rutaceae

样地名称（Plot name）= JH
个体数（Individual number/1 hm²）= 11
最大胸径（Max DBH）= 7.2 cm
重要值排序（Importance value rank）= 76

灌木或小乔木，高约 3 m。叶有小叶（3～）5～7 片；小叶狭披针形，两端渐尖，长 5～9 cm，宽 1.5～2.5 cm，边全缘，干后浅波浪状起伏，叶面灰榄绿色，叶背淡黄色，幼嫩叶背沿中脉两侧被稀疏的鳞秕状锈色微柔毛，叶面中脉下半段稍凹陷。花序腋生，1～5 朵。子房球形；花柱极短。花期 7～10 月，果期 1～3 月。

Shrubs or small trees, to 3 m tall. Leaves (3 or) 5- or 7-foliolate. Leaflet blades narrowly lanceolate, 5–9 × 1.5–2.5 cm, base attenuate, margin entire or repand, apex acuminate. Inflorescences axillary, 1–5-flowered. Ovary globose; style extremely short. Fl. Jul.–Oct., fr. Jan.–Mar..

果枝　Fruiting branches
摄影：李健星　Photo by: Li Jianxing

花序　Inflorescences
摄影：李健星　Photo by: Li Jianxing

果　Fruits
摄影：李健星　Photo by: Li Jianxing

径级分布表 DBH class

胸径区间 Diameter class (cm)	个体数 No. of individuals in the plot	比例 Proportion (%)
1～2	9	81.82
2～5	1	9.09
5～10	1	9.09
10～20	0	0.00
20～35	0	0.00
35～50	0	0.00
≥50	0	0.00

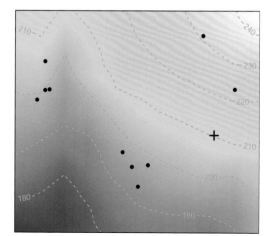

● 1～5 cm DBH　＋ 5～20 cm DBH　○ ≥20 cm DBH
个体分布图 Distribution of individuals

76 三桠苦 | sān yā kǔ （密茱萸）

蜜茱萸属

Melicope pteleifolia **(Champ. ex Benth.) T. G. Hartley**

芸香科 Rutaceae

样地名称（Plot name）= JH
个体数（Individual number/1 hm^2）= 16
最大胸径（Max DBH）= 3.0 cm
重要值排序（Importance value rank）= 62

灌木或乔木，高 1～14 m。3 小叶，有时偶有 2 小叶或单小叶同时存在，叶柄基部稍增粗，小叶长椭圆形，两端尖，有时倒卵状椭圆形，长 6～22 cm，宽 2～8 cm。花序腋生。分果瓣淡黄或茶褐色，散生肉眼可见的透明油点。花期 4～6 月，果期 7～10 月。

Shrubs or trees, 1–14 m tall. Leaves 3-foliolate (occasional leaves 1-foliolate). Leaflet blades ovate-elliptic, elliptic, elliptic-obovate, or narrowly so, in terminal leaflet 6–22 × 2–8 cm, apex acuminate or subcaudate. Inflorescences axillary. Fruit follicles subglobose to ellipsoid to obovoid. Fl. Apr.–Jun., fr. Jul.–Oct..

果枝　Fruiting branch
摄影：丁涛　Photo by: Ding Tao

叶　Leaves
摄影：丁涛　Photo by: Ding Tao

叶背　Leaf backs
摄影：丁涛　Photo by: Ding Tao

径级分布表　DBH class

胸径区间 Diameter class (cm)	个体数 No. of individuals in the plot	比例 Proportion (%)
1～2	7	43.75
2～5	9	56.25
5～10	0	0.00
10～20	0	0.00
20～35	0	0.00
35～50	0	0.00
≥50	0	0.00

● 1～5 cm DBH　＋ 5～20 cm DBH　○ ≥20 cm DBH
个体分布图　Distribution of individuals

77　九里香 | jiǔ lǐ xiāng　（千里香）

Murraya exotica L.

芸香科 Rutaceae

样地名称（Plot name）＝ JH
个体数（Individual number/1 hm^2）＝ 1
最大胸径（Max DBH）＝ 5.5 cm
重要值排序（Importance value rank）＝ 129

乔木，高可达 8 m。枝白灰或淡黄灰色，但当年生枝绿色。叶有小叶 3~7 片，小叶倒卵形成倒卵状椭圆形，两侧常不对称，长 1~6 cm，宽 0.5~3 cm，顶端圆或钝，有时微凹，基部短尖，一侧略偏斜，边全缘，平展；小叶柄甚短。花序通常顶生，或顶生兼腋生。果橙黄至朱红色，阔卵形，长 8~12 mm，横径 6~10 mm。花期 4~8 月，果期 9~12 月。

Trees, to 8 m tall. Older branchlets grayish white to pale yellowish gray. Leaves 3–7-foliolate; petiolules rather short. Leaflet blades elliptic-obovate or obovate, 1–6 × 0.5–3 cm, margin entire, apex rounded or obtuse. Inflorescences terminal or terminal and axillary. Fruit orange to vermilion, broadly ovoid, 8–12 × 6–10 mm. Fl. Apr.–Aug., fr. Sep.–Dec..

枝叶　Branch and leaves
摄影：丁涛　Photo by：Ding Tao

叶背　Leaf backs
摄影：丁涛　Photo by：Ding Tao

果枝　Fruiting branch
摄影：丁涛　Photo by：Ding Tao

径级分布表　DBH class

胸径区间 Diameter class (cm)	个体数 No. of individuals in the plot	比例 Proportion (%)
1~2	0	0.00
2~5	0	0.00
5~10	1	100.00
10~20	0	0.00
20~35	0	0.00
35~50	0	0.00
≥50	0	0.00

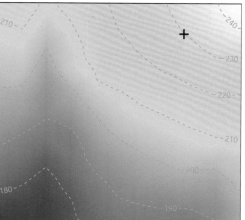

● 1~5 cm DBH　＋ 5~20 cm DBH　○ ≥20 cm DBH
个体分布图　Distribution of individuals

78 簕欓花椒 | lè dǎng huā jiāo （簕档花椒）

花椒属

Zanthoxylum avicennae (Lam.) DC.
芸香科 Rutaceae

样地名称（Plot name）= JH
个体数（Individual number/1 hm²）= 6
最大胸径（Max DBH）= 4.5 cm
重要值排序（Importance value rank）= 84

落叶乔木，高稀达 15 m。叶有小叶 11～21 片，稀较少；小叶通常对生或偶有不整齐对生，斜卵形，斜长方形或呈镰刀状，有时倒卵形，顶部短尖或钝，全缘，或中部以上有疏裂齿，叶轴常呈狭翼状。花序顶生，花多。分果瓣淡紫红色，油点大且多。花期 6～8 月，果期 10～12 月。

Deciduous trees, to 15 m tall. Leaves 11–21-foliolate; rachis winged; leaflet blades opposite or rarely subopposite, obliquely ovate, rhomboidal, obovate, or falcate, margin entire or apically crenate, apex mucronate to blunt. Inflorescences terminal, many flowered. Fruit pedicel; follicles pale purplish red, oil glands numerous, large. Fl. Jun.–Aug., fr. Oct.–Dec..

树干 　　Trunk
摄影：李健星　　Photo by：Li Jianxing

花序 　　Inflorescence
摄影：李健星　　Photo by：Li Jianxing

果序 　　Infructescence
摄影：李健星　　Photo by：Li Jianxing

径级分布表 DBH class

胸径区间 Diameter class (cm)	个体数 No. of individuals in the plot	比例 Proportion (%)
1～2	0	0.00
2～5	6	100.00
5～10	0	0.00
10～20	0	0.00
20～35	0	0.00
35～50	0	0.00
≥50	0	0.00

● 1～5 cm DBH　　+ 5～20 cm DBH　　○ ≥20 cm DBH
个体分布图 Distribution of individuals

79 米仔兰 | mǐ zǎi lán （小叶米仔兰）

Aglaia odorata Lour.
楝科 Meliaceae

样地名称（Plot name）＝ JH
个体数（Individual number/1 hm^2）＝ 1
最大胸径（Max DBH）＝ 10.0 cm
重要值排序（Importance value rank）＝ 123

灌木或小乔木；茎多小枝。幼枝顶部被星状锈色的鳞片。叶长5～12（～16）cm，叶轴和叶柄具狭翅，有小叶3～7（～9）片；小叶对生，厚纸质，长1～7（～11）cm，宽0.5～3.5（～5）cm，顶端1片最大，先端钝，基部楔形。聚伞圆锥花序腋生。果为浆果，卵形或近球形。花期5～12月，果期7月至翌年3月。

Shrubs or small trees, much branching. Young branches apically with stellate or lepidote trichomes. Leaves 5–12 (–16) cm; petiole and rachis narrowly winged. Leaflets 3–7 (or 9), opposite. Leaflet blades usually obovate, sometimes elliptic, 1–7 (–11) × 0.5–3.5 (–5) cm with apical one biggest, base cuneate, apex obtuse. Thyrses axillary. Fruit indehiscent, ovoid to subglobose. Fl. May–Dec., fr. Jul.–Mar. of next year.

树干　Trunk
摄影：李健星　Photo by：Li Jianxing

复叶　Compound leaves
摄影：丁涛　Photo by：Ding Tao

果序　Infructescences
摄影：李健星　Photo by：Li Jianxing

径级分布表　DBH class

胸径区间 Diameter class (cm)	个体数 No. of individuals in the plot	比例 Proportion (%)
1～2	0	0.00
2～5	0	0.00
5～10	0	0.00
10～20	1	100.00
20～35	0	0.00
35～50	0	0.00
≥50	0	0.00

● 1～5 cm DBH　＋ 5～20 cm DBH　○ ≥20 cm DBH
个体分布图　Distribution of individuals

80 破布叶 | pò bù yè （布渣叶）

***Microcos paniculata* L.**

锦葵科 Malvaceae

样地名称（Plot name）= JH
个体数（Individual number/1 hm^2）= 4
最大胸径（Max DBH）= 3.6 cm
重要值排序（Importance value rank）= 109

灌木或小乔木，高 3～12 m。树皮粗糙。叶薄革质，卵状长圆形，长 8～18 cm，宽 4～8 cm，先端渐尖，基部圆形，两面初时有极稀疏星状柔毛，以后变秃净，三出脉的两侧脉从基部发出，向上行超过叶片中部，边缘有细钝齿。顶生圆锥花序长 4～10 cm。核果近球形或倒卵形，果柄短。花期 6～7 月，果期冬季。

Shrubs or small trees, 3–12 m tall. Bark rough. Leaf blade ovate or oblong, 8–18 × 4–8 cm, thinly leathery, very sparsely stellate at first and glabrescent both abaxially and adaxially, basal veins 3, laterals more than 1/2 as long as leaf blade, base rounded, margin finely crenate, apex acuminate. Panicles terminal, 4–10 cm. Drupe nearly globose or obovoid, stipe short. Fl. Jun.–Jul., fr. winter.

叶背　　Leaf backs
摄影：丁涛　　Photo by: Ding Tao

果　　Fruits
摄影：丁涛　　Photo by: Ding Tao

果枝　　Fruiting branch
摄影：丁涛　　Photo by: Ding Tao

径级分布表　DBH class

胸径区间 Diameter class (cm)	个体数 No. of individuals in the plot	比例 Proportion (%)
1～2	2	50.00
2～5	2	50.00
5～10	0	0.00
10～20	0	0.00
20～35	0	0.00
35～50	0	0.00
≥50	0	0.00

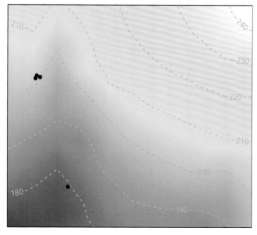

● 1～5 cm DBH　　+ 5～20 cm DBH　　○ ≥20 cm DBH
个体分布图　Distribution of individuals

81 翻白叶树 | fān bái yè shù （异叶翅子木）

翅子树属

***Pterospermum heterophyllum* Hance**

锦葵科 Malvaceae

样地名称（Plot name）= JH
个体数（Individual number/1 hm²）= 60
最大胸径（Max DBH）= 35.7 cm
重要值排序（Importance value rank）= 21

乔木，高达 20 m。叶二形，生于幼树或萌蘖枝上的叶盾形，直径约 15 cm，掌状 3~5 裂，基部截形而略近半圆形；生于成长的树上的叶矩圆形至卵状矩圆形，长 7~15 cm，宽 3~10 cm，顶端钝、急尖或渐尖，基部钝、截形或斜心形。花单生或 2~4 朵组成聚伞花序。蒴果木质。花期秋季，果期 6~11 月。

Trees, to 20 m tall. Leaves dimorphic, juvenile and coppice leaves, leaf blade palmately 3–5-lobed, ca. 15 cm in diam., base conspicuously peltate, truncate or slightly rounded. Leaf blade oblong-ovate to oblong, 7–15 × 3–10 cm, base obtuse, truncate or obliquely cordate, apex acute or acuminate; juvenile and coppice leaves. Flowers solitary or in cymes of 2–4. Capsule woody. Fl. autumn, fr. Jun.–Nov..

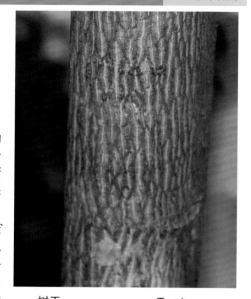

树干　　　Trunk
摄影：李健星　　Photo by: Li Jianxing

叶背　　　Leaf backs
摄影：李健星　　Photo by: Li Jianxing

枝叶　　　Branch and leaves
摄影：李健星　　Photo by: Li Jianxing

径级分布表　DBH class

胸径区间 Diameter class (cm)	个体数 No. of individuals in the plot	比例 Proportion (%)
1~2	3	5.00
2~5	15	25.00
5~10	31	51.67
10~20	9	15.00
20~35	1	1.67
35~50	1	1.66
≥50	0	0.00

● 1~5 cm DBH　＋ 5~20 cm DBH　○ ≥20 cm DBH
个体分布图　Distribution of individuals

82 梭罗树 | suō luó shù （两广梭罗）

Reevesia pubescens Mast.

锦葵科 Malvaceae

样地名称（Plot name）= JH
个体数（Individual number/1 hm^2）= 3
最大胸径（Max DBH）= 3.7 cm
重要值排序（Importance value rank）= 112

常绿乔木。树皮灰褐色。叶革质，长圆形到椭圆形，长 5~7 cm，宽 2.5~3 cm，顶端急尖或渐尖，基部圆形或钝，两面均无毛；叶柄长 1~3 cm，两端膨大。聚伞状伞房花序顶生，被毛，花密集。蒴果矩圆状梨形，有 5 棱，长约 3 cm，被短柔毛；种子连翅长约 2 cm。花期 3~4 月，果期 4~12 月。

Evergreen trees. Bark gray-brown. Petiole 1–3 cm, swollen at both ends. Leaf blade oblong to elliptic, 5–7 × 2.5–3 cm, leathery, both surfaces glabrous, base rounded or obtuse, apex acute or acuminate. Inflorescence cymose-corymbose, densely flowered, puberulent. Capsule on ca. 2 cm stipe, oblong-pyriform, 5-angular, ca. 3 cm, puberulent. Seeds ca.2 cm including wing. Fl. Mar.–Apr., fr. Apr.–Dec..

果枝　Fruiting branches
摄影：丁涛　Photo by：Ding Tao

叶背　Leaf back
摄影：丁涛　Photo by：Ding Tao

果　Fruit
摄影：丁涛　Photo by：Ding Tao

径级分布表　DBH class

胸径区间 Diameter class (cm)	个体数 No. of individuals in the plot	比例 Proportion (%)
1~2	0	0.00
2~5	3	100.00
5~10	0	0.00
10~20	0	0.00
20~35	0	0.00
35~50	0	0.00
≥50	0	0.00

● 1~5 cm DBH　+ 5~20 cm DBH　○ ≥20 cm DBH
个体分布图　Distribution of individuals

83 假苹婆 | jiǎ píng pó

苹婆属

***Sterculia lanceolata* Cav.**

锦葵科 Malvaceae

样地名称（Plot name）= JH
个体数（Individual number/1 hm²）= 140
最大胸径（Max DBH）= 21.7 cm
重要值排序（Importance value rank）= 11

乔木。小枝幼时被毛。叶椭圆形、披针形或椭圆状披针形，长 9~20 cm，宽 3.5~8 cm，顶端急尖，基部钝形或近圆形。圆锥花序腋生，蓇葖果鲜红色，长卵形或长椭圆形，顶端有喙，基部渐狭，密被短柔毛。种子黑褐色，椭圆状卵形，直径约 1 cm。每果有种子 2~4 个。花期 4~6 月，果期 6~8 月。

Trees. Branchlets at first pilose. Leaf blade elliptic, lanceolate, or elliptic-lanceolate, 9–20 × 3.5–8 cm, base obtuse or nearly rounded, apex acute. Inflorescence paniculate. Follicle red when fresh, narrowly ovoid or ellipsoid, 2–4-seeded, densely puberulent, base attenuate, apex beaked. Seeds black-brown, ellipsoid-ovoid, ca. 1 cm. Fl. Apr.–Jun., fr. Jun.–Aug..

果枝　Fruiting branch
摄影：丁涛　Photo by：Ding Tao

花序　Inflorescence
摄影：丁涛　Photo by：Ding Tao

蓇葖果　Follicle
摄影：丁涛　Photo by：Ding Tao

径级分布表　DBH class

胸径区间 Diameter class (cm)	个体数 No. of individuals in the plot	比例 Proportion (%)
1~2	26	18.57
2~5	35	25.00
5~10	41	29.29
10~20	37	26.43
20~35	1	0.71
35~50	0	0.00
≥50	0	0.00

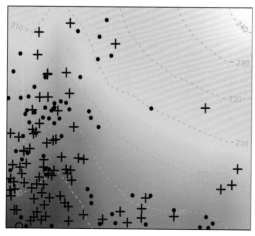

● 1~5 cm DBH　＋ 5~20 cm DBH　○ ≥20 cm DBH
个体分布图　Distribution of individuals

84 八角枫 | bā jiǎo fēng

八角枫属

Alangium chinense (Lour.) Harms
山茱萸科 Cornaceae

样地名称（Plot name）＝ JH
个体数（Individual number/1 hm²）＝ 2
最大胸径（Max DBH）＝ 7.4 cm
重要值排序（Importance value rank）＝ 113

灌木或小乔木，高 3～5 m。叶缘卵形或圆形到心形，长 8～20 cm，宽 5～12 cm，叶脉腋部背面丛生短柔毛，正面无毛，基部 3～5 脉，基部通常斜，偶尔圆形或近圆形，或三角形，边缘全缘或具少量浅裂片，先端渐尖。聚伞花序腋生，核果卵圆形。花期 5～7 月和 9～10 月，果期 7～11 月。

Shrubs or small trees, 3–5 m tall. Leaf blade ovate or orbicular to cordate, 8–20 × 5–12 cm, abaxially tufted pubescent at axils of veins, adaxially glabrous, strongly 3–5-veined at base, base usually oblique, occasionally rounded or subrounded, or triangular, margin entire or with few shallow lobes, apex acuminate. Inflorescences axillary cymes. Drupe ovoid. Fl. May–Jul., Sep.–Oct., fr. Jul.–Nov..

叶　　Leaves
摄影：丁涛　　Photo by：Ding Tao

叶背　　Leaf backs
摄影：丁涛　　Photo by：Ding Tao

花　　Flowers
摄影：丁涛　　Photo by：Ding Tao

径级分布表　DBH class

胸径区间 Diameter class (cm)	个体数 No. of individuals in the plot	比例 Proportion (%)
1～2	1	50.00
2～5	0	0.00
5～10	1	50.00
10～20	0	0.00
20～35	0	0.00
35～50	0	0.00
≥50	0	0.00

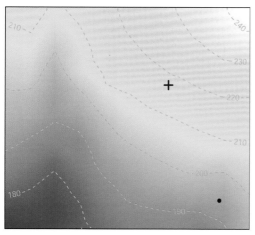

● 1～5 cm DBH　＋ 5～20 cm DBH　○ ≥20 cm DBH
个体分布图　Distribution of individuals

85 黑柃 | hēi líng

柃属

Eurya macartneyi Champion

五列木科 Pentaphylacaceae

样地名称（Plot name）= JH
个体数（Individual number/1 hm²）= 84
最大胸径（Max DBH）= 23.3 cm
重要值排序（Importance value rank）= 23

灌木或小乔木，高 2~7 m。嫩枝粗壮，圆柱形，淡红褐色，无毛，小枝灰褐色或褐色。叶革质，长圆状椭圆形或椭圆形，长 6~14 cm，宽 2~4.5 cm，顶端短渐尖，基部近钝形或阔楔形，边缘几全缘，或上半部密生细微锯齿。花 1~4 朵簇生于叶腋。果实圆球形，成熟时黑色。花期 11 月至翌年 1 月，果期 6~8 月。

Shrubs or small trees, 2–7 m tall. Young branches grayish brown to brown; current year branchlets pale reddish brown, stout, terete, glabrous. Leaf blade oblong-elliptic to elliptic, 6–14 × 2–4.5 cm, leathery, base broadly cuneate to obtuse, margin subentire to apically serrulate, apex shortly acuminate. Flowers axillary, solitary or to 4 in a cluster. Fruit purplish black when mature, globose. Fl. Nov.–Jan. of next year, fr. Jun.–Aug..

枝叶　Branch and leaves
摄影：杨平　Photo by: Yang Ping

叶　Leaves
摄影：杨平　Photo by: Yang Ping

果　Fruits
摄影：杨平　Photo by: Yang Ping

径级分布表　DBH class

胸径区间 Diameter class (cm)	个体数 No. of individuals in the plot	比例 Proportion (%)
1~2	12	14.29
2~5	37	44.05
5~10	29	34.52
10~20	4	4.76
20~35	2	2.38
35~50	0	0.00
≥50	0	0.00

● 1~5 cm DBH　+ 5~20 cm DBH　○ ≥20 cm DBH
个体分布图　Distribution of individuals

86 大果毛柃 | dà guǒ máo líng （大毛果柃）

Eurya megatrichocarpa Hung T. Chang

五列木科 Pentaphylacaceae

样地名称（Plot name）＝ JH
个体数（Individual number/1 hm²）＝ 9
最大胸径（Max DBH）＝ 28.7 cm
重要值排序（Importance value rank）＝ 58

灌木或乔木，高 6~8 m。叶薄革质，长圆形或长圆状椭圆形，长 7~11 cm，宽 2~3 cm，顶端渐尖或锐尖，基部楔形或阔楔形，边缘密生细锯齿，侧脉 8~10 对，纤细，在离叶缘颇远处即连结，在上面不甚明显，下面明显隆起。花通常 2~5 朵簇生于叶腋。果实圆球形，成熟时暗紫色。花期 11~12 月，果期翌年 7~8 月。

Shrubs or trees, 6–8 m tall. Leaf blade oblong to oblong-elliptic, 7–11 × 2–3 cm, thinly leathery, secondary veins 8–10 on each side of midvein, slender, abaxially raised, and adaxially obscure, base cuneate to broadly cuneate, margin closely serrulate, apex acuminate to acute. Flowers axillary, usually 2–5 in a cluster. Fruit dark purple when mature, globose. Fl. Nov.–Dec., fr. Jul.–Aug. of next year.

枝叶　Branch and leaves
摄影：向悟生　Photo by: Xiang Wusheng

叶　Leaves
摄影：向悟生　Photo by: Xiang Wusheng

叶背　Leaf backs
摄影：向悟生　Photo by: Xiang Wusheng

径级分布表　DBH class

胸径区间 Diameter class (cm)	个体数 No. of individuals in the plot	比例 Proportion (%)
1~2	1	11.11
2~5	3	33.34
5~10	2	22.22
10~20	2	22.22
20~35	1	11.11
35~50	0	0.00
≥50	0	0.00

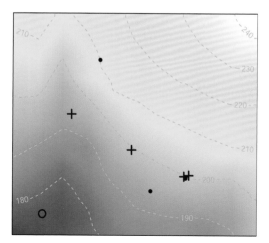

● 1~5 cm DBH　＋ 5~20 cm DBH　○ ≥20 cm DBH
个体分布图　Distribution of individuals

87 细齿叶柃 | xì chǐ yè líng （台湾柃）　柃属

Eurya nitida Korth.
五列木科 Pentaphylacaceae

样地名称（Plot name）= JH
个体数（Individual number/1 hm²）= 38
最大胸径（Max DBH）= 19.8 cm
重要值排序（Importance value rank）= 27

灌木或乔木，高 2~5 m。嫩枝稍纤细，具 2 棱。叶薄革质，椭圆形、长圆状椭圆形或倒卵状长圆形，长（3~）4~6（~7）cm，宽 1.5~2.5 cm，顶端渐尖或短渐尖，尖头钝，基部楔形，边缘密生锯齿或细钝齿，侧脉 9~12 对。花生于叶腋。果实圆球形，成熟时蓝黑色。花期 11 月至翌年 1 月，果期翌年 7~9 月。

Shrubs or trees, 2–5 m tall. Young branches slender, 2-ribbed. Leaf blade elliptic, oblong-elliptic, or obovate-oblong, (3–) 4–6 (–7) × 1.5–2.5 cm, thinly leathery to leathery, secondary veins 9–12 on each side of midvein, base cuneate to rounded, margin closely serrulate, crenulate, or subentire, apex obtuse, acute. Flowers axillary. Fruit bluish black when mature, globose. Fl. Nov.–Jan. of next year, fr. Jul.–Sep. of next year.

枝叶　Branch and leaves
摄影：丁涛　Photo by: Ding Tao

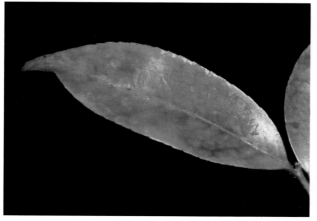
叶　Leaf
摄影：丁涛　Photo by: Ding Tao

叶背　Leaf backs
摄影：丁涛　Photo by: Ding Tao

径级分布表　DBH class

胸径区间 Diameter class (cm)	个体数 No. of individuals in the plot	比例 Proportion (%)
1~2	10	26.32
2~5	12	31.58
5~10	6	15.79
10~20	10	26.31
20~35	0	0.00
35~50	0	0.00
≥50	0	0.00

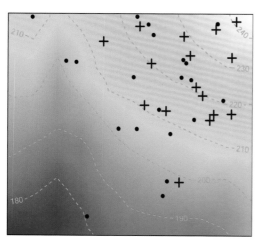

● 1~5 cm DBH　+ 5~20 cm DBH　○ ≥20 cm DBH
个体分布图　Distribution of individuals

88 大叶五室柃 | dà yè wǔ shì líng （细叶子）

Eurya quinquelocularis Kobuski

五列木科 Pentaphylacaceae

样地名称（Plot name）＝ JH
个体数（Individual number/1 hm²）＝ 20
最大胸径（Max DBH）＝ 19.5 cm
重要值排序（Importance value rank）＝ 45

灌木或小乔木，高 3～10 （～20）m。叶近膜质或薄纸质，长圆形或长圆状卵形，长 7～13 cm，宽 2～3.5 cm，顶端渐尖或长渐尖而呈尾状，基部楔形或阔楔形，边缘密生细锯齿，侧脉 12～14 对，网脉两面均甚明显。花 1 至数朵簇生于叶腋。果圆球形，成熟时黑色。花期 11～12 月，果期翌年 6～7 月。

Shrubs or small trees, 3–10 (–20) m tall. Leaf blade oblong to oblong-ovate, 7–13 × 2–3.5 cm, membranous to papery, secondary veins 12–14 on each side of midvein, secondary and reticulate veins visible on both surfaces, base cuneate to broadly cuneate, margin closely serrulate, apex acuminate to caudate-acuminate. Flowers axillary. Fruit black when mature, globose. Fl. Nov.–Dec., fr. Jun.–Jul. of next year.

果枝 Fruiting branches
摄影：杨平 Photo by: Yang Ping

叶背 Leaf backs
摄影：杨平 Photo by: Yang Ping

果 Fruits
摄影：杨平 Photo by: Yang Ping

径级分布表 DBH class

胸径区间 Diameter class (cm)	个体数 No. of individuals in the plot	比例 Proportion (%)
1～2	3	15.00
2～5	9	45.00
5～10	4	20.00
10～20	4	20.00
20～35	0	0.00
35～50	0	0.00
≥50	0	0.00

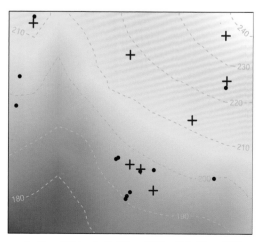

● 1～5 cm DBH　＋ 5～20 cm DBH　○ ≥20 cm DBH
个体分布图 Distribution of individuals

89 锈毛梭子果 | xiù máo suō zǐ guǒ

***Eberhardtia aurata* (Pierre ex Dubard) Lecomte**

山榄科 Sapotaceae

样地名称（Plot name）= JH
个体数（Individual number/1 hm^2）= 15
最大胸径（Max DBH）= 17.6 cm
重要值排序（Importance value rank）= 48

乔木，高 7~15 m，胸径约 20~40 cm。树皮暗灰色；嫩枝被锈色绒毛，叶近革质，长圆形、倒卵状长圆形或椭圆形，长 12~24 cm，宽 4.5~9.5 cm，先端骤然渐尖，基部楔形或近圆形。花具香味。果核果状，近球形，绿色转锈褐色，下垂，被锈色绒毛，干时现 5 棱，基部具宿存萼。花期 3 月，果期 9~12 月。

Trees, 7–15 m tall. Trunk 20–40 cm d.b.h., bark dark gray. Branchlets rust colored tomentose. Leaf blade oblong, obovate-oblong, or elliptic, 12–24×4.5–9.5 cm, base cuneate to subrounded, apex abruptly acuminate. Flowers aromatic. Fruit with persistent calyx, subglobose, 5-ribbed, rust colored tomentose. Fl. Mar., fr. Feb.–Dec..

枝叶　Branch and leaves
摄影：丁涛　Photo by: Ding Tao

花　Flower
摄影：杨平　Photo by: Yang Ping

果　Fruit
摄影：椰子　Photo by: Ye Zi

径级分布表　DBH class

胸径区间 Diameter class (cm)	个体数 No. of individuals in the plot	比例 Proportion (%)
1~2	5	33.33
2~5	4	26.67
5~10	2	13.33
10~20	4	26.67
20~35	0	0.00
35~50	0	0.00
≥50	0	0.00

● 1~5 cm DBH　+ 5~20 cm DBH　○ ≥20 cm DBH
个体分布图　Distribution of individuals

90 紫荆木 | zǐ jīng mù （铁色）

Madhuca pasquieri (Dubard) H. J. Lam
山榄科 Sapotaceae

样地名称（Plot name）= JH
个体数（Individual number/1 hm^2）= 5
最大胸径（Max DBH）= 9.0 cm
重要值排序（Importance value rank）= 103

乔木，高达30 m，胸径达60 cm。树皮灰黑色。叶星散或密聚于分枝顶端，革质，倒卵形或倒卵状长圆形，长6~16 cm，宽2~6 cm，先端阔渐尖而钝头或骤然收缩，基部阔渐尖或尖楔形。花数朵簇生叶腋。果椭圆形或小球形，先端具宿存、花后延长的花柱，果皮肥厚，被锈色绒毛。花期7~9月，果期10月至翌年1月。

Trees, to 30 m tall. Trunk to 60 cm d.b.h.. Bark blackish. Leaves scattered or more often closely clustered at end of branchlets. Leaf blade obovate to obovate-oblong, 6–16×2–6 cm, base broadly acuminate to cuneate, apex broadly acuminate to abruptly acute. Flowers several, axillary, fascicled. Fruit ellipsoid to globose, with lengthened style, rust colored tomentose but glabrescent; pericarp fleshy. Fl. Jul.–Sep., fr. Oct.–Jan. of next year.

树干　　Trunk
摄影：李健星　　Photo by：Li Jianxing

叶背　　Leaf backs
摄影：李健星　　Photo by：Li Jianxing

枝叶　　Branch and leaves
摄影：向悟生　　Photo by：Xiang Wusheng

径级分布表　DBH class

胸径区间 Diameter class (cm)	个体数 No. of individuals in the plot	比例 Proportion (%)
1~2	2	40.00
2~5	1	20.00
5~10	2	40.00
10~20	0	0.00
20~35	0	0.00
35~50	0	0.00
≥50	0	0.00

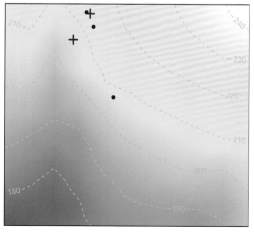
● 1~5 cm DBH　　+ 5~20 cm DBH　　○ ≥20 cm DBH
个体分布图　Distribution of individuals

91 肉实树 | ròu shí shù （水石梓）

Sarcosperma laurinum (Benth.) Hook. f.
山榄科 Sapotaceae

样地名称（Plot name）= JH
个体数（Individual number/1 hm²）= 328
最大胸径（Max DBH）= 27.5 cm
重要值排序（Importance value rank）= 3

乔木，高 6~15（~26）m。板根显著；小枝具棱，无毛。叶近革质，通常倒卵形或倒披针形，长 7~16（~19）cm，宽 3~6 cm，先端通常骤然急尖，有时钝至钝渐尖，基部楔形。总状花序或为圆锥花序腋生。核果长圆形或椭圆形，由绿至红至紫红转黑色。花期 8~9 月，果期 12 月至翌年 1 月。

Trees, 6–15 (–26) m tall. Conspicuously buttressed. Branchlets angulate, glabrous. leaf blade usually obovate to oblanceolate, 7–16 (–19)×3–6 cm, almost leathery, adaxially dark and shiny green, base cuneate, apex usually acute but sometimes obtuse to obtuse-acuminate. Racemes or panicles axillary. Drupe green, becoming red and finally black, oblong to ellipsoid. Fl. Aug.–Sep., fr. Dec.–Jan. of next year.

树干　Trunk
摄影：丁涛　Photo by: Ding Tao

叶　Leaves
摄影：丁涛　Photo by: Ding Tao

花序　Inflorescence
摄影：刘冰　Photo by: Liu Bing

径级分布表　DBH class

胸径区间 Diameter class (cm)	个体数 No. of individuals in the plot	比例 Proportion (%)
1~2	61	18.60
2~5	89	27.13
5~10	91	27.74
10~20	79	24.09
20~35	8	2.44
35~50	0	0.00
≥50	0	0.00

● 1~5 cm DBH　+ 5~20 cm DBH　○ ≥20 cm DBH
个体分布图　Distribution of individuals

92 罗浮柿 | luó fú shì

Diospyros morrisiana Hance

柿科 Ebenaceae

样地名称（Plot name）= JH
个体数（Individual number/1 hm²）= 88
最大胸径（Max DBH）= 20.0 cm
重要值排序（Importance value rank）= 15

落叶乔木或灌木，高可达 20 m，胸径可达 30 cm。树皮呈片状剥落，表面黑色。叶薄革质，长椭圆形或下部的为卵形，长 5~11.5 cm，宽 2.5~4.3 cm，先端短渐尖或钝，基部楔形。雄花密集，下弯，聚伞花序式，雌花单生。果球形，黄色。花期 5~6 月，果期 11 月。

Deciduous shrubs or trees, up to 20 m tall. Trunk to 30 cm d.b.h.. Bark peeling off in thin pieces, surface black. Leaf blade elliptic, 5–11.5×2.5–4.3 cm, thinly leathery, base cuneate or obtuse, apex acuminate to obtuse. Male flowers congested, cymose, nodding; female flowers solitary. Berries yellow, globose. Fl. May–Jun., fr. Nov..

树干　　　　Trunk
摄影：丁涛　　Photo by：Ding Tao

枝叶　　　　Branch and leaves
摄影：丁涛　　Photo by：Ding Tao

果　　　　Fruits
摄影：丁涛　　Photo by：Ding Tao

径级分布表　DBH class

胸径区间 Diameter class (cm)	个体数 No. of individuals in the plot	比例 Proportion (%)
1~2	8	9.09
2~5	20	22.73
5~10	33	37.50
10~20	26	29.54
20~35	1	1.14
35~50	0	0.00
≥50	0	0.00

● 1~5 cm DBH　　+ 5~20 cm DBH　　○ ≥20 cm DBH
个体分布图　Distribution of individuals

93 大罗伞树 | dà luó sǎn shù （郎伞树）

紫金牛属

Ardisia hanceana Mez

报春花科 Primulaceae

样地名称（Plot name）＝ JH
个体数（Individual number/1 hm²）＝ 6
最大胸径（Max DBH）＝ 2.9 cm
重要值排序（Importance value rank）＝ 86

灌木，高 1～3 m。叶片坚纸质，椭圆状或长圆状披针形，顶端长急尖或渐尖，基部楔形，长 9～12（～15）cm，宽 2.5～4 cm，近全缘或具边缘反卷的疏突尖锯齿，齿尖具边缘腺点，背面近边缘通常具隆起的疏腺点，侧脉 12～15 对，近边缘连成边缘脉。复伞房状伞形花序。果球形，深红色。花期 6～7 月，果期 12 至翌年 4 月。

Shrubs, 1–3 m tall. Leaf blade elliptic, oblanceolate, or rarely ovate, 9–12 (–15) × 2.5–4 cm, leathery, punctate, base cuneate and decurrent, margin crenate or subentire, apex acute or acuminate; lateral veins 12–15 on each side of midrib, marginal vein obscure. Inflorescences terminal corymbose panicles of umbels. Fruit dull red or black, globose. Fl. Jun.–Jul., fr. Dec.–Apr. of next year.

果枝　Fruiting branch
摄影：丁涛　Photo by: Ding Tao

叶背　Leaf backs
摄影：丁涛　Photo by: Ding Tao

果　Fruits
摄影：丁涛　Photo by: Ding Tao

径级分布表　DBH class

胸径区间 Diameter class (cm)	个体数 No. of individuals in the plot	比例 Proportion (%)
1～2	3	50.00
2～5	3	50.00
5～10	0	0.00
10～20	0	0.00
20～35	0	0.00
35～50	0	0.00
≥50	0	0.00

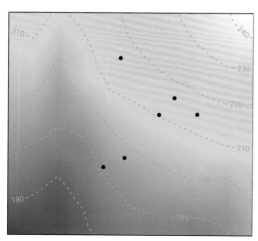

● 1～5 cm DBH　＋ 5～20 cm DBH　○ ≥20 cm DBH
个体分布图　Distribution of individuals

94 罗伞树 | luó sǎn shù （海南罗伞树）

紫金牛属

Ardisia quinquegona Blume
报春花科 Primulaceae

样地名称（Plot name）= JH
个体数（Individual number/1 hm²）= 2
最大胸径（Max DBH）= 2.8 cm
重要值排序（Importance value rank）= 125

灌木，高约 2 m，可达 6 m 以上。小枝细，无毛，有纵纹，嫩时被锈色鳞片。叶片坚纸质，长圆状披针形、椭圆状披针形至倒披针形，顶端渐尖，基部楔形，长 8～16 cm，宽 2～4 cm，全缘。聚伞花序腋生。果扁球形，具钝 5 棱。花期 3～7 月，果期 8 月至翌年 2 月。

Shrubs, 2 (–6) m tall. Branchlets angular, brown scaly, glabrescent, longitudinally ridged. Leaf blade oblong, elliptic, or oblanceolate, 8–16 × 2–4 cm, membranous, base cuneate, margin flat, entire, apex narrowly acute to acuminate. Cymes axillary. Fruit depressed, obtusely 5-angled. Fl. Mar.–Jul., fr. Aug.–Feb. of next year.

果枝　Fruiting branches
摄影：丁涛　Photo by: Ding Tao

叶背　Leaf backs
摄影：丁涛　Photo by: Ding Tao

果　Fruits
摄影：丁涛　Photo by: Ding Tao

径级分布表　DBH class

胸径区间 Diameter class (cm)	个体数 No. of individuals in the plot	比例 Proportion (%)
1～2	0	0.00
2～5	2	100.00
5～10	0	0.00
10～20	0	0.00
20～35	0	0.00
35～50	0	0.00
≥50	0	0.00

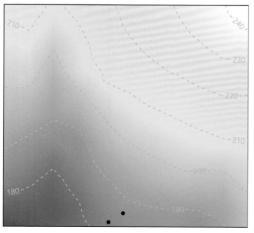

● 1～5 cm DBH　+ 5～20 cm DBH　○ ≥20 cm DBH
个体分布图　Distribution of individuals

95 米珍果 | mǐ zhēn guǒ （尖叶杜茎山） 杜茎山属

***Maesa acuminatissima* Merr.**
报春花科 Primulaceae

样地名称（Plot name）= JH
个体数（Individual number/1 hm²）= 27
最大胸径（Max DBH）= 13.1 cm
重要值排序（Importance value rank）= 42

灌木，高1~2（~4）m。小枝纤细，无毛。叶片膜质或略厚，披针形或广披针形，顶端渐尖或尾状渐尖，常镰状，基部钝或近圆形，长9~17 cm，宽2~5 cm，全缘或具极不明显的疏离浅波状齿。金字塔形圆锥花序，顶生及腋生。果球形或近卵圆形。花期1~2月，果期11~12月。

Shrubs, 1–2 (–4) m tall. Branchlets terete, slender, glabrous; pith hollow. Leaf blade lanceolate or broadly so, 9–17 × 2–5 cm, papery, glabrous, inconspicuously pellucid punctate, glossy adaxially, dull abaxially, base obtuse or subrounded, margin entire to obscurely undulate, apex acuminate or caudate-acuminate. Inflorescences terminal or subterminal, paniculate. Fruit globose or subovoid. Fl. Jan.–Feb., fr. Nov.–Dec..

树干　　Trunk
摄影：李健星　　Photo by: Li Jianxing

叶　　Leaves
摄影：李健星　　Photo by: Li Jianxing

果序　　Infructescences
摄影：丁涛　　Photo by: Ding Tao

径级分布表　DBH class

胸径区间 Diameter class (cm)	个体数 No. of individuals in the plot	比例 Proportion (%)
1~2	7	25.93
2~5	8	29.63
5~10	9	33.33
10~20	3	11.11
20~35	0	0.00
35~50	0	0.00
≥50	0	0.00

● 1~5 cm DBH　＋ 5~20 cm DBH　○ ≥20 cm DBH
个体分布图　Distribution of individuals

96 密花树 | mì huā shù

Myrsine seguinii H. Lév.

报春花科 Primulaceae

样地名称（Plot name）= JH
个体数（Individual number/1 hm²）= 207
最大胸径（Max DBH）= 24.5 cm
重要值排序（Importance value rank）= 8

灌木或小乔木，高 2~7 m，可达 12 m。小枝圆柱形，具皱纹，有皮孔。叶片革质，长圆状倒披针形至倒披针形，顶端急尖或钝，基部楔形，多少下延，长 7~17 cm，宽 1.3~6 cm，全缘。花序 3~10 朵花簇生。果球形或近卵形，灰绿色或紫黑色。花期 4~5 月，果期 10~12 月。

Shrubs or small trees, 2–7 (–12) m tall. Branchlets terete, white lenticellate, rugose. Leaf blade elliptic to narrowly linear-oblanceolate 7–17 × 1.3–6 cm, leathery, glabrous, base cuneate, decurrent, margin entire, apex acute. Inflorescences 3–10-flowered. Fruit grayish green or purplish black, globose or subovate. Fl. Apr.–May, fr. Oct.–Dec..

树干　　Trunk
摄影：李健星　Photo by: Li Jianxing

叶　　Leaves
摄影：李健星　Photo by: Li Jianxing

果　　Fruits
摄影：丁涛　Photo by: Ding Tao

径级分布表　DBH class

胸径区间 Diameter class (cm)	个体数 No. of individuals in the plot	比例 Proportion (%)
1~2	57	27.54
2~5	79	38.16
5~10	38	18.36
10~20	31	14.97
20~35	2	0.97
35~50	0	0.00
≥50	0	0.00

• 1~5 cm DBH　+ 5~20 cm DBH　○ ≥20 cm DBH
个体分布图　Distribution of individuals

97 长尾毛蕊茶 | cháng wěi máo ruǐ chá （膜叶连蕊茶） 山茶属

***Camellia caudata* Wall.**

山茶科 Theaceae

样地名称（Plot name）＝ JH
个体数（Individual number/1 hm^2）＝ 158
最大胸径（Max DBH）＝ 17.6 cm
重要值排序（Importance value rank）＝ 22

灌木至小乔木，高 2～8 m。叶薄革质，长圆形、披针形或椭圆形，长 7～12 cm，宽 2.5～4（～5）cm，先端尾状渐尖，基部楔形，边缘有细锯齿，叶柄长 3～7 mm，有柔毛或茸毛。花腋生，单生或 3 朵一簇。蒴果圆球形，果爿薄。花期 10 月至翌年 1 月，果期 9～10 月。

Shrubs to small trees, 2–8 m tall. Petiole 3–7 mm, villous or pubescent; Leaf blade elliptic, oblong-elliptic, or oblong, 7–12 × 2.5–4 (–5) cm, thinly leathery, base cuneate to broadly cuneate, margin serrulate to crenulate-serrulate, apex long caudate to caudate. Flowers axillary, solitary or to 3 in a cluster. Capsule ellipsoid-globose; pericarp thinly leathery. Fl. Oct.–Jan. of next year, fr. Sep.–Oct..

枝叶　Branches and leaves
摄影：丁涛　Photo by：Ding Tao

果　Fruit
摄影：丁涛　Photo by：Ding Tao

果枝　Fruiting branches
摄影：丁涛　Photo by：Ding Tao

径级分布表　DBH class

胸径区间 Diameter class (cm)	个体数 No. of individuals in the plot	比例 Proportion (%)
1～2	84	53.16
2～5	66	41.77
5～10	6	3.80
10～20	2	1.27
20～35	0	0.00
35～50	0	0.00
≥50	0	0.00

● 1～5 cm DBH　＋ 5～20 cm DBH　○ ≥20 cm DBH
个体分布图　Distribution of individuals

98 显脉金花茶 | xiǎn mài jīn huā chá

Camellia euphlebia Merr. ex Sealy
山茶科 Theaceae

样地名称（Plot name）= JH
个体数（Individual number/1 hm²）= 13
最大胸径（Max DBH）= 2.7 cm
重要值排序（Importance value rank）= 43

灌木或小乔木，高 2~5 m。叶革质，椭圆形，长（11~）14~20（~25）cm，宽（4.5~）5~8（~15）cm，先端急短尖，基部钝或近于圆，背面淡绿色，有棕色腺点，正面深绿色，侧脉 11~13 对，在上面稍下陷，在下面显著突起。花单生于叶腋。花期 12 月，果期翌年 10 月。

Shrubs or small trees, 2–5 m tall. Leaf blade elliptic, (11–) 14–20 (–25) × (4.5-) 5–8 (–15) cm, leathery, abaxially pale green, brown glandular punctate, adaxially dark green, midvein abaxially elevated and adaxially impressed, secondary veins 11–13, abaxially raised, and adaxially slightly impressed, base obtuse to subrounded, margin serrulate. Flowers axillary, solitary. Fl. Dec., fr. Oct. of next year.

植株　Whole plant
摄影：丁涛　Photo by: Ding Tao

叶背　Leaf backs
摄影：丁涛　Photo by: Ding Tao

花　Flower
摄影：李健星　Photo by: Li Jianxing

径级分布表　DBH class

胸径区间 Diameter class (cm)	个体数 No. of individuals in the plot	比例 Proportion (%)
1~2	12	92.31
2~5	1	7.69
5~10	0	0.00
10~20	0	0.00
20~35	0	0.00
35~50	0	0.00
≥50	0	0.00

● 1~5 cm DBH　+ 5~20 cm DBH　○ ≥20 cm DBH
个体分布图　Distribution of individuals

99 硬叶糙果茶 | yìng yè cāo guǒ chá （山油茶）

***Camellia gaudichaudii* (Gagnep.) Sealy**

山茶科 Theaceae

样地名称（Plot name）= JH
个体数（Individual number/1 hm^2）= 1
最大胸径（Max DBH）= 1.6 cm
重要值排序（Importance value rank）= 136

灌木或乔木，高 3~5 m。叶硬革质，椭圆形，长 5~7 cm，宽 2.5~3 cm，先端急短尖，尖头钝，基部楔形，上面干后暗褐色，稍发亮，无毛，下面浅褐色，无毛，有腺点，边缘上半部有疏而小的浅钝齿，或近全缘。花顶生，或近枝顶叶腋生。蒴果圆球形，3 室，3 片裂开。花期 12 月至翌年 1 月，果期翌年 8~9 月。

Shrubs or trees, 3–5 m tall. Leaf blade elliptic to ovate-elliptic, 5–7 × 2.5–3 cm, rigidly leathery, abaxially pale green and brown glandular punctate, adaxially dark green, both surfaces glabrous, base broadly cuneate, margin widely and shallowly denticulate, apex obtuse to rounded. Flowers axillary or subterminal. Capsule grayish, subglobose, 3-loculed; splitting into 3 valves. Fl. Dec.–Jan. of next year, fr. Aug.–Sep. of next year.

果枝　Fruiting branch
摄影：丁涛　Photo by: Ding Tao

叶　Leaves
摄影：丁涛　Photo by: Ding Tao

果　Fruit
摄影：丁涛　Photo by: Ding Tao

径级分布表　DBH class

胸径区间 Diameter class (cm)	个体数 No. of individuals in the plot	比例 Proportion (%)
1~2	1	100.00
2~5	0	0.00
5~10	0	0.00
10~20	0	0.00
20~35	0	0.00
35~50	0	0.00
≥50	0	0.00

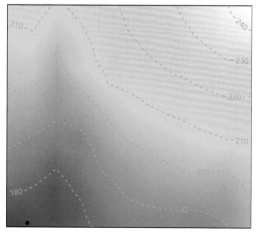

● 1~5 cm DBH　+ 5~20 cm DBH　○ ≥20 cm DBH
个体分布图　Distribution of individuals

100 东兴金花茶 | dōng xīng jīn huā chá

Camellia indochinensis var. *tunghinensis* (Hung T. Chang) T. L. Ming W. J. Zhang
山茶科 Theaceae

样地名称（Plot name）= JH
个体数（Individual number/1 hm²）= 60
最大胸径（Max DBH）= 3.9 cm
重要值排序（Importance value rank）= 39

灌木，高1～4 m。叶薄革质，椭圆形，长6～10.5 cm，宽（2.5～）3～4.5 cm，先端锐尖到渐尖，基部阔楔形，上面深绿色，干后变灰绿色，两面无毛，有棕色腺点，侧脉6～7对，边缘有细锯齿。花腋生，单生，金黄色。蒴果球形。花期3～4月，果期10～11月。

Shrubs, 1–4 m tall. Leaf blade elliptic, 6–10.5 × (2.5-) 3–4.5 cm, thinly leathery, abaxially pale green, brown glandular punctate, adaxially dark green and becoming grayish green when dry, both surfaces glabrous, secondary veins 6 or 7 on each side of midvein, base broadly cuneate, margin serrulate, apex acute to acuminate. Flowers axillary, solitary, pale yellow to yellowish white. Capsule oblate. Fl. Mar.–Apr., fr. Oct.–Nov..

花　　Flower
摄影：丁涛　　Photo by：Ding Tao

枝叶　　Branch and leaves
摄影：丁涛　　Photo by：Ding Tao

果　　Fruit
摄影：李健星　　Photo by：Li Jianxing

径级分布表　DBH class

胸径区间 Diameter class (cm)	个体数 No. of individuals in the plot	比例 Proportion (%)
1～2	36	60.00
2～5	24	40.00
5～10	0	0.00
10～20	0	0.00
20～35	0	0.00
35～50	0	0.00
≥50	0	0.00

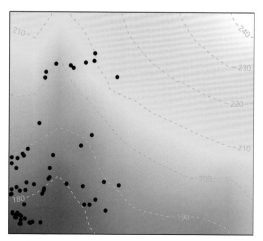

● 1～5 cm DBH　　+ 5～20 cm DBH　　○ ≥20 cm DBH
个体分布图　Distribution of individuals

101 柔毛紫茎 | róu máo zǐ jīng （毛折柄茶）

Stewartia villosa Merr.
山茶科 Theaceae

样地名称（Plot name）= JH
个体数（Individual number/1 hm^2）= 7
最大胸径（Max DBH）= 9.7 cm
重要值排序（Importance value rank）= 87

常绿乔木，高 8~20 m。叶长圆形至椭圆形，革质，长（4.5~）6~13（~21）cm，宽（2.4~）3~5.5（~7）cm，先端锐尖，基部圆形，不发亮，下面密被灰褐色柔毛，侧脉 10~16 对。叶柄长 10~20 mm，有翅，翅宽 2 mm。花单生于叶腋。蒴果球形。花期 6~7 月，果期 10~11 月。

Evergreen trees, 8–20 m tall. Petiole 10–20 mm, wings ca. 2 mm wide. Leaf blade elliptic-oblong, oblong, or lanceolate, (4.5–) 6–13 (–21) × (2.4–) 3–5.5 (–7) cm, both surfaces villous or pubescent but sometimes glabrescent, secondary veins 10–16 on each side of midvein, base rounded to obtuse, margin sparsely serrate, apex abruptly acute to shortly acuminate. Flowers solitary. Capsule conic. Fl. Jun.–Jul., fr. Oct.–Nov..

枝叶　Branch and leaves
摄影：杨平　Photo by: Yang Ping

叶背　Leaf back
摄影：杨平　Photo by: Yang Ping

花　Flower
摄影：唐忠炳　Photo by: Tang Zhongbing

径级分布表　DBH class

胸径区间 Diameter class (cm)	个体数 No. of individuals in the plot	比例 Proportion (%)
1~2	1	14.28
2~5	3	42.86
5~10	3	42.86
10~20	0	0.00
20~35	0	0.00
35~50	0	0.00
≥50	0	0.00

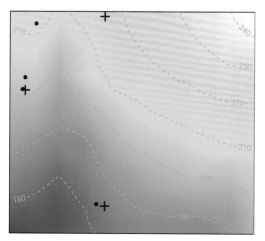

● 1~5 cm DBH　+ 5~20 cm DBH　○ ≥20 cm DBH
个体分布图　Distribution of individuals

102 薄叶山矾 | báo yè shān fán

Symplocos anomala Brand
山矾科 Symplocaceae

样地名称（Plot name）= JH
个体数（Individual number/1 hm²）= 4
最大胸径（Max DBH）= 6.6 cm
重要值排序（Importance value rank）= 89

灌木或小乔木。老枝通常黑褐色。叶薄革质，狭椭圆形、椭圆形或卵形，长2~7（~11）cm，宽1.2~3 cm，先端渐尖，基部楔形，全缘或具锐锯齿，叶面有光泽，中脉和侧脉在叶面均凸起，侧脉每边7~10条。总状花序腋生。核果褐色，长圆形，被短柔毛，有明显的纵棱，顶端宿萼裂片直立或向内伏。花果期4~12月。

Shrubs or small trees. Leaf blade narrowly elliptic, elliptic, or ovate, 2–7 (–11)×1.2–3 cm, thinly leathery, base attenuate-cuneate, margin entire or finely glandular dentate, apex acuminate, midvein and lateral veins adaxially prominent, lateral veins 7–10 pairs. Racemes axillary. Drupes brown, oblong-globose, apex with persistent erect calyx lobes. Fl. and fr. Apr.–Dec..

枝叶　　　　Branch and leaves
摄影：杨平　　Photo by: Yang Ping

叶背　　　　Leaf backs
摄影：杨平　　Photo by: Yang Ping

果　　　　　Fruits
摄影：杨平　　Photo by: Yang Ping

径级分布表　DBH class

胸径区间 Diameter class (cm)	个体数 No. of individuals in the plot	比例 Proportion (%)
1~2	0	0.00
2~5	3	75.00
5~10	1	25.00
10~20	0	0.00
20~35	0	0.00
35~50	0	0.00
≥50	0	0.00

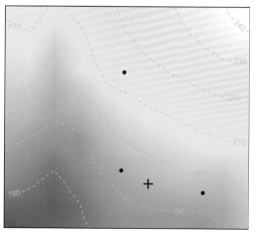

● 1~5 cm DBH　+ 5~20 cm DBH　○ ≥20 cm DBH
个体分布图　Distribution of individuals

103 光叶山矾 | guāng yè shān fán

Symplocos lancifolia Siebold Zucc.
山矾科 Symplocaceae

样地名称（Plot name）= JH
个体数（Individual number/1 hm²）= 50
最大胸径（Max DBH）= 18.2 cm
重要值排序（Importance value rank）= 30

灌木或乔木，高达 20 m。芽、嫩枝、嫩叶背面脉上、花序均被毛。叶纸质或近膜质，干后有时呈红褐色，卵形至阔披针形，长 2～10 cm，宽 1.5～4.3 cm，先端尾状渐尖，基部阔楔形或稍圆。核果近球形，顶端宿萼裂片。花期 3～11 月，果期 6～12 月。

Shrubs or trees, to 20 m. Buds, young branchlets, and inflorescence axes appressed to hairy. Leaf blade ovate, elliptic, narrowly ovate, or narrowly elliptic, 2–10 × 1.5–4.3 cm, submembranous to papery, base attenuate-cuneate, apex caudate-acuminate. Drupes subglobose, apex with persistent calyx lobes. Fl. Mar.–Nov., fr. Jun.–Dec..

果枝　Fruiting branch
摄影：唐忠炳　Photo by: Tang Zhongbing

枝叶　Branch and leaves
摄影：蒋裕良　Photo by: Jiang Yuliang

花序　Inflorescences
摄影：椰子　Photo by: Ye Zi

径级分布表　DBH class

胸径区间 Diameter class (cm)	个体数 No. of individuals in the plot	比例 Proportion (%)
1～2	4	8.00
2～5	18	36.00
5～10	21	42.00
10～20	7	14.00
20～35	0	0.00
35～50	0	0.00
≥50	0	0.00

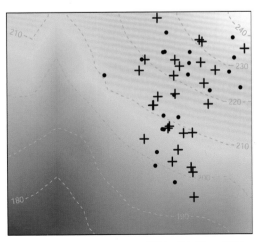

● 1～5 cm DBH　+ 5～20 cm DBH　○ ≥20 cm DBH
个体分布图　Distribution of individuals

104 老鼠屎 | lǎo shǔ shǐ （老鼠矢） 山矾属

***Symplocos stellaris* Brand**

山矾科 Symplocaceae

样地名称（Plot name）= JH
个体数（Individual number/1 hm²）= 2
最大胸径（Max DBH）= 1.9 cm
重要值排序（Importance value rank）= 117

灌木或小乔木。叶厚革质，叶面有光泽，叶背粉褐色，披针状椭圆形或狭长圆状椭圆形，长 6～23 cm，宽 1.8～5 cm，先端急尖或短渐尖，基部阔楔形或圆，通常全缘，很少有细齿；侧脉每边 7～15 条。团伞花序着生于 2 年生枝的叶痕之上。核果狭卵状圆柱形，顶端宿萼裂片直立。花期 2～5 月，果期 6～9 月。

Shrubs or small trees. Leaf blade narrowly oblong-elliptic to narrowly obovate, 6–23 × 1.8–5 cm, thickly leathery, glabrous, abaxially smooth, often light colored, and subglaucous, base broadly cuneate to rarely subrounded, apex mucronate-acuminate to acute, lateral veins 7–15 pairs. Inflorescences a glomerule or condensed spike. Drupes narrowly ovoid-cylindric, apex with persistent erect calyx lobes. Fl. Feb.–May, fr. Jun.–Sep..

枝叶　Branch and leaves
摄影：蒋裕良　Photo by: Jiang Yuliang

叶背　Leaf backs
摄影：蒋裕良　Photo by: Jiang Yuliang

花序　Inflorescence
摄影：杨平　Photo by: Yang Ping

径级分布表　DBH class

胸径区间 Diameter class (cm)	个体数 No. of individuals in the plot	比例 Proportion (%)
1～2	2	100.00
2～5	0	0.00
5～10	0	0.00
10～20	0	0.00
20～35	0	0.00
35～50	0	0.00
≥50	0	0.00

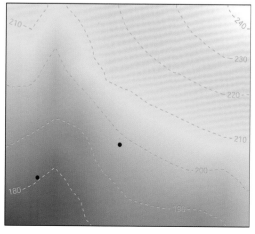
● 1～5 cm DBH　＋ 5～20 cm DBH　○ ≥20 cm DBH
个体分布图　Distribution of individuals

105 赤杨叶 | chì yáng yè （水冬瓜）

Alniphyllum fortunei (Hemsl.) Makino
安息香科 Styracaceae

样地名称（Plot name）= JH
个体数（Individual number/1 hm²）= 8
最大胸径（Max DBH）= 9.5 cm
重要值排序（Importance value rank）= 79

乔木，高 20 m。叶纸质，椭圆形、宽椭圆形或倒卵状椭圆形，长 5～15（～20）cm，宽 4～7（～11）cm，顶端急尖至渐尖，基部宽楔形或楔形，边缘具疏离硬质锯齿，两面疏生至密被褐色星状短柔毛或星状绒毛。总状花序或圆锥花序，顶生或腋生。果实长圆形或长椭圆形，疏被白色星状柔毛或无毛。花期 4～7 月，果期 8～10 月。

Trees, to 20 m tall. Leaf blade, elliptic, broadly elliptic, or obovate-elliptic, 5–15 (–20)×4–7 (–11) cm, papery, sparsely to densely brown stellate pubescent to tomentose, base broadly cuneate to cuneate, margin remotely serrate, apex acute to acuminate. Inflorescences terminal or axillary, racemes or panicles. Fruit oblong to ellipsoid, sparsely white stellate pubescent or glabrous. Fl. Apr.–Jul., fr. Aug.–Oct..

树干　Trunk
摄影：李健星　Photo by: Li Jianxing

叶　Leaf
摄影：唐文秀　Photo by: Tang Wenxiu

花　Flowers
摄影：丁涛　Photo by: Ding Tao

径级分布表　DBH class

胸径区间 Diameter class (cm)	个体数 No. of individuals in the plot	比例 Proportion (%)
1～2	1	12.50
2～5	1	12.50
5～10	6	75.00
10～20	0	0.00
20～35	0	0.00
35～50	0	0.00
≥50	0	0.00

● 1～5 cm DBH　+ 5～20 cm DBH　○ ≥20 cm DBH
个体分布图　Distribution of individuals

106 茜树 | qiàn shù （越南香楠）

茜树属

***Aidia cochinchinensis* Lour.**

茜草科 Rubiaceae

样地名称（Plot name）= JH
个体数（Individual number/1 hm²）= 134
最大胸径（Max DBH）= 9.9 cm
重要值排序（Importance value rank）= 19

灌木或乔木，高2~15 m。叶革质或纸质，对生，椭圆形到披针形，长9~15 cm，宽3~5 cm，顶端渐尖至尾状渐尖，有时短尖，基部楔形，两面无毛，上面稍光亮，下面脉腋内的小窝孔中常簇生短柔毛；侧脉5~10对。聚伞花序与叶对生或生于无叶的节上，多花。浆果球形，紫黑色。花期3~6月，果期5月至翌年2月。

Shrubs or trees, 2–15 m tall. Leaf blade drying leathery or papery, elliptic to lanceolate, 9–15 × 3–5 cm, both surfaces glabrous, base acute to obtuse, apex acute to acuminate; secondary veins 5–10 pairs, in abaxial axils usually with pilosulous and/or foveolate domatia; apex acuminate. Inflorescences cymose. Berry globose, purple black. Fl. Mar.–Jun., fr. May–Feb. of next year.

树干　　　Trunk
摄影：李健星　　Photo by：Li Jianxing

枝叶　　　Branch and leaves
摄影：蒋裕良　　Photo by：Jiang Yuliang

果　　　Fruit
摄影：李健星　　Photo by：Li Jianxing

径级分布表　DBH class

胸径区间 Diameter class (cm)	个体数 No. of individuals in the plot	比例 Proportion (%)
1~2	63	47.02
2~5	50	37.31
5~10	21	15.67
10~20	0	0.00
20~35	0	0.00
35~50	0	0.00
≥50	0	0.00

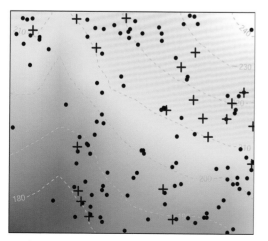

● 1~5 cm DBH　　+ 5~20 cm DBH　　○ ≥20 cm DBH
个体分布图　Distribution of individuals

107 猪肚木 | zhū dù mù （刺鱼骨木）

Canthium horridum Blume

茜草科 Rubiaceae

样地名称（Plot name）= JH
个体数（Individual number/1 hm²）= 15
最大胸径（Max DBH）= 9.2 cm
重要值排序（Importance value rank）= 53

灌木，高 2～3 m，具刺。小枝纤细；刺长 3～30 mm，劲直，锐尖。叶纸质，卵形，椭圆形或长卵形，长 2～6 cm，宽 1～3.5 cm，顶端钝、急尖或近渐尖，基部圆或阔楔形。花序簇生。核果黄色，卵形到近球形。花期 4～6 月，果期 7～11 月。

Shrubs, 2–3 m tall, spined. Branches rather slender; thorns slender to stout, 3–30 mm, straight. Leaves blade drying papery, ovate to lanceolate or elliptic, 2–6 × 1–3.5 cm, base rounded or obtuse, apex obtuse, acute, or weakly acuminate. Inflorescences fasciculate. Drupes yellow, ovoid to subglobose. Fl. Apr.–Jun., fr. Jul.–Nov..

叶背　　Leaf backs
摄影：丁涛　　Photo by：Ding Tao

枝叶　　Branch and leaves
摄影：杨平　　Photo by：Yang Ping

果　　Fruit
摄影：杨平　　Photo by：Yang Ping

径级分布表 DBH class

胸径区间 Diameter class (cm)	个体数 No. of individuals in the plot	比例 Proportion (%)
1～2	3	20.00
2～5	11	73.33
5～10	1	6.67
10～20	0	0.00
20～35	0	0.00
35～50	0	0.00
≥50	0	0.00

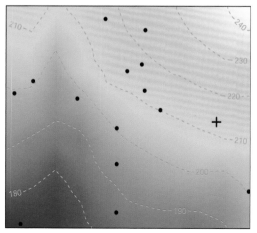

● 1～5 cm DBH　　+ 5～20 cm DBH　　○ ≥20 cm DBH
个体分布图　Distribution of individuals

108 大叶猪肚木 | dà yè zhū dù mù （大叶鱼骨木） 猪肚木属

Canthium simile Merr. Chun

茜草科 Rubiaceae

样地名称（Plot name）＝ JH
个体数（Individual number/1 hm^2）＝ 4
最大胸径（Max DBH）＝ 4.8 cm
重要值排序（Importance value rank）＝ 98

直立灌木至小乔木，高 4~10 m，有时高达 18 m，无刺，无毛。叶纸质，卵状长圆形，长 9~13 cm，宽 4.5~6.5 cm，顶端短渐尖，基部阔而急尖，两面无毛，微有光泽。伞房花序式的聚伞花序；总花梗长 10~14 mm。核果倒卵形。花期 1~3 月，果期 6~7 月。

Erect shrubs to small trees, 4–10 (–18) m tall, unarmed. Branches glabrous. Leaves blade papery, ovate to ovate-oblong, 9–13 × 4.5–6.5 cm, glabrous, base obtuse to rounded, apex shortly acuminate. Inflorescences corymbiform to cymose, peduncle 10–14 mm. Drupes obovoid. Fl. Jan.–Mar., fr. Jun.–Jul..

树干　Trunk
摄影：李健星　Photo by: Li Jianxing

叶背　Leaf backs
摄影：李健星　Photo by: Li Jianxing

枝叶　Branch and leaves
摄影：李健星　Photo by: Li Jianxing

径级分布表　DBH class

胸径区间 Diameter class (cm)	个体数 No. of individuals in the plot	比例 Proportion (%)
1~2	1	25.00
2~5	3	75.00
5~10	0	0.00
10~20	0	0.00
20~35	0	0.00
35~50	0	0.00
≥50	0	0.00

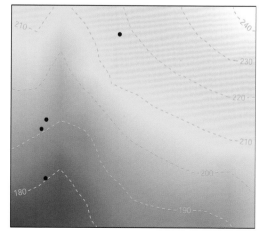

● 1~5 cm DBH　＋ 5~20 cm DBH　○ ≥20 cm DBH
个体分布图　Distribution of individuals

109 山石榴 | shān shí liú

Catunaregam spinosa (Thunb.) Tirveng.
茜草科 Rubiaceae

样地名称（Plot name）= JH
个体数（Individual number/1 hm²）= 19
最大胸径（Max DBH）= 5.9 cm
重要值排序（Importance value rank）= 60

有刺灌木或小乔木，高 1~10 m。刺腋生，对生，粗壮，长 1~5 cm。叶纸质或近革质，倒卵形或长圆状倒卵形，少为卵形至匙形，长 1.8~11 cm，宽 1~5.7 cm，顶端钝或短尖，基部楔形或下延。花单生或簇生于具叶、抑发的侧生短枝的顶部。浆果球形。花期 3~6 月，果期 5 月至翌年 1 月。

Shrubs or small trees, spined, 1–10 m tall. Armed with axillary stout paired thorns 1–5 cm. Leaf blade drying papery or subleathery, obovate or oblong-obovate or rarely ovate to spatulate, 1.8–11 × 1–5.7 cm, base cuneate and sometimes decurrent, apex acute. Inflorescences terminal on lateral short shoots together with tufted leaves. Berry globose. Fl. Mar.–Jun., fr. May–Jan. of next year.

树干　　　Trunk
摄影：李健星　　Photo by：Li Jianxing

枝叶　　　Branches and leaves
摄影：李健星　　Photo by：Li Jianxing

果　　　Fruit
摄影：丁涛　　Photo by：Ding Tao

径级分布表　DBH class

胸径区间 Diameter class (cm)	个体数 No. of individuals in the plot	比例 Proportion (%)
1~2	6	31.58
2~5	11	57.89
5~10	2	10.53
10~20	0	0.00
20~35	0	0.00
35~50	0	0.00
≥50	0	0.00

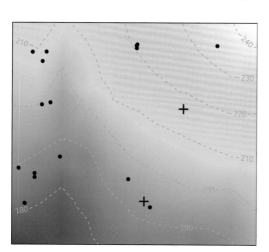

● 1~5 cm DBH　　+ 5~20 cm DBH　　○ ≥20 cm DBH
个体分布图　Distribution of individuals

110 白花龙船花 | bái huā lóng chuán huā

Ixora henryi H. Lév.

茜草科 Rubiaceae

样地名称（Plot name）= JH
个体数（Individual number/1 hm²）= 15
最大胸径（Max DBH）= 2.4 cm
重要值排序（Importance value rank）= 81

灌木，高 1~3 m，全部无毛。叶对生，纸质，长圆形或披针形，长 5~15 cm，宽 1.5~4 cm，顶端长渐尖或渐尖，基部楔形至阔楔形；侧脉每边 7~8 条，在叶片两面稍明显。花序顶生，多花，排成三歧伞房式的聚伞花序。果球形，直径 0.8~1 cm。花期 8~12 月，果期翌年 5~7 月。

Shrubs, 1–3 m tall. Branches glabrous. Leaves opposite; blade drying papery, elliptic-oblong, lanceolate, 5–15 × 1.5–4 cm, glabrous on both surfaces, base cuneate, obtuse, or rounded, apex sharply acute to usually acuminate; secondary veins 7 or 8 pairs. Inflorescence terminal, corymbiform to congested-cymose, many flowered, glabrous. Drupe subglobose, 8–10 mm in diam.. Fl. Apr.–Dec., fr. May–Jul. of next year.

果枝　Fruiting branch
摄影：丁涛　Photo by: Ding Tao

叶背　Leaf back
摄影：丁涛　Photo by: Ding Tao

果序　Infructescence
摄影：丁涛　Photo by: Ding Tao

径级分布表　DBH class

胸径区间 Diameter class (cm)	个体数 No. of individuals in the plot	比例 Proportion (%)
1~2	13	86.67
2~5	2	13.33
5~10	0	0.00
10~20	0	0.00
20~35	0	0.00
35~50	0	0.00
≥50	0	0.00

● 1~5 cm DBH　＋ 5~20 cm DBH　○ ≥20 cm DBH
个体分布图　Distribution of individuals

111 斜基粗叶木 | xié jī cū yè mù

粗叶木属

***Lasianthus attenuatus* Jack**

茜草科 Rubiaceae

样地名称（Plot name）= JH
个体数（Individual number/1 hm²）= 4
最大胸径（Max DBH）= 1.1 cm
重要值排序（Importance value rank）= 111

灌木，高通常 1~2 m，除叶上面和花冠外密被多细胞长硬毛或长柔毛。叶片纸质或近革质，椭圆状卵形或长圆状卵形，长 5~12 cm，宽 2.5~5 cm，顶端骤然渐尖，基部心形，两侧稍不对称。聚伞花序。核果近球形。花期 4 月，果期 8~9 月。

Shrubs, 1–2 m tall. Branches and branchlets densely tomentose to hirsute. Leaf blade leathery or subleathery, oblong, elliptic-lanceolate, or oblong-ovate, 5–12 × 2.5–5 cm, base slightly cordate or rarely rounded, slightly to markedly oblique, apex acute to cuspidate-acuminate. Inflorescence cymose. Fruit globose or ovoid. Fl. Apr., fr. Aug.–Sep..

果枝　Fruiting branch
摄影：丁涛　Photo by: Ding Tao

叶背　Leaf backs
摄影：丁涛　Photo by: Ding Tao

果序　Infructescence
摄影：丁涛　Photo by: Ding Tao

径级分布表　DBH class

胸径区间 Diameter class (cm)	个体数 No. of individuals in the plot	比例 Proportion (%)
1~2	4	100.00
2~5	0	0.00
5~10	0	0.00
10~20	0	0.00
20~35	0	0.00
35~50	0	0.00
≥50	0	0.00

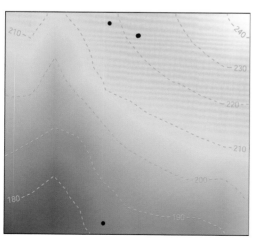

● 1~5 cm DBH　　+ 5~20 cm DBH　　○ ≥20 cm DBH
个体分布图　Distribution of individuals

112 粗叶木 | cū yè mù

Lasianthus chinensis (Champ. ex Benth.) Benth.

茜草科 Rubiaceae

样地名称（Plot name）= JH
个体数（Individual number/1 hm^2）= 2
最大胸径（Max DBH）= 1.3 cm
重要值排序（Importance value rank）= 118

灌木，高通常 1~4 m。枝和小枝均粗壮，被褐色短柔毛。叶薄近革质，通常为长圆形或长圆状披针形，很少椭圆形，长 11~25 cm，宽 2.5~7 cm，顶端常骤尖或有时近短尖，基部阔楔形或钝，上面无毛。花近头状到聚伞花序。核果近卵球形。花期 5~6 月，果期 9~10 月。

Shrubs, 1–4 m tall. Branches and branchlets terete, densely pubescent. Leaf blade subleathery, oblong to elliptic, 11–25 × 2.5–7 cm, glabrous adaxially, base acute or obtuse, apex acute or acuminate. Inflorescences subcapitate to congested-cymose. Fruit globose. Fl. May–Jun., fr. Sep.–Oct..

树干　Trunk
摄影：李健星　Photo by: Li Jianxing

枝叶　Branch and leaves
摄影：李健星　Photo by: Li Jianxing

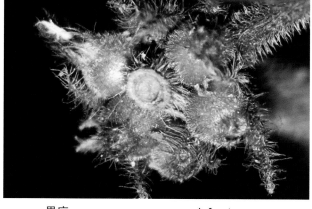

果序　Infructescence
摄影：李健星　Photo by: Li Jianxing

径级分布表　DBH class

胸径区间 Diameter class (cm)	个体数 No. of individuals in the plot	比例 Proportion (%)
1~2	2	100.00
2~5	0	0.00
5~10	0	0.00
10~20	0	0.00
20~35	0	0.00
35~50	0	0.00
≥50	0	0.00

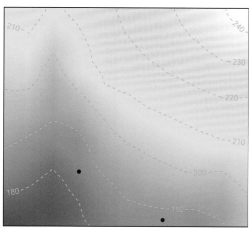

● 1~5 cm DBH　＋ 5~20 cm DBH　○ ≥20 cm DBH
个体分布图　Distribution of individuals

113 西南粗叶木 | xī nán cū yè mù

粗叶木属

***Lasianthus henryi* Hutch. in Sargent**

茜草科 Rubiaceae

样地名称（Plot name）= JH
个体数（Individual number/1 hm²）= 1
最大胸径（Max DBH）= 1.2 cm
重要值排序（Importance value rank）= 138

灌木，高 1~1.5 m；小枝常近长，密被贴伏的绒毛。叶纸质，广披针形，长 6~12 cm，宽 1.5~3.5 cm，顶端渐尖或短尾状渐尖，基部钝或圆，有缘毛，上面无毛，下面中脉、侧脉和横行小脉上被贴伏或稍伸展的硬毛状柔毛；侧脉 7~8 对。花序近头状。核果近球形，成熟时蓝色，无毛。花期 6 月，果期 7~10 月。

Shrubs, 1–1.5 m tall. Branches and branchlets densely appressed pubescent or strigillose. Leaf blade subleathery or papery, oblong-lanceolate, 6–12 × 1.5–3.5 cm, appressed pubescent abaxially on nerves or strigillose on veins, base acute or cuneate, apex cuspidate-acuminate; lateral veins 7 or 8 pairs. Inflorescences glomerulate to subcapitate. Drupes subglobose, blue at maturity. Fl. Jun., fr. Jul.–Oct..

枝叶　Branch and leaves
摄影：蒋裕良　Photo by：Jiang Yuliang

叶　Leaves
摄影：蒋裕良　Photo by：Jiang Yuliang

果和叶背　Fruits and leaf backs
摄影：蒋裕良　Photo by：Jiang Yuliang

径级分布表　DBH class

胸径区间 Diameter class (cm)	个体数 No. of individuals in the plot	比例 Proportion (%)
1~2	1	100.00
2~5	0	0.00
5~10	0	0.00
10~20	0	0.00
20~35	0	0.00
35~50	0	0.00
≥50	0	0.00

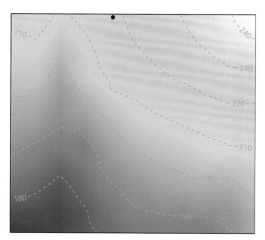

● 1~5 cm DBH　＋ 5~20 cm DBH　○ ≥20 cm DBH
个体分布图　Distribution of individuals

114 玉叶金花 | yù yè jīn huā

Mussaenda pubescens W. T. Aiton

茜草科 Rubiaceae

样地名称（Plot name）= JH
个体数（Individual number/1 hm²）= 8
最大胸径（Max DBH）= 9.0 cm
重要值排序（Importance value rank）= 91

攀援灌木。叶对生或轮生，膜质或薄纸质，卵状长圆形或卵状披针形，长 2~9 cm，宽 1~4 cm，顶端渐尖，基部楔形。聚伞花序顶生，密花。浆果近球形。花期 4~7 月，果期 6~12 月。

Climbing shrubs. Leaves opposite or perhaps rarely whorled; blade drying membranous or thinly papery, ovate-oblong, ovate-lanceolate, elliptic, lanceolate, or oblanceolate, 2–9 × 1–4 cm, base acute to obtuse. Inflorescences terminal, subcapitate to congested-cymose. Berry subglobose. Fl. Apr.–Jul., fr. Jun.–Dec..

树干　Trunk
摄影：李健星　Photo by: Li Jianxing

叶　Leaves
摄影：李健星　Photo by: Li Jianxing

枝叶　Branch and leaves
摄影：李健星　Photo by: Li Jianxing

径级分布表　DBH class

胸径区间 Diameter class (cm)	个体数 No. of individuals in the plot	比例 Proportion (%)
1~2	5	62.50
2~5	2	25.00
5~10	1	12.50
10~20	0	0.00
20~35	0	0.00
35~50	0	0.00
≥50	0	0.00

● 1~5 cm DBH　＋ 5~20 cm DBH　○ ≥20 cm DBH
个体分布图　Distribution of individuals

115 香港大沙叶 | xiāng gǎng dà shā yè

Pavetta hongkongensis Bremek.

茜草科 Rubiaceae

样地名称（Plot name）= JH
个体数（Individual number/1 hm²）= 25
最大胸径（Max DBH）= 7.3 cm
重要值排序（Importance value rank）= 46

灌木或小乔木，高1~4 m。叶对生，膜质，长圆形至椭圆状倒卵形，长8~15 cm，宽3~6.5 cm，顶端渐尖，基部楔形，花序生于侧枝顶部。果球形。花期3~7月，果期7~11月。

Shrubs or small trees, 1–4 m tall. Leaf blade drying membranous, elliptic-oblong to elliptic-oblanceolate, 8–15 × 3–6.5 cm, base cuneate to acute, apex acuminate to acute. Inflorescences terminal on lateral branches. Drupes globose. Fl. Mar.–Jul., fr. Jul.–Nov..

果枝　Fruiting branch
摄影：丁涛　Photo by: Ding Tao

叶　Leaves
摄影：丁涛　Photo by: Ding Tao

花序　Inflorescence
摄影：丁涛　Photo by: Ding Tao

径级分布表　DBH class

胸径区间 Diameter class (cm)	个体数 No. of individuals in the plot	比例 Proportion (%)
1~2	14	56.00
2~5	7	28.00
5~10	4	16.00
10~20	0	0.00
20~35	0	0.00
35~50	0	0.00
≥50	0	0.00

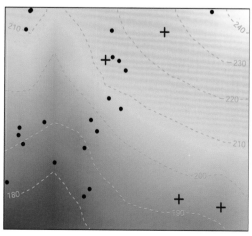

● 1~5 cm DBH　+ 5~20 cm DBH　○ ≥20 cm DBH
个体分布图　Distribution of individuals

116 南山花 | nán shān huā （四蕊三角瓣花）

Prismatomeris tetrandra (Roxb.) K. Schum. in Engler Prantl

茜草科 Rubiaceae

样地名称（Plot name）= JH
个体数（Individual number/1 hm^2）= 359
最大胸径（Max DBH）= 6.6 cm
重要值排序（Importance value rank）= 12

灌木至小乔木，高 2~8 m。小枝四棱柱形。叶长圆形至披针形，近革质，有时卵形或倒卵形，长 4~18 cm，宽 2~5 cm，全缘，顶端渐尖或钝，基部狭楔形；侧脉每边 5~9 条，两面凸起。伞形花序顶生. 核果近球形，无毛，光滑。花期 5~9 月，果期 9~12 月。

Shrubs or small trees, to 8 m tall. Branches quadrangular. Leaf blade drying leathery, elliptic, ovate, obovate, 4–18 × 2–5 cm, base cuneate to acute, apex acuminate or acute to obtuse; secondary veins 5–9 pairs; fascicled or umbellate. Drupes subglobose, glabrous, smooth. Fl. May–Sep., fr. Sep.–Dec..

果枝　Fruiting branch
摄影：丁涛　Photo by: Ding Tao

花序　Inflorescence
摄影：丁涛　Photo by: Ding Tao

果　Fruits
摄影：丁涛　Photo by: Ding Tao

径级分布表　DBH class

胸径区间 Diameter class (cm)	个体数 No. of individuals in the plot	比例 Proportion (%)
1~2	247	68.80
2~5	110	30.64
5~10	2	0.56
10~20	0	0.00
20~35	0	0.00
35~50	0	0.00
≥50	0	0.00

● 1~5 cm DBH　+ 5~20 cm DBH　○ ≥20 cm DBH
个体分布图　Distribution of individuals

117 九节 | jiǔ jié

Psychotria asiatica L.
茜草科 Rubiaceae

样地名称（Plot name）= JH
个体数（Individual number/1 hm^2）= 1050
最大胸径（Max DBH）= 18.9 cm
重要值排序（Importance value rank）= 2

灌木或小乔木，高 0.5~5 m。叶对生，纸质或革质，椭圆状长圆形或倒披针状长圆形，稀长圆状倒卵形，长 5~23.5 cm，宽 2~9 cm，顶端渐尖、急渐尖或短尖，基部锐尖至钝。聚伞花序顶生。核果球形或宽椭圆形，红色。花果期全年。

Shrubs or small trees, 0.5–5 m tall. Leaf blade drying papery to leathery, elliptic-oblong, lanceolate-oblong, or rarely oblong-ovate, 5–23.5 × 2–9 cm, base acute to obtuse, apex acute to acuminate or obtuse. Inflorescences terminal, cymose to paniculiform. Drupes red, subglobose to broadly ellipsoid. Fl. and fr. throughout year.

枝叶　Branch and leaves
摄影：李健星　Photo by: Li Jianxing

花序　Inflorescence
摄影：丁涛　Photo by: Ding Tao

果序　Infructescence
摄影：丁涛　Photo by: Ding Tao

径级分布表　DBH class

胸径区间 Diameter class (cm)	个体数 No. of individuals in the plot	比例 Proportion (%)
1~2	490	46.67
2~5	517	49.24
5~10	38	3.62
10~20	5	0.47
20~35	0	0.00
35~50	0	0.00
≥50	0	0.00

● 1~5 cm DBH　＋ 5~20 cm DBH　○ ≥20 cm DBH
个体分布图　Distribution of individuals

118 黄脉九节 | huáng mài jiǔ jié

***Psychotria straminea* Hutch. in Sargent**

茜草科 Rubiaceae

样地名称（Plot name）= JH
个体数（Individual number/1 hm²）= 3
最大胸径（Max DBH）= 5.8 cm
重要值排序（Importance value rank）= 108

灌木，高 0.5~3 m，无毛。叶对生，纸质或膜质，椭圆状披针形、长圆形、倒卵状长圆形，少为椭圆形或披针形，长 5.5~29 cm，宽 0.8~10.5 cm，顶端渐尖或短尖，基部楔形或稍圆，有时不等侧，全缘，干时黄绿色。聚伞花序顶生，少花。浆果状核果近球形或椭圆形，成熟时黑色。花期 1~7 月，果期 6 月至翌年 1 月。

Shrubs, 0.5–3 m tall. Stems glabrous. Leaf blade drying papery to membranous, yellowish green, often pale below, elliptic, elliptic-oblong, oblanceolate, or obovate, 5.5–29 × 0.8–10.5 cm, glabrous on both surfaces, base cuneate to obtuse, margins flat, apex acuminate or acute. Inflorescences terminal, few flowered, cymose. Drupes red or perhaps ultimately black, subglobose or ellipsoid. Fl. Jan.–Jul., fr. Jun.–Jan. of next year.

果序　　Infructescence
摄影：丁涛　　Photo by: Ding Tao

叶　　Leaves
摄影：丁涛　　Photo by: Ding Tao

花序　　Inflorescence
摄影：朱鑫鑫　　Photo by: Zhu Xinxin

径级分布表　DBH class

胸径区间 Diameter class (cm)	个体数 No. of individuals in the plot	比例 Proportion (%)
1~2	0	0.00
2~5	1	33.33
5~10	2	66.67
10~20	0	0.00
20~35	0	0.00
35~50	0	0.00
≥50	0	0.00

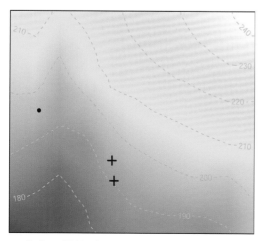

● 1~5 cm DBH　　+ 5~20 cm DBH　　○ ≥20 cm DBH
个体分布图　Distribution of individuals

119 鱼骨木 | yú gǔ mù （铁屎米） 鱼骨木属

Psydrax dicocca Gaertn.

茜草科 Rubiaceae

样地名称（Plot name）＝ JH
个体数（Individual number/1 hm²）＝ 8
最大胸径（Max DBH）＝ 21.5 cm
重要值排序（Importance value rank）＝ 77

无刺灌木至小乔木，高 4~10（~18）m。小枝初时呈压扁形或四棱柱形，后变圆柱形，黑褐色。叶薄纸质到厚纸质，卵形，椭圆形至卵状披针形，长 9~19 cm，宽 4~8.5 cm，顶端长渐尖或钝或钝急尖，基部楔形。聚伞花序。核果倒卵形，略扁，多少近孪生，花期 1~5 月，果期 6~7 月。

Erect shrubs to small trees, 4–10 (–18) m tall, unarmed. Branches somewhat compressed becoming terete. Leaves blade drying thinly to thickly papery, ovate to ovate-oblong or elliptic-oblong, 9–19 × 4–8.5 cm, base obtuse to rounded, apex shortly acuminate. Inflorescences corymbiform to cymose. Drupes obovoid, laterally compressed, often somewhat dicoccous. Fl. Jan.–Mar., fr. Jun.–Jul..

枝叶　Branch and leaves
摄影：丁涛　Photo by: Ding Tao

叶　Leaves
摄影：丁涛　Photo by: Ding Tao

托叶　Stipule
摄影：丁涛　Photo by: Ding Tao

径级分布表 DBH class

胸径区间 Diameter class (cm)	个体数 No. of individuals in the plot	比例 Proportion (%)
1~2	4	50.00
2~5	2	25.00
5~10	0	0.00
10~20	1	12.50
20~35	1	12.50
35~50	0	0.00
≥50	0	0.00

● 1~5 cm DBH　＋ 5~20 cm DBH　○ ≥20 cm DBH
个体分布图 Distribution of individuals

120 白皮乌口树 | bái pí wū kǒu shù （白骨木）

Tarenna depauperata Hutch. in Sargent

茜草科 Rubiaceae

样地名称（Plot name）= JH
个体数（Individual number/1 hm²）= 1
最大胸径（Max DBH）= 6.0 cm
重要值排序（Importance value rank）= 128

灌木或小乔木，高 1~6 m。叶纸质或革质，椭圆状倒卵形、椭圆形或近卵形，长 4~15 cm，宽 2~6.5 cm，顶端短渐尖或骤然渐尖，尖端常稍钝，基部楔形或短尖；中脉在两面凸起，侧脉 5~11 对。伞房状聚伞花序顶生。浆果球形，光亮。花期 4~11 月，果期 4 月至翌年 1 月。

Shrubs or small trees, 1–6 m tall. Leaf blade drying papery or leathery, and somewhat shiny adaxially, elliptic-obovate, elliptic, or subovate, 4–15 × 2–6.5 cm, both surfaces glabrous, base cuneate or acute, apex shortly acuminate often abruptly acuminate with tip often slightly obtuse; secondary veins 5–11 pairs. Inflorescences corymbiform to pyramidal. Berries globose, shiny. Fl. Apr.–Nov., fr. Apr.–Jan. of next year.

果枝　Fruiting branch
摄影：陆昭岑　Photo by：Lu Zhaochen

叶背　Leaf backs
摄影：蒋裕良　Photo by：Jiang Yuliang

花序　Inflorescences
摄影：陆昭岑　Photo by：Lu Zhaochen

径级分布表　DBH class

胸径区间 Diameter class (cm)	个体数 No. of individuals in the plot	比例 Proportion (%)
1~2	0	0.00
2~5	0	0.00
5~10	1	100.00
10~20	0	0.00
20~35	0	0.00
35~50	0	0.00
≥50	0	0.00

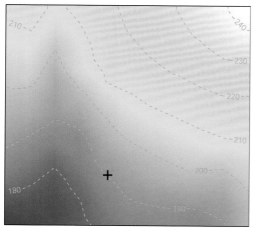

● 1~5 cm DBH　+ 5~20 cm DBH　○ ≥20 cm DBH
个体分布图　Distribution of individuals

121 短花水金京 | duǎn huā shuǐ jīn jīng （虾飞木）

水锦树属

***Wendlandia formosana* subsp. *breviflora* F. C. How**

茜草科 Rubiaceae

样地名称（Plot name）= JH
个体数（Individual number/1 hm²）= 4
最大胸径（Max DBH）= 6.0 cm
重要值排序（Importance value rank）= 90

灌木或乔木，高 2~8 m。叶纸质，椭圆形或椭圆状披针形，长 6~14 cm，宽 2~6.5 cm，顶端骤然渐尖或短渐尖，有时短尖，基部渐狭；叶侧脉 7~10 对，较密，常在叶下面凸起；托叶阔三角形，直立。圆锥状聚伞花序顶生。蒴果球形，无毛。花期 4~6 月，果期 5~12 月。

Shrubs or trees, 2–8 m tall. Leaves opposite, blade drying papery, elliptic or elliptic-lanceolate, 6–14 × 2–6.5 cm, base acute to obtuse, apex acute to acuminate; secondary veins 7–10 pairs, closely set, usually prominent abaxially; stipules generally persistent, broadly triangular. Inflorescences paniculate. Capsules globose, glabrous. Fl. Apr.–Jun., fr. May–Dec..

花序　　Inflorescences
摄影：林春蕊　　Photo by：Lin Chunrui

叶背　　Leaf backs
摄影：林春蕊　　Photo by：Lin Chunrui

枝叶　　Branch and leaves
摄影：林春蕊　　Photo by：Lin Chunrui

径级分布表　DBH class

胸径区间 Diameter class (cm)	个体数 No. of individuals in the plot	比例 Proportion (%)
1~2	1	25.00
2~5	2	50.00
5~10	1	25.00
10~20	0	0.00
20~35	0	0.00
35~50	0	0.00
≥50	0	0.00

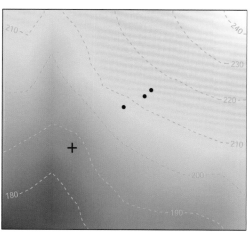

● 1~5 cm DBH　+ 5~20 cm DBH　○ ≥20 cm DBH
个体分布图　Distribution of individuals

122 水锦树 | shuǐ jǐn shù （中华水锦树）

Wendlandia uvariifolia Hance

茜草科 Rubiaceae

样地名称（Plot name）= JH
个体数（Individual number/1 hm²）= 218
最大胸径（Max DBH）= 21.3 cm
重要值排序（Importance value rank）= 9

灌木或乔木，高 2~15 m。叶对生，纸质，宽椭圆形、长圆形、卵形或长圆状披针形，长 7~26 cm，宽 4~14 cm，顶端短渐尖或骤然渐尖，基部楔形或短尖，圆锥状聚伞花序顶生。蒴果球形。花期 1~5 月，果期 4~10 月。

Shrubs or trees, 2–15 m tall. Leaves opposite; blade drying papery, broadly elliptic, elliptic-oblong, ovate, or oblong-lanceolate, 7–26 × 4–14 cm, base acute to obtuse, apex shortly to abruptly acuminate. Inflorescences paniculate, pyramidal in outline. Capsules subglobose. Fl. Jan.–May, fr. Apr.–Oct..

花序　Inflorescence
摄影：丁涛　Photo by: Ding Tao

叶　Leaves
摄影：丁涛　Photo by: Ding Tao

叶背　Leaf backs
摄影：丁涛　Photo by: Ding Tao

径级分布表　DBH class

胸径区间 Diameter class (cm)	个体数 No. of individuals in the plot	比例 Proportion (%)
1~2	26	11.93
2~5	84	38.53
5~10	88	40.37
10~20	19	8.71
20~35	1	0.46
35~50	0	0.00
≥50	0	0.00

● 1~5 cm DBH　+ 5~20 cm DBH　○ ≥20 cm DBH
个体分布图　Distribution of individuals

123 思茅山橙 | sī máo shān chéng

Melodinus cochinchinensis (Lour.) Merr.
夹竹桃科 Apocynaceae

样地名称（Plot name）= JH
个体数（Individual number/1 hm²）= 1
最大胸径（Max DBH）= 13.1 cm
重要值排序（Importance value rank）= 120

攀援木质藤本，长达 10 m，除花序被稀疏的柔毛外，其余无毛。叶柄长约 1.2 cm。叶近革质，椭圆形或卵圆形，长 5~10 cm，宽 1.8~5 cm，顶端短渐尖，基部渐尖或圆形。聚伞花序顶生和腋生。浆果球形，直径 5~8 cm。花期 5~11 月，果期 8~12 月。

Lianas, to 10 m, glabrous except for inflorescences. Petiole to 1.2 cm. Leaf blade elliptic or ovate, 5–10 × 1.8–5 cm, leathery, base attenuate to rounded, apex short acuminate. Cymes terminal and axillary. Berries globose, 5–8 cm in diam. Fl. May–Nov., fr. Aug.–Dec..

果枝　Fruiting branch
摄影：丁涛　Photo by: Ding Tao

枝叶　Branch and leaves
摄影：丁涛　Photo by: Ding Tao

果　Fruit
摄影：丁涛　Photo by: Ding Tao

径级分布表 DBH class

胸径区间 Diameter class (cm)	个体数 No. of individuals in the plot	比例 Proportion (%)
1~2	0	0.00
2~5	0	0.00
5~10	0	0.00
10~20	1	100.00
20~35	0	0.00
35~50	0	0.00
≥50	0	0.00

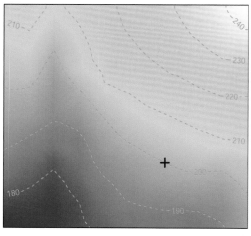

● 1~5 cm DBH　＋ 5~20 cm DBH　○ ≥20 cm DBH
个体分布图 Distribution of individuals

124 长花厚壳树 | cháng huā hòu qiào shù

厚壳树属

Ehretia longiflora Champ. ex Benth.
紫草科 Boraginaceae

样地名称（Plot name）＝ JH
个体数（Individual number/1 hm^2）＝ 1
最大胸径（Max DBH）＝ 1.8 cm
重要值排序（Importance value rank）＝ 133

乔木，高 5~10 m。树皮深灰色至暗褐色，片状剥落。叶椭圆形、长圆形或长圆状倒披针形，长 3~12 cm，宽 2~6 cm，先端急尖，基部楔形，稀圆形，全缘，无毛，侧脉 4~7 对，小脉不明显；叶柄长 1~2 cm，无毛。聚伞花序生侧枝顶端。核果淡黄色或红色，核具棱，分裂成 4 个具单种子的分核。花期 4 月，果期 6~7 月。

Trees, 5–10 m tall. Bark dark gray to dark brown, scaly. Leaf blade elliptic to oblong or oblong-oblanceolate, 3–12 × 2–6 cm, glabrous, base cuneate, rarely rounded, margin entire, apex acute; lateral veins 4–7 pairs, reticulate veins inconspicuous. Petiole 1–2 cm. Cymes terminating lateral branches, flat topped. Drupes pale yellow or red; endocarp ribbed, divided at maturity into 4 1-seeded pyrenes. Fl. Apr., fr. Jun.–Jul..

果枝　　　　Fruiting branch
摄影：杨平　　Photo by：Yang Ping

叶背　　　Leaf back
摄影：杨平　Photo by：Yang Ping

果序　　　Infructescence
摄影：杨平　Photo by：Yang Ping

径级分布表　DBH class

胸径区间 Diameter class (cm)	个体数 No. of individuals in the plot	比例 Proportion (%)
1~2	1	100.00
2~5	0	0.00
5~10	0	0.00
10~20	0	0.00
20~35	0	0.00
35~50	0	0.00
≥50	0	0.00

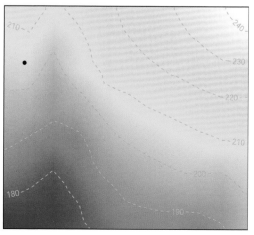

● 1~5 cm DBH　＋ 5~20 cm DBH　○ ≥20 cm DBH
个体分布图　Distribution of individuals

125 牛屎果 | niú shǐ guǒ （牛矢果）

木樨属

Osmanthus matsumuranus Hayata

木樨科 Oleaceae

样地名称（Plot name）= JH
个体数（Individual number/1 hm²）= 2
最大胸径（Max DBH）= 16.0 cm
重要值排序（Importance value rank）= 102

灌木或乔木，高 2.5~10 m，无毛。小枝扁平。叶片薄革质或厚纸质，倒披针形，稀为倒卵形，长 8~14（~19）cm，宽 2.5~4.5（~6）cm，先端渐尖，具尖头，基部狭楔形。聚伞花序组成短小圆锥花序。果椭圆形，成熟时紫色至黑色。花期 5~6 月，果期 11~12 月。

Shrubs or trees, 2.5–10 m tall glabrous. Branchlets compressed. Leaf blade oblanceolate, rarely obovate, 8–14 (–19) × 2.5–4.5 (–6) cm, thin leathery to thick papery, base attenuate and decurrent, margin entire or serrate along distal half, apex acuminate and mucronate. Cymes in short panicles, axillary. Drupe ripening purple to black, ellipsoid. Fl. May–Jun., fr. Nov.–Dec..

叶 Leaves
摄影：丁涛 Photo by: Ding Tao

花序 Inflorescence
摄影：叶喜阳 Photo by: Ye Xiyang

果枝 Fruiting branch
摄影：叶喜阳 Photo by: Ye Xiyang

径级分布表 DBH class

胸径区间 Diameter class (cm)	个体数 No. of individuals in the plot	比例 Proportion (%)
1~2	1	50.00
2~5	0	0.00
5~10	0	0.00
10~20	1	50.00
20~35	0	0.00
35~50	0	0.00
≥50	0	0.00

• 1~5 cm DBH   + 5~20 cm DBH   ○ ≥20 cm DBH
个体分布图 Distribution of individuals

126 大青 | dà qīng （鸡屎青）

***Clerodendrum cyrtophyllum* Turca.**

唇形科 Lamiaceae

样地名称（Plot name）= JH
个体数（Individual number/1 hm²）= 22
最大胸径（Max DBH）= 4.0 cm
重要值排序（Importance value rank）= 59

灌木或小乔木，高1～10 m。叶片纸质，椭圆形、卵状椭圆形、长圆形或长圆状披针形，长6～20 cm，宽3～9 cm，顶端渐尖或急尖，基部圆形或宽楔形，通常全缘。伞房状聚伞花序。果实球形或倒卵形，蓝紫色。花果期6月至翌年2月。

Shrubs or small trees, 1–10 m tall. Leaf blade oblong, oblong-lanceolate, elliptic, or ovate-elliptic, 6–20 × 3–9 cm, papery, base rounded to cuneate, margin entire or rounded serrate, apex acuminate to acute. Inflorescences corymbose. Drupes blue-purple, obovate to globose. Fl. and fr. Jun.–Feb. of next year.

树干　　　Trunk
摄影：李健星　　Photo by：Li Jianxing

叶背　　　Leaf backs
摄影：李健星　　Photo by：Li Jianxing

枝叶　　　Branch and leaves
摄影：李健星　　Photo by：Li Jianxing

径级分布表　DBH class

胸径区间 Diameter class (cm)	个体数 No. of individuals in the plot	比例 Proportion (%)
1～2	12	54.55
2～5	10	45.45
5～10	0	0.00
10～20	0	0.00
20～35	0	0.00
35～50	0	0.00
≥50	0	0.00

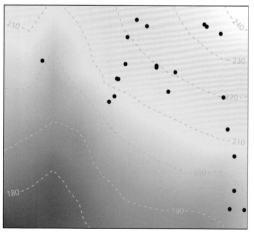

● 1～5 cm DBH　　+ 5～20 cm DBH　　○ ≥20 cm DBH
个体分布图　Distribution of individuals

127 垂茉莉 | chuí mò lì （长花龙吐珠） 大青属

***Clerodendrum wallichii* Merr.**

唇形科 Lamiaceae

样地名称（Plot name）= JH
个体数（Individual number/1 hm²）= 6
最大胸径（Max DBH）= 5.4 cm
重要值排序（Importance value rank）= 95

直立灌木或小乔木，高 2~4 m。小枝锐四棱形或呈翅状，无毛。叶片近革质，长圆形或长圆状披针形，长 11~18 cm，宽 2.5~4 cm，顶端渐尖或长渐尖，基部狭楔形，全缘。聚伞花序排列成圆锥状。核果球形，初时黄绿色，成熟后紫黑色。花果期 10 月至翌年 4 月。

Shrubs or small trees, 2–4 m tall, erect. Branchlets 4-angled, ± winged, glabrous. Leaf blade oblong to oblong-lanceolate, 11–18 × 2.5–4 cm, sub-leathery, base narrowly cuneate, margin entire, apex acuminate to acute. Inflorescences pendent thyrses. Drupes yellow-green when young, black and shiny at maturity. Fl. and fr. Oct.–Apr. of next year.

树干　Trunk
摄影：李健星　Photo by: Li Jianxing

叶背　Leaf backs
摄影：李健星　Photo by: Li Jianxing

枝叶　Branch and leaves
摄影：李健星　Photo by: Li Jianxing

径级分布表　DBH class

胸径区间 Diameter class (cm)	个体数 No. of individuals in the plot	比例 Proportion (%)
1~2	2	33.33
2~5	3	50.00
5~10	1	16.67
10~20	0	0.00
20~35	0	0.00
35~50	0	0.00
≥50	0	0.00

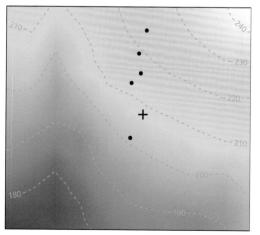

● 1~5 cm DBH　＋ 5~20 cm DBH　○ ≥20 cm DBH
个体分布图　Distribution of individuals

128 山牡荆 | shān mǔ jīng

***Vitex quinata* (Lour.) F. W. Williams**

唇形科 Lamiaceae

样地名称（Plot name）= JH
个体数（Individual number/1 hm²）= 2
最大胸径（Max DBH）= 2.9 cm
重要值排序（Importance value rank）= 126

常绿乔木，高 4~12 m。树皮褐色。掌状复叶，对生，有 3~5 小叶，小叶片倒卵形至倒卵状椭圆形，顶端渐尖至短尾状，基部楔形至阔楔形，通常全缘。圆锥花序顶生。核果球形或倒卵形，黑色。花期 5~7 月，果期 8~9 月。

Evergreen trees, 4–12 m tall. Bark brown. Leaves 3–5-foliolate; Leaflets obovate-elliptic to obovate or oblong to elliptic, thickly papery, abaxially yellow glandular, base cuneate, apex acuminate, acute, or obtuse. Panicles terminal. Drupes black, obovoid to globose. Fl. May–Jul., fr. Aug.–Sep..

树干　　Trunk
摄影：李健星　　Photo by：Li Jianxing

叶　　Leaves
摄影：李健星　　Photo by：Li Jianxing

叶背　　Leaf backs
摄影：李健星　　Photo by：Li Jianxing

径级分布表　DBH class

胸径区间 Diameter class (cm)	个体数 No. of individuals in the plot	比例 Proportion (%)
1~2	1	50.00
2~5	1	50.00
5~10	0	0.00
10~20	0	0.00
20~35	0	0.00
35~50	0	0.00
≥50	0	0.00

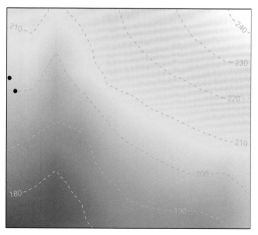

● 1~5 cm DBH　＋ 5~20 cm DBH　○ ≥20 cm DBH
个体分布图　Distribution of individuals

129 微毛布惊 | wēi máo bù jīng

Vitex quinata var. *puberula* (H. J. Lam) Moldenke

唇形科 Lamiaceae

样地名称（Plot name）= JH
个体数（Individual number/1 hm^2）= 12
最大胸径（Max DBH）= 14.9 cm
重要值排序（Importance value rank）= 63

常绿乔木，高 4~12 m。树皮褐色。掌状复叶，对生，叶柄长 2.5~6 cm，有 3~5 小叶，通常 5，小叶片倒卵形至倒卵状椭圆形，顶端渐尖至短尾状，基部楔形至阔楔形，通常全缘，表面通常有灰白色小窝点；中间的 1 枚小叶较大，长 15~20 cm，宽 5~8.5 cm。圆锥花序顶生。核果近球形，成熟后呈黑色。花期 5~7 月，果期 8~9 月。

Evergreen trees, 4–12 m tall. Bark brown. Leaves 3–5-foliolate, usually 5; petiole 2.5–6 cm. Leaflets obovate-elliptic to obovate or oblong to elliptic, thickly papery, both surfaces shiny, base cuneate, margin entire or sometimes apically crenulate dentate, apex acuminate, acute, or obtuse. Central leaflet, 15–20×5–8.5 cm. Panicles terminal. Drupes subglobose, black at maturity. Fl. May–Jul., fr. Aug.–Sep..

果枝　　　　　　　　　　Fruiting branch
摄影：丁涛　　　　　　　Photo by：Ding Tao

叶　　　　　　　　　　　Leaves
摄影：丁涛　　　　　　　Photo by：Ding Tao

叶背　　　　　　　　　　Leaf backs
摄影：丁涛　　　　　　　Photo by：Ding Tao

径级分布表　DBH class

胸径区间 Diameter class (cm)	个体数 No. of individuals in the plot	比例 Proportion (%)
1~2	2	16.67
2~5	7	58.33
5~10	1	8.33
10~20	2	16.67
20~35	0	0.00
35~50	0	0.00
≥50	0	0.00

● 1~5 cm DBH　+ 5~20 cm DBH　○ ≥20 cm DBH
个体分布图　Distribution of individuals

130 粗丝木 | cū sī mù （海南粗丝木）

Gomphandra tetrandra (Wall.) Sleumer

粗丝木科 Stemonuraceae

样地名称（Plot name）= JH
个体数（Individual number/1 hm²）= 9
最大胸径（Max DBH）= 3.1 cm
重要值排序（Importance value rank）= 70

灌木或小乔木，高 2~10 m。树皮灰色。叶纸质，幼时膜质，狭披针形、长椭圆形或阔椭圆形，长 6~15 cm，宽 2~6 cm，先端渐尖或成尾状，基部楔形。聚伞花序与叶对生。核果椭圆形，由青转黄，成熟时白色，浆果状。花果期全年。

Shrubs or small trees, 2–10 m tall. Bark gray. Leaf blade shiny, narrowly lanceolate or narrowly or broadly elliptic, 6–15 × 2–6 cm, membranous when young, base cuneate, apex acuminate or caudate. Cymes opposite leaves. Drupe berrylike, changing from green to yellow to white, ellipsoid. Fl. and fr. throughout year.

枝叶　Branch and leaves
摄影：丁涛　Photo by: Ding Tao

叶背　Leaf backs
摄影：丁涛　Photo by: Ding Tao

果　Fruit
摄影：丁涛　Photo by: Ding Tao

径级分布表　DBH class

胸径区间 Diameter class (cm)	个体数 No. of individuals in the plot	比例 Proportion (%)
1~2	5	55.56
2~5	4	44.44
5~10	0	0.00
10~20	0	0.00
20~35	0	0.00
35~50	0	0.00
≥50	0	0.00

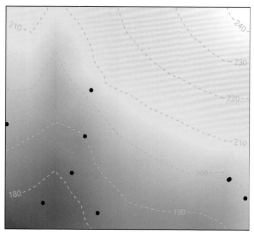

● 1~5 cm DBH　　+ 5~20 cm DBH　　○ ≥20 cm DBH
个体分布图　Distribution of individuals

131 棱枝冬青 | léng zhī dōng qīng

Ilex angulata **Merr. Chun**

冬青科 Aquifoliaceae

样地名称（Plot name）＝ JH
个体数（Individual number/1 hm²）＝ 150
最大胸径（Max DBH）＝ 23.2 cm
重要值排序（Importance value rank）＝ 17

常绿灌木或小乔木，高 4～10 m。树皮灰白色。小枝纤细，"之"字形，具纵棱脊。叶生于 1～3 年生枝，叶片纸质或幼时膜质，椭圆形或阔椭圆形，长 3.5～5 cm，宽 1.5～2 cm。聚伞花序单生。果椭圆形，成熟时红色，具纵棱。花期 4 月，果期 7～10 月。

Evergreen shrubs or small trees, 4–10 m tall. Bark gray-white. Branchlets zigzag, slender, ridged, narrowly sulcate. Leaf blade elliptic or broadly elliptic, 3.5–5 × 1.5–2 cm, papery or membranous when young. Inflorescences cymes solitary. Fruit ellipsoidal, red at maturity, longitudinally angular. Fl. Apr., fr. Jul.–Oct..

枝叶　　Branch and leaves
摄影：李健星　　Photo by：Li Jianxing

叶背　　Leaf backs
摄影：丁涛　　Photo by：Ding Tao

花　　Flowers
摄影：周欣欣　　Photo by：Ding Tao

径级分布表　DBH class

胸径区间 Diameter class (cm)	个体数 No. of individuals in the plot	比例 Proportion (%)
1～2	44	29.33
2～5	77	51.33
5～10	25	16.67
10～20	3	2.00
20～35	1	0.67
35～50	0	0.00
≥50	0	0.00

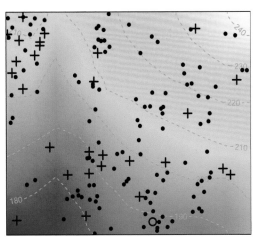

● 1～5 cm DBH　＋ 5～20 cm DBH　○ ≥20 cm DBH
个体分布图　Distribution of individuals

132 榕叶冬青 | róng yè dōng qīng （白粘）

***Ilex ficoidea* Hemsl.**

冬青科 Aquifoliaceae

样地名称（Plot name）= JH
个体数（Individual number/1 hm²）= 4
最大胸径（Max DBH）= 7.2 cm
重要值排序（Importance value rank）= 107

常绿乔木，高 2~12 m。叶片革质，长圆状椭圆形、卵状或稀倒卵状椭圆形，长 4.5~10 cm，宽 1.5~3.5 cm，先端骤然尾状渐尖，渐尖头长可达 15 mm，基部钝、楔形或近圆形，边缘具不规则的细圆齿状锯齿，齿尖变黑色，干后稍反卷。聚伞花序或单花簇生于当年生枝的叶腋内。果球形或近球形，成熟后红色。花期 3~5 月，果期 8~11 月。

Evergreen tree, 2–12 m tall. Leaf blade elliptic, oblong, ovate, or obovate-elliptic, 4.5–10 × 1.5–3.5 cm, leathery, base obtuse, cuneate, or sub-rounded, margin irregularly crenate-serrate, apices of teeth black and slightly recurved when dry, apex abruptly long caudate, acumen ca. 15 mm. Inflorescences: fasciculate, axillary on current year's branchlets. Fruit red, globose or subglobose. Fl. Mar.-May, fr. Aug.-Nov..

枝叶　Branch and leaves
摄影：蒋裕良　Photo by: Jiang Yuliang

花序　Inflorescences
摄影：椰子　Photo by: Ye Zi

果　Fruits
摄影：唐忠炳　Photo by: Tang Zhongbing

径级分布表 DBH class

胸径区间 Diameter class (cm)	个体数 No. of individuals in the plot	比例 Proportion (%)
1~2	1	25.00
2~5	2	50.00
5~10	1	25.00
10~20	0	0.00
20~35	0	0.00
35~50	0	0.00
≥50	0	0.00

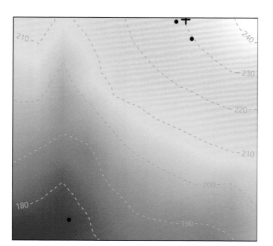

● 1~5 cm DBH　＋ 5~20 cm DBH　○ ≥20 cm DBH
个体分布图 Distribution of individuals

133 铁冬青 | tiě dōng qīng （小果铁冬青） 冬青属

Ilex rotunda Thunb.

冬青科 Aquifoliaceae

样地名称（Plot name）= JH
个体数（Individual number/1 hm²）= 2
最大胸径（Max DBH）= 12.1 cm
重要值排序（Importance value rank）= 106

常绿灌木或乔木，高可达 20 m，胸径达 1 m。树皮灰色至灰黑色。叶片薄革质或纸质，卵形、倒卵形或椭圆形，长 4~9 cm，宽 1.8~4 cm。聚伞花序或伞状花序，单生于当年生枝的叶腋内。果近球形或稀椭圆形，成熟时红色。花期 4~6 月，果期 8~12 月。

Evergreen shrubs or trees, to 20 m tall, trunk to 1 m d.b.h.. Bark gray to gray-black. Leaf blade ovate, obovate, or elliptic, 4–9 × 1.8–4 cm, thinly leathery or papery. Inflorescences cymes, umbelliform, solitary, axillary on current year's branchlets. Fruit red, subglobose, rarely ellipsoid. Fl. Apr.–Jun., fr. Aug.–Dec..

果枝 Fruiting branch
摄影：丁涛 Photo by: Ding Tao

叶 Leaves
摄影：丁涛 Photo by: Ding Tao

叶背 Leaf backs
摄影：丁涛 Photo by: Ding Tao

径级分布表 DBH class

胸径区间 Diameter class (cm)	个体数 No. of individuals in the plot	比例 Proportion (%)
1~2	0	0.00
2~5	1	50.00
5~10	0	0.00
10~20	1	50.00
20~35	0	0.00
35~50	0	0.00
≥50	0	0.00

● 1~5 cm DBH + 5~20 cm DBH ○ ≥20 cm DBH
个体分布图 Distribution of individuals

134 三花冬青 | sān huā dōng qīng

Ilex triflora Blume
冬青科 Aquifoliaceae

样地名称（Plot name）= JH
个体数（Individual number/1 hm^2）= 1
最大胸径（Max DBH）= 21.1 cm
重要值排序（Importance value rank）= 105

常绿灌木或乔木，高 2~10 m。叶片近革质，椭圆形，长圆形或卵状椭圆形，长 2.5~10 cm，宽 1.5~4 cm，先端急尖至渐尖，渐尖头长 3~4 mm，基部圆形或钝，边缘具近波状线齿，叶面深绿色，干时呈褐色或橄榄绿色，背面具腺点，疏被短柔毛。聚伞花序，簇生，腋生。果球形，成熟后黑色。花期 5~7 月，果期 8~11 月。

Evergreen shrubs or trees, 2–10 m tall. Leaf blade brown or olivaceous when dry, elliptic, oblong, ovate-elliptic, obovate, or oblong-elliptic, 2.5–10 × 1.5–4 cm, subleathery, abaxially sparsely puberulent, punctate, base rounded or obtuse, margin ± undulate, serrate, apex acute to acuminate, acumen 3-4 mm. Inflorescences: cymes, fasciculate, axillary. Fruit black, globose. Fl. May–Jul., fr. Aug.–Nov..

枝叶　　Branches and leaves
摄影：杨平　　Photo by：Yang Ping

叶　　Leaves
摄影：杨平　　Photo by：Yang Ping

果序　　Infructescences
摄影：杨平　　Photo by：Yang Ping

径级分布表　DBH class

胸径区间 Diameter class (cm)	个体数 No. of individuals in the plot	比例 Proportion (%)
1~2	0	0.00
2~5	0	0.00
5~10	0	0.00
10~20	0	0.00
20~35	1	100.00
35~50	0	0.00
≥50	0	0.00

● 1~5 cm DBH　　＋ 5~20 cm DBH　　○ ≥20 cm DBH
个体分布图　Distribution of individuals

135 南方荚蒾 | nán fāng jiá mí

荚蒾属

***Viburnum fordiae* Hance**

五福花科 Adoxaceae

样地名称（Plot name）＝ JH
个体数（Individual number/1 hm²）＝ 1
最大胸径（Max DBH）＝ 1.9 cm
重要值排序（Importance value rank）＝ 132

落叶灌木，高达 5 m。当年小枝密被土黄色或黄绿色开展的小刚毛状粗毛及簇状短毛，2 年生小枝暗紫褐色。叶纸质，宽倒卵形或宽卵形，长 3～10（～13）cm，宽 2～7（～11）cm。复伞形聚伞花序稠密。果实红色，椭圆状卵圆形。花期 5～6 月，果熟期 9～11 月。

Deciduous shrubs, to 5 m tall. Branchlets of current year, densely bristle-like hairy and stellate-pubescent; branchlets of previous year dark purple-brown. Leaf blade broadly obovate, obovate, or broadly ovate, 3–10 (–13) × 2–7 (–11) cm, papery. Inflorescence a compound umbel-like cyme. Fruit maturing red, ellipsoid-ovoid. Fl. May–Jul., fr. Sep.–Nov..

果枝　Fruiting branches
摄影：杨平　Photo by: Yang Ping

花序　Inflorescence
摄影：杨平　Photo by: Yang Ping

果序　Infructescence
摄影：杨平　Photo by: Yang Ping

径级分布表 DBH class

胸径区间 Diameter class (cm)	个体数 No. of individuals in the plot	比例 Proportion (%)
1～2	1	100.00
2～5	0	0.00
5～10	0	0.00
10～20	0	0.00
20～35	0	0.00
35～50	0	0.00
≥50	0	0.00

● 1～5 cm DBH　＋ 5～20 cm DBH　○ ≥20 cm DBH
个体分布图 Distribution of individuals

136 常绿荚蒾 | cháng lǜ jiá mí

Viburnum sempervirens K. Koch
五福花科 Adoxaceae

样地名称（Plot name）= JH
个体数（Individual number/1 hm²）= 3
最大胸径（Max DBH）= 2.6 cm
重要值排序（Importance value rank）= 114

常绿灌木，高可达 4 m。叶革质，干后上面变黑色至黑褐色或灰黑色，椭圆形至椭圆状卵形，较少宽卵形，有时矩圆形或倒披针形，长 4～12（～16）cm，宽 2.5～5（～6.5）cm，顶端尖或短渐尖，基部渐狭至钝形，有时近圆形。复伞形聚伞花序顶生。果实红色，卵圆形。花期 4～5 月，果熟期 7～12 月。

Evergreen shrubs, to 4 m tall. Leaf blade green when young, black to black-brown or gray-black when dry, elliptic to elliptic-ovate, rarely broadly ovate, sometimes oblong or oblanceolate, 4–12 (–16) × 2.5–5 (–6.5) cm, leathery, abaxially with tiny brown glandular dots throughout, apex acute or shortly acuminate. Inflorescence a compound umbel-like cyme. Fruit maturing red, ovoid. Fl. Apr.–May, fr. Jul.–Dec..

果枝　　Fruiting branches
摄影：唐忠炳　　Photo by: Tang Zhongbing

叶背　　Leaf back
摄影：唐忠炳　　Photo by: Tang Zhongbing

果序　　Infructescence
摄影：丁涛　　Photo by: Ding Tao

径级分布表　DBH class

胸径区间 Diameter class (cm)	个体数 No. of individuals in the plot	比例 Proportion (%)
1～2	2	66.67
2～5	1	33.33
5～10	0	0.00
10～20	0	0.00
20～35	0	0.00
35～50	0	0.00
≥50	0	0.00

● 1～5 cm DBH　　＋ 5～20 cm DBH　　○ ≥20 cm DBH
个体分布图　Distribution of individuals

137 广西海桐 | guǎng xī hǎi tóng

Pittosporum kwangsiense H. T. Chang S. Z. Yan

海桐科 Pittosporaceae

样地名称（Plot name）= JH
个体数（Individual number/1 hm²）= 34
最大胸径（Max DBH）= 11.2 cm
重要值排序（Importance value rank）= 37

灌木或小乔木。小枝无毛，灰白色，多皮孔。叶簇生于枝顶，2年生，革质，倒卵状矩圆形，长 10~15 cm，宽 4~6 cm，先端尖锐，基部楔形。伞房花序。蒴果短圆形，稍压扁。花期 3~5 月，果期 5~11 月。

Shrubs or small trees. Young branchlets gray-white, glabrous, many lenticellate. Leaves clustered at branchlet apex, biennial, 10–15× 4–6cm. Leaf blade green and shiny adaxially, base cuneate. Inflorescence. Capsule oblate, slightly compressed. Fl. Mar.–May, fr. May–Nov..

枝叶　Branch and leaves
摄影：丁涛　Photo by: Ding Tao

果　Fruit
摄影：丁涛　Photo by: Ding Tao

果枝　Fruiting branch
摄影：丁涛　Photo by: Ding Tao

径级分布表　DBH class

胸径区间 Diameter class (cm)	个体数 No. of individuals in the plot	比例 Proportion (%)
1~2	16	47.06
2~5	14	41.18
5~10	3	8.82
10~20	1	2.94
20~35	0	0.00
35~50	0	0.00
≥50	0	0.00

● 1~5 cm DBH　＋ 5~20 cm DBH　○ ≥20 cm DBH
个体分布图　Distribution of individuals

138 罗伞 | luó sǎn （短梗罗伞）

Brassaiopsis glomerulata (Blume) Regel
五加科 Araliaceae

样地名称（Plot name）= JH
个体数（Individual number/1 hm²）= 376
最大胸径（Max DBH）= 14.5 cm
重要值排序（Importance value rank）= 7

乔木，高 3～20 m。树皮灰棕色，上部的枝有刺，新枝有红锈色绒毛。叶有小叶 5～9；小叶片纸质或薄革质，椭圆形至阔披针形，或卵状长圆形，长 15～35 cm，宽 6～15 cm，先端渐尖，基部通常楔形，边缘全缘或疏生细锯齿。圆锥花序。果实阔扁球形或球形，紫黑色。花期 6～8 月，果期翌年 1～2 月。

Trees, to 3–20 m tall. Branches prickly, ferruginous red tomentose when young. Leaves palmately compound, with 5–9 Leaflets. Leaflets oblong, ovate-elliptic, or broadly lanceolate, 15–35 × 6–15 cm, papery or subleathery, margin entire or sparsely serrulate, apex acuminate. Inflorescence terminal, pendent, unarmed. Fruit globose or compressed-globose to didymo-globose, purple-black. Fl. Jun.–Aug., fr. Jan.–Feb. of next year.

叶背　Compound leaf back
摄影：蒋裕良　Photo by: Jiang Yuliang

复叶　Compound leaf
摄影：蒋裕良　Photo by: Jiang Yuliang

枝干　Branch
摄影：蒋裕良　Photo by: Jiang Yuliang

径级分布表 DBH class

胸径区间 Diameter class (cm)	个体数 No. of individuals in the plot	比例 Proportion (%)
1～2	185	49.20
2～5	177	47.07
5～10	11	2.93
10～20	3	0.80
20～35	0	0.00
35～50	0	0.00
≥50	0	0.00

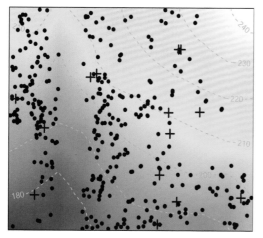

● 1～5 cm DBH　＋ 5～20 cm DBH　○ ≥20 cm DBH
个体分布图 Distribution of individuals

139 鹅掌柴 | é zhǎng chái

Schefflera heptaphylla (L.) Frodin
五加科 Araliaceae

样地名称（Plot name）= JH
个体数（Individual number/1 hm²）= 331
最大胸径（Max DBH）= 28.8 cm
重要值排序（Importance value rank）= 1

乔木，高达 15 m。叶柄长 10~30 cm。叶有小叶 6~9，最多至 11，椭圆形至倒卵形椭圆形，7~18×3~5 cm，纸质至革质，密被幼时星状短柔毛，先端急尖或短渐尖，基部渐狭或楔形到钝或圆形，边缘全缘。圆锥花序顶生，呈球形。花期 9~12 月，果期 12 月至翌年 2 月。

Trees, to 15 m tall. Petiole 10–30 cm. Leaflets 6–9 (–11). elliptic to obovate-elliptic, 7–18 × 3–5 cm, papery to leathery, densely stellate pubescent when young, apex abruptly acute to acuminate, base attenuate or cuneate to obtuse or rounded, margin entire. Inflorescence a terminal panicle of umbels. Fruit globose. Fl. Sep.–Dec., fr. Dec.–Feb. of next year.

树干　Trunk
摄影：李健星　Photo by: Li Jianxing

复叶　Compound leaf
摄影：李健星　Photo by: Li Jianxing

果序　Infructescence
摄影：向悟生　Photo by: Xiang Wusheng

径级分布表　DBH class

胸径区间 Diameter class (cm)	个体数 No. of individuals in the plot	比例 Proportion (%)
1~2	40	12.09
2~5	39	11.78
5~10	87	26.28
10~20	148	44.71
20~35	17	5.14
35~50	0	0.00
≥50	0	0.00

● 1~5 cm DBH　＋ 5~20 cm DBH　○ ≥20 cm DBH
个体分布图　Distribution of individuals

广西防城季节性雨林物种及其分布格局
GUANGXI FANGCHENG SEASONAL RAIN FORESTS: SPECIES AND THEIR DISTRIBUTION PATTERNS

4 广西防城十万山 1 hm² 样地（SW）的树种及其分布格局

Tree Species and their Distribution Patterns in the Guangxi Fangcheng SW 1 hm² Plot

1 杉木 | shān mù （沙木） 杉木属

***Cunninghamia lanceolata* (Lamb.) Hook.**

柏科 Cupressaceae

样地名称（Plot name）= SW
个体数（Individual number/1 hm²）= 2
最大胸径（Max DBH）= 7.1 cm
重要值排序（Importance value rank）= 100

乔木或灌木，高达 50 m，胸径可达 3 m。树皮灰褐色，裂成长条片脱落，内皮淡红色。叶披针形或条状披针形，呈镰状，竖硬，长 0.8~6.5（~7）cm，宽 1.5~5 mm，上面深绿色，下面淡绿色，沿中脉两侧各有 1 条白粉气孔带。圆锥花序簇生。球果卵圆形；种子深褐色。花期 1~5 月，球果 8~11 月。

Trees or shrubs, to 50 m tall, trunk to 3 m d.b.h.. Bark dark gray to dark brown, or reddish brown, longitudinally fissured. Leaves glossy deep green adaxially, narrowly linear-lanceolate, straight or slightly falcate, 0.8–6.5 (–7) × 0.15–0.5 cm, midvein green abaxially, stomatal bands present on both surfaces. Pollen cone fascicles terminal. Seed ovoid, seeds dark brown. Pollination Jan.–May, seed maturity Aug.–Nov..

树干　　　　Trunk
摄影：王斌　　Photo by: Wang Bin

叶　　　　Leaves
摄影：王斌　　Photo by: Wang Bin

枝叶　　　　Branch and leaves
摄影：王斌　　Photo by: Wang Bin

径级分布表　DBH class

胸径区间 Diameter class (cm)	个体数 No. of individuals in the plot	比例 Proportion (%)
1~2	0	0.00
2~5	0	0.00
5~10	2	100.00
10~20	0	0.00
20~35	0	0.00
35~50	0	0.00
≥50	0	0.00

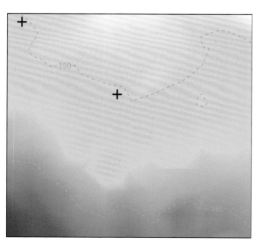

● 1~5 cm DBH　　+ 5~20 cm DBH　　○ ≥20 cm DBH
个体分布图　Distribution of individuals

2 黄丹木姜子 | huáng dān mù jiāng zǐ （黄壳楠）

Litsea elongata (Wall. ex Ness) Benth. Hook. f.

樟科 Lauraceae

样地名称（Plot name）= SW
个体数（Individual number/1 hm^2）= 1
最大胸径（Max DBH）= 7.3 cm
重要值排序（Importance value rank）= 117

常绿小乔木或中乔木，高达 10 m。叶互生，长圆形、长圆状披针形至倒披针形，长 6~22 cm，宽 2~6 cm，先端钝或短渐尖，基部楔形或近圆，革质，羽状脉，侧脉 6~10 对，中脉及侧脉在叶上面平或稍下陷，在下面突起。伞形花序单生。果长圆形，成熟时黑紫色。花期 5~11 月，果期翌年 2~6 月。

Evergreen small or medium-sized trees, up to 10 m tall. Leaves alternate. Leaf blade oblong or oblong-lanceolate to oblanceolate, 6–22 × 2–6 cm, apex obtuse or shortly acuminate, lateral veins 6–10 pairs, base cuneate, apex abruptly acute. Umbels solitary. Fruit ellipsoid, seated on cup-shaped perianth tube, red becoming deep purple-black at maturity. Fl. May–Nov., fr. Feb.–Jun. of next year.

花枝　　Flowering branches
摄影：蒋裕良　Photo by: Jiang Yuliang

叶背　　Leaf backs
摄影：杨平　Photo by: Yang Ping

果　　Fruits
摄影：杨平　Photo by: Yang Ping

径级分布表 DBH class

胸径区间 Diameter class (cm)	个体数 No. of individuals in the plot	比例 Proportion (%)
1~2	0	0.00
2~5	0	0.00
5~10	1	100.00
10~20	0	0.00
20~35	0	0.00
35~50	0	0.00
≥50	0	0.00

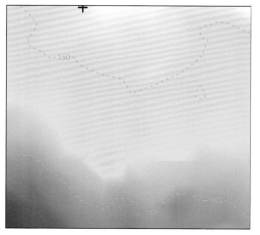

● 1~5 cm DBH　＋ 5~20 cm DBH　○ ≥20 cm DBH
个体分布图 Distribution of individuals

3 短序润楠 | duǎn xù rùn nán

Machilus breviflora (Benth.) Hemsl.

樟科 Lauraceae

样地名称（Plot name）= SW
个体数（Individual number/1 hm²）= 1
最大胸径（Max DBH）= 18.9 cm
重要值排序（Importance value rank）= 105

乔木，高约 8 m。树皮灰褐色。小枝无毛。叶倒卵形至倒卵状披针形，长 4~5 cm，宽 1.5~2 cm，先端钝，基部渐狭，革质，两面无毛，干时下面稍粉绿或带褐色，中脉上面凹入，下面凸起，侧脉和网脉纤细。圆锥花序 3~5 个，顶生。果球形，直径约 8~10 mm。花期 7~8 月，果期 10~12 月。

Trees, ca. 8 m tall. Bark gray-brown. Branchlets glabrous. Leaf blade obovate to obovate-lanceolate, 4–5 × 1.5–2 cm, glabrous on both surfaces, midrib raised abaxially, concave adaxially, lateral veins and veinlets slender, almost invisible, base attenuate, apex obtuse. Panicles terminal in cluster of 3–5. Fruit globose, 8–10 mm in diam.. Fl. Jul.–Aug., fr. Oct.–Dec..

树干　　　Trunk
摄影：李健星　　Photo by: Li Jianxing

叶　　　Leaves
摄影：李健星　　Photo by: Li Jianxing

叶背　　　Leaf backs
摄影：李健星　　Photo by: Li Jianxing

径级分布表　DBH class

胸径区间 Diameter class (cm)	个体数 No. of individuals in the plot	比例 Proportion (%)
1~2	0	0.00
2~5	0	0.00
5~10	0	0.00
10~20	1	100.00
20~35	0	0.00
35~50	0	0.00
≥50	0	0.00

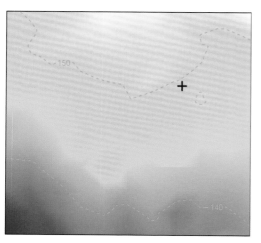

● 1~5 cm DBH　　+ 5~20 cm DBH　　O ≥20 cm DBH
个体分布图　Distribution of individuals

4 狭叶润楠 | xiá yè rùn nán

Machilus rehderi C. K. Allen

樟科 Lauraceae

样地名称（Plot name）= SW
个体数（Individual number/1 hm²）= 6
最大胸径（Max DBH）= 20.0 cm
重要值排序（Importance value rank）= 69

小乔木，高 4~15 m。枝无毛，紫黑色，有皱纹。叶聚生于小枝上部，披针形至倒披针形，长 7~14.5 cm，宽 1.5~3 cm，先端长渐尖，尖头钝，向基部渐狭，革质，两面无毛，光亮，上面黄绿色，下面较淡。圆锥或总状花序。果球形，直径 7~8 mm。花期 4 月，果期 7 月。

Small trees, 4–15 m tall. Branchlets purple-black, glabrous, striate when dry. Leaves clustered at upper part of branchlet. Leaf blade shiny, adaxially yellowish green, lanceolate to oblanceolate, 7–14.5 × 1.5–3 cm, leathery, glabrous on both surfaces, midrib slightly impressed, base cuneate, apex acuminate, summit obtuse. Inflorescences in panicle or raceme. Fruit globose, 7–8 mm in diam.. Fl. Apr., fr. Jun..

枝叶 Branch and leaves
摄影：蒋裕良 Photo by：Jiang Yuliang

叶 Leaves
摄影：蒋裕良 Photo by：Jiang Yuliang

叶背 Leaf backs
摄影：蒋裕良 Photo by：Jiang Yuliang

径级分布表 DBH class

胸径区间 Diameter class (cm)	个体数 No. of individuals in the plot	比例 Proportion (%)
1~2	2	33.33
2~5	2	33.33
5~10	1	16.67
10~20	0	0.00
20~35	1	16.67
35~50	0	0.00
≥50	0	0.00

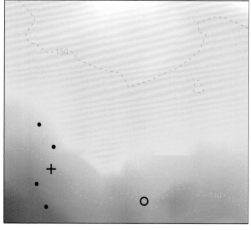

● 1~5 cm DBH   + 5~20 cm DBH   ○ ≥20 cm DBH

个体分布图 Distribution of individuals

5 长圆叶新木姜子 | cháng yuán yè xīn mù jiāng zǐ

新木姜子属

***Neolitsea oblongifolia* Merr. Chun**

樟科 Lauraceae

样地名称（Plot name）＝ SW
个体数（Individual number/1 hm²）＝ 1
最大胸径（Max DBH）＝ 1.6 cm
重要值排序（Importance value rank）＝ 137

乔木，高 8~12 m。叶互生，有时 3~5 片簇生呈近轮生状，长圆形或长圆状披针形，长 4~10 cm，宽 0.8~2.3 cm，先端钝或急尖或略渐尖，基部急尖，薄革质，上面深绿色，光亮，下面淡绿色或灰绿，羽状脉，中脉在两面均突起，侧脉每边 4~5 条，纤细。伞形花序常 5~7 个簇生叶腋。果球形，直径 8~10 mm。花期 8~11 月，果期 9~12 月。

Trees, 8–12 m tall. Leaves 3–5 subverticillate. Leaf blade oblong-lanceolate, oblong-elliptic, lanceolate, obovate, or elliptic, 4–10 × 0.8–2.3 cm, glabrate abaxially, pinninerved or subtriplinerved, lateral veins 4 or 5 pairs, base narrow or cuneate, apex acuminate, subcaudate, or abruptly acute. Umbels 5–7 clustered in axil. Fruit globose, 8–10 mm in diam.. Fl. Aug.–Nov., fr. Sep.–Dec..

果枝 Fruiting branches
摄影：杨平 Photo by：Yang Ping

枝叶 Branch and leaves
摄影：杨平 Photo by：Yang Ping

花序 Inflorescences
摄影：杨平 Photo by：Yang Ping

径级分布表 DBH class

胸径区间 Diameter class (cm)	个体数 No. of individuals in the plot	比例 Proportion (%)
1~2	1	100.00
2~5	0	0.00
5~10	0	0.00
10~20	0	0.00
20~35	0	0.00
35~50	0	0.00
≥50	0	0.00

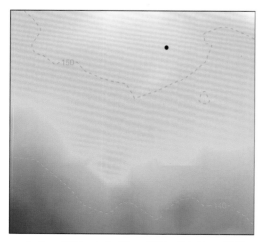

● 1~5 cm DBH ＋ 5~20 cm DBH ○ ≥20 cm DBH
个体分布图 Distribution of individuals

6 枇杷叶山龙眼 | pí pá yè shān lóng yǎn （野乌榄）

Helicia obovatifolia var. *mixta* (H. L. Li) Sleumer

山龙眼科 **Proteaceae**

样地名称（Plot name）= SW
个体数（Individual number/1 hm²）= 1
最大胸径（Max DBH）= 1.4 cm
重要值排序（Importance value rank）= 138

乔木，高 6~14 m。叶倒卵状长圆形或阔倒披针形，长 9~28 cm，宽 5~15 cm，顶端急尖或短渐尖，基部楔形或阔楔形，边具疏粗锯齿或上半部具细锯齿，有时全缘，成长叶上面无毛，下面的绒毛逐渐稀疏；侧脉 9~12 对。总状花序长 10~16 cm。果椭圆状或橄榄状，黑色。花期 6~7 月，果期 10~12 月。

Trees, 6–14 m tall. Leaf blade obovate-oblong to broadly oblanceolate, 9–28 × 5–15 cm, margin coarsely serrate or serrulate in upper half, sometimes entire, apex acute to shortly acuminate; secondary veins 9–12 on each side of midvein. Inflorescences 10–16 cm. Fruit ellipsoidal, black. Fl. Jun.–Jul., fr. Oct.–Dec..

树干　Trunk
摄影：李健星　Photo by: Li Jianxing

叶背　Leaf backs
摄影：李健星　Photo by: Li Jianxing

枝叶　Branch and leaves
摄影：李健星　Photo by: Li Jianxing

径级分布表　DBH class

胸径区间 Diameter class (cm)	个体数 No. of individuals in the plot	比例 Proportion (%)
1~2	1	100.00
2~5	0	0.00
5~10	0	0.00
10~20	0	0.00
20~35	0	0.00
35~50	0	0.00
≥50	0	0.00

● 1~5 cm DBH　+ 5~20 cm DBH　○ ≥20 cm DBH
个体分布图　Distribution of individuals

7 调羹树 | diào gēng shù （叉腮树）

假山龙眼属

***Heliciopsis lobata* (Merr.) Sleumer**

山龙眼科 **Proteaceae**

样地名称（Plot name）= SW
个体数（Individual number/1 hm²）= 1
最大胸径（Max DBH）= 4.6 cm
重要值排序（Importance value rank）= 123

乔木，高 5～20 m。叶二型，革质，全缘叶长圆形，长 10～25 cm，宽 5～7 cm，顶端短渐尖，基部楔形；分裂叶轮廓近椭圆形，长 20～60 cm，宽 20～40 cm，通常具 2～8 对羽状深裂片，有时为 3 裂叶。花序生于小枝已落叶腋部。果椭圆状或卵状椭圆形，黄绿色。花期 5～7 月，果期 11～12 月。

Trees, 5–20 m tall. Leaves dimorphic, simple or pinnatipartite. Leaf blade leathery, base cuneate, margin entire, apex shortly acuminate; reticulate veins raised, conspicuous. Simple leaf blade oblong, 10–25 × 5–7 cm. Pinnatipartite leaf blade 20–60 × 20–40 cm, lobes 2–8 pairs, sinuses somewhat rounded. Inflorescences ramiflorous, pilose. Fruit yellowish green, ellipsoid to ovoid-ellipsoid. Fl. May–Jul., fr. Nov.–Dec..

树干　　Trunk
摄影：李健星　　Photo by: Li Jianxing

叶背　　Leaf backs
摄影：李健星　　Photo by: Li Jianxing

枝叶　　Branch and leaves
摄影：李健星　　Photo by: Li Jianxing

径级分布表　DBH class

胸径区间 Diameter class (cm)	个体数 No. of individuals in the plot	比例 Proportion (%)
1～2	0	0.00
2～5	1	100.00
5～10	0	0.00
10～20	0	0.00
20～35	0	0.00
35～50	0	0.00
≥50	0	0.00

● 1～5 cm DBH　　+ 5～20 cm DBH　　○ ≥20 cm DBH
个体分布图　Distribution of individuals

8 牛耳枫 | niú ěr fēng （南岭虎皮楠）

***Daphniphyllum calycinum* Benth.**
虎皮楠科 **Daphniphyllaceae**

样地名称（Plot name）= SW
个体数（Individual number/1 hm²）= 1
最大胸径（Max DBH）= 2.0 cm
重要值排序（Importance value rank）= 134

灌木，高 1.5~4 m。小枝灰褐色，具稀疏皮孔。叶纸质，阔椭圆形或倒卵形，长 12~16 cm，宽 4~9 cm，先端钝或圆形，具短尖头，基部阔楔形，全缘，略反卷，干后两面绿色，叶面具光泽，叶背多少被白粉，具细小乳突体。总状花序腋生。果卵圆形，被白粉，具小疣状突起。花期 4~6 月，果期 8~11 月。

Shrubs, 1.5–4 m tall. Branchlets grayish brown, sparsely lenticellate. Leaf blade obovate or obovate-elliptic, 12–16 × 4–9 cm, chartaceous, glaucous and inconspicuously papillate abaxially, green and shining adaxially, base broadly cuneate, margin slightly reflexed, apex obtuse or rounded, mucronate. Inflorescences axillary, racemose. Drupe ovoid-ellipsoidal, tuberculate. Fl. Apr.–Jun., fr. Aug.–Nov..

果枝　Fruiting branches
摄影：丁涛　Photo by：Ding Tao

叶　Leaves
摄影：丁涛　Photo by：Ding Tao

叶背　Leaf backs
摄影：丁涛　Photo by：Ding Tao

径级分布表　DBH class

胸径区间 Diameter class (cm)	个体数 No. of individuals in the plot	比例 Proportion (%)
1~2	0	0.00
2~5	1	100.00
5~10	0	0.00
10~20	0	0.00
20~35	0	0.00
35~50	0	0.00
≥50	0	0.00

● 1~5 cm DBH　＋ 5~20 cm DBH　○ ≥20 cm DBH
个体分布图　Distribution of individuals

9 海红豆 | hǎi hóng dòu （小籽海红豆）

Adenanthera pavonina var. *microsperma* (Teijsm. Binn.) I. C. Nielsen

豆科 Fabaceae

样地名称（Plot name）= SW
个体数（Individual number/1 hm²）= 5
最大胸径（Max DBH）= 5.4 cm
重要值排序（Importance value rank）= 84

落叶乔木，高 5~20 余米。二回羽状复叶；叶柄和叶轴被微柔毛，无腺体；羽片 3~5 对，小叶 4~7 对，互生，长圆形或卵形，长 2.5~3.5 cm，宽 1.5~2.5 cm，两端圆钝，两面均被微柔毛。总状花序单生于叶腋或在枝顶排成圆锥花序。荚果狭长圆形，盘旋，开裂后果瓣旋卷。花期 4~7 月，果期 7~10 月。

Deciduous trees, 5–20 m tall. Pinnae 3–5 pairs. Leaflets 4–7 pairs, alternate, oblong or ovate, 2.5–3.5 × 1.5–2.5 cm, both surfaces puberulent, both ends rounded-obtuse. Racemes simple, axillary or arranged in panicles at apices of branchlets. Legume narrowly oblong, valves contorted after de-hiscence. Fl. Apr.–Jul., fr. Jul.–Oct..

树干　Trunk
摄影：李健星　Photo by: Li Jianxing

复叶　Compound leaf
摄影：李健星　Photo by: Li Jianxing

荚果　Legume
摄影：李健星　Photo by: Li Jianxing

径级分布表　DBH class

胸径区间 Diameter class (cm)	个体数 No. of individuals in the plot	比例 Proportion (%)
1~2	2	40.00
2~5	2	40.00
5~10	1	20.00
10~20	0	0.00
20~35	0	0.00
35~50	0	0.00
≥50	0	0.00

● 1~5 cm DBH　+ 5~20 cm DBH　○ ≥20 cm DBH
个体分布图　Distribution of individuals

10 楹树 | yíng shù

Albizia chinensis (Osbeck) Merr.
豆科 Fabaceae

样地名称（Plot name）= SW
个体数（Individual number/1 hm²）= 1
最大胸径（Max DBH）= 4.1 cm
重要值排序（Importance value rank）= 124

落叶乔木，高达 30 m。小枝被黄色柔毛。托叶大，膜质，心形，先端有小尖头，早落。二回羽状复叶，，羽片 6~12 对，小叶 20~35（~40）对，无柄，长椭圆形，长 6~10 mm，宽 2~3 mm，先端渐尖，基部近截平，具缘毛，下面被长柔毛；中脉紧靠上边缘。头状花序有花 10~20 朵。荚果扁平。花期 3~5 月，果期 6~12 月。

Deciduous trees, to 30 m tall. Branchlets yellow pubes-cent. Stipules deciduous, cordate, large, membranous, apex api-culate. Bipinnate compound leaf, pinnae 6-12 pairs, leaflets 20-35(-40) pairs, sessile, oblong-linear, falcate, 6–10 × 2–3 mm, abaxially villous, main vein close to upper margin, base subtruncate, margin ciliate, apex acuminate. Heads 10–20-flowered. Seeds elliptic, flat. Fl. Mar.–May, fr. Jun.–Dec..

复叶 Compound leaves
摄影：杨平 Photo by: Yang Ping

花序 Inflorescences
摄影：杨平 Photo by: Yang Ping

花枝 Flowering branches
摄影：杨平 Photo by: Yang Ping

径级分布表 DBH class

胸径区间 Diameter class (cm)	个体数 No. of individuals in the plot	比例 Proportion (%)
1~2	0	0.00
2~5	1	100.00
5~10	0	0.00
10~20	0	0.00
20~35	0	0.00
35~50	0	0.00
≥50	0	0.00

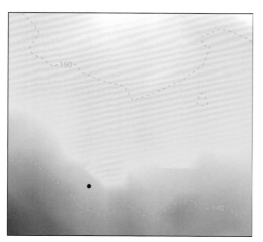

● 1~5 cm DBH ＋ 5~20 cm DBH ○ ≥20 cm DBH
个体分布图 Distribution of individuals

11 薄叶猴耳环 | báo yè hóu ěr huán （薄叶围涎树） 猴耳环属

***Archidendron utile* (Chun F. C. How) I. C. Nielsen**

豆科 **Fabaceae**

样地名称（Plot name）＝ SW
个体数（Individual number/1 hm^2）＝ 2
最大胸径（Max DBH）＝ 13.4 cm
重要值排序（Importance value rank）＝ 112

灌木，高 1~2 m，稀小乔木。羽片 2~3 对，长 10~18 cm，总叶柄和顶端 1~2 对小叶着生处稍下的叶轴上有腺体；小叶膜质，4~7 对，对生，长方菱形，长 2~9 cm，宽 1.5~4 cm，顶部的较大，往下渐小，顶端钝，有小凸头，基部钝或急尖。头状花序直径约 1 cm。荚果红褐色，弯卷或镰刀状。花期 3~8 月；果期 4~12 月。

Shrubs, 1–2 m tall, rarely small trees. Pinnae 2 or 3 pairs, 10–18 cm; glands on petiole and rachis of apical 1 or 2 leaflets at places of insertion, glands circular, sessile. Leaflets 4–7 pairs, opposite, oblong-rhombic, 2–9 × 1.5–4 cm, apical ones larger, downward smaller, membranous, base ob-tuse or acute, apex obtuse, mucronate. Heads ca. 15-flowered, ca.1 cm in diam.. Legume red-brown, falcate. Fl. Mar.–Aug., fr. Apr.–Dec..

果枝　Fruiting branches
摄影：杨平　Photo by: Yang Ping

复叶　Compound leaf
摄影：杨平　Photo by: Yang Ping

荚果　Legume
摄影：杨平　Photo by: Yang Ping

径级分布表　DBH class

胸径区间 Diameter class (cm)	个体数 No. of individuals in the plot	比例 Proportion (%)
1~2	1	50.00
2~5	0	0.00
5~10	0	0.00
10~20	1	50.00
20~35	0	0.00
35~50	0	0.00
≥50	0	0.00

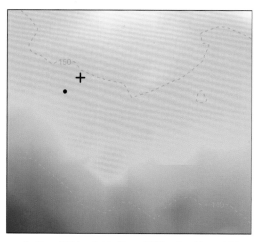

● 1~5 cm DBH　＋ 5~20 cm DBH　○ ≥20 cm DBH
个体分布图　Distribution of individuals

12 老虎刺 | lǎo hǔ cì （黄牛筋）

老虎刺属

***Pterolobium punctatum* Hemsl.**

豆科 Fabaceae

样地名称（Plot name）= SW
个体数（Individual number/1 hm²）= 1
最大胸径（Max DBH）= 3.5 cm
重要值排序（Importance value rank）= 126

木质藤本或攀援性灌木，高 3~10 m。叶轴长 12~20 cm；亦有成对黑色托叶刺；羽片 9~14 对，狭长；羽轴长 5~8 cm，上面具槽，小叶片 19~30 对，对生，狭长圆形，顶端圆钝具凸尖或微凹，基部微偏斜。总状花序。荚果长 4~6 cm。花期 6~8 月，果期 9 月至翌年 1 月。

Climbers, woody, or climbing shrubs, 3–10 m tall. Leaf rachis 12–20 cm; with paired blackish stip-ulaceous spines; pinnae 9–14 pairs, long and narrow; rachis of pinnae 5–8 cm, sulcate; petiolules short, articulate. Leaflets 19–30 pairs, opposite, narrowly oblong, base slightly oblique, apex rounded-cuspidate or emarginate. Racemes. Legume shiny, 4–6 cm. Fl. Jun.–Aug., fr. Apr., Sep.–Jan. of nexr year.

树干　Trunk
摄影：向悟生　Photo by: Xiang Wusheng

枝叶　Branch and leaves
摄影：向悟生　Photo by: Xiang Wusheng

荚果　Legumes
摄影：杨平　Photo by: Yang Ping

径级分布表　DBH class

胸径区间 Diameter class (cm)	个体数 No. of individuals in the plot	比例 Proportion (%)
1~2	0	0.00
2~5	1	100.00
5~10	0	0.00
10~20	0	0.00
20~35	0	0.00
35~50	0	0.00
≥50	0	0.00

● 1~5 cm DBH　+ 5~20 cm DBH　○ ≥20 cm DBH
个体分布图　Distribution of individuals

13 黄叶树 | huáng yè shù 黄叶树属

***Xanthophyllum hainanense* Hu**
远志科 Polygalaceae

样地名称（Plot name）= SW
个体数（Individual number/1 hm²）= 5
最大胸径（Max DBH）= 2.5 cm
重要值排序（Importance value rank）= 79

乔木，高 5~20 m。树皮暗灰色，具细纵裂。叶片革质，卵状椭圆形至长圆状披针形，长 4~12 cm，宽 1.5~5 cm，先端长渐尖，基部楔形至钝，全缘，有时波状，两面均无毛，干时黄绿色。总状花序或小型圆锥花序腋生或顶生。核果球形，淡黄色，直径 1.5~2 cm。花期 3~5 月，果期 4~7 月。

Trees, 5–20 m tall. Bark gray, longitudinally thinly fissured. Leaf blade yellow-green when dry, ovate-elliptic to oblong-lanceolate, 4–12 × 1.5–5 cm, leathery, both surfaces glabrous, midvein and lateral veins raised on both surfaces, base cuneate or obtuse, margin sometimes undulate, apex long acuminate. Racemes or small panicles axillary or terminal. Drupe yellowish, 1.5–2 cm in diam.. Fl. Mar.–May, fr. Apr.–Jul..

枝叶　Branch and leaves
摄影：杨平　Photo by: Yang Ping

叶背　Leaf backs
摄影：杨平　Photo by: Yang Ping

花序　Inflorescences
摄影：孙观灵　Photo by: Sun Guanling

径级分布表 DBH class

胸径区间 Diameter class (cm)	个体数 No. of individuals in the plot	比例 Proportion (%)
1~2	4	80.00
2~5	1	20.00
5~10	0	0.00
10~20	0	0.00
20~35	0	0.00
35~50	0	0.00
≥50	0	0.00

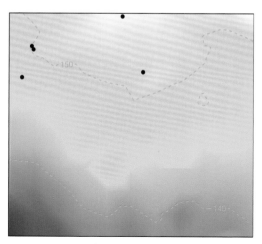

● 1~5 cm DBH　＋ 5~20 cm DBH　○ ≥20 cm DBH
个体分布图 Distribution of individuals

14 朴树 | pò shù

***Celtis sinensis* Pers.**
大麻科 Cannabaceae

样地名称（Plot name）= SW
个体数（Individual number/1 hm^2）= 1
最大胸径（Max DBH）= 1.3 cm
重要值排序（Importance value rank）= 141

落叶乔木，高达 20 m。树皮灰白色。叶厚纸质，通常卵形或卵状椭圆形，但不带菱形，长 3~10 cm，宽 3.5~6 cm，基部几乎不偏斜或仅稍偏斜，先端尖至渐尖，但不为尾状渐尖。花簇生在叶腋和茎基部。果近球形，直径约 5~7 (~8) mm。花期 3~4 月，果期 9~10 月。

Deciduous trees, to 20 m tall. Bark gray. Leaf blade ovate to ovate-elliptic, 3–10 × 3.5–6 cm, thickly papery, base rounded, obtuse, or obliquely truncate, ± symmetric to moderately oblique, margin subentire to crenate on apical half, apex acute to shortly acuminate. Flowers fascicled in leaf axils and stem bases. Drupe ± globose, 5–7 (–8) mm in diam.. Fl. Mar–Apr., fr. Sep.–Oct..

树干　Trunk
摄影：李健星　Photo by: Li Jianxing

叶　Leaves
摄影：李健星　Photo by: Li Jianxing

枝叶　Branch and leaves
摄影：李健星　Photo by: Li Jianxing

径级分布表　DBH class

胸径区间 Diameter class (cm)	个体数 No. of individuals in the plot	比例 Proportion (%)
1~2	1	100.00
2~5	0	0.00
5~10	0	0.00
10~20	0	0.00
20~35	0	0.00
35~50	0	0.00
≥50	0	0.00

● 1~5 cm DBH　＋ 5~20 cm DBH　○ ≥20 cm DBH
个体分布图　Distribution of individuals

15 二色波罗蜜 | èr sè bō luó mì （红山梅）

波罗蜜属

***Artocarpus styracifolius* Pierre**

桑科 Moraceae

样地名称（Plot name）＝ SW
个体数（Individual number/1 hm^2）＝ 1
最大胸径（Max DBH）＝ 21.9 cm
重要值排序（Importance value rank）＝ 97

乔木，高达 20 m。叶互生排为 2 列，皮纸质，长圆形或倒卵状披针形，有时椭圆形，长 4～8 cm，宽 2.5～3 cm，先端渐尖为尾状，基部楔形，略下延至叶柄，全缘，背面被苍白色粉沫状毛，脉上更密，侧脉 4～7 对。花雌雄同株，花序单生叶腋。核果球形。花期秋初，果期秋末冬初。

Trees, to 20 m tall. Leaves distichous. Leaf blade oblong, obovate-lanceolate, or sometimes elliptic, 4–8 × 2.5–3 cm, leathery to papery, abaxially densely covered with white farinaceous hairs especially on veins, base cuneate to decurrent on petiole, margin entire, apex acuminate to caudate; secondary veins 4–7 on each side of midvein. Inflorescences axillary, solitary. Drupes globose. Fl. early autumn, fr. late autumn to early winter.

果枝　Fruiting branches
摄影：陆昭岑　Photo by: Lu Zhaochen

叶　Leaves
摄影：陆昭岑　Photo by: Lu Zhaochen

果　Fruit
摄影：陆昭岑　Photo by: Lu Zhaochen

径级分布表　DBH class

胸径区间 Diameter class (cm)	个体数 No. of individuals in the plot	比例 Proportion (%)
1～2	0	0.00
2～5	0	0.00
5～10	0	0.00
10～20	0	0.00
20～35	1	100.00
35～50	0	0.00
≥50	0	0.00

● 1～5 cm DBH　＋ 5～20 cm DBH　○ ≥20 cm DBH
个体分布图 Distribution of individuals

16 垂叶榕 | chuí yè róng

Ficus benjamina L.

桑科 Moraceae

样地名称（Plot name）= SW
个体数（Individual number/1 hm²）= 1
最大胸径（Max DBH）= 40.0 cm
重要值排序（Importance value rank）= 65

乔木，高达 20 m。叶薄革质，卵形至卵状椭圆形，长 4~8 cm，宽 2~4 cm，先端短渐尖，基部圆形或楔形，全缘，一级侧脉与二级侧脉难于区分，平行展出，直达近叶边缘，网结成边脉，两面光滑无毛。榕果成对或单生叶腋，基部缢缩成柄，球形或扁球形，光滑，成熟时红色至黄色。花期 8~11 月，果期全年。

Trees, to 20 m tall. Leaf blade ovate to broadly elliptic, 4–8× 2–4 cm, ± leathery leathery, glabrous, base rounded to cuneate, margin entire, apex shortly acuminate; secondary veins, parallel, anastomosing near margin, indistinct from tertiary veins. Figs axillary on leafy branchlets, paired or solitary, red-yellow at maturity. Fl. Aug.–Nov., fr. throughout year.

果枝　　Fruiting branches
摄影：杨平　　Photo by：Yang Ping

叶　　Leaves
摄影：杨平　　Photo by：Yang Ping

果　　Fruit
摄影：杨平　　Photo by：Yang Ping

径级分布表 DBH class

胸径区间 Diameter class (cm)	个体数 No. of individuals in the plot	比例 Proportion (%)
1~2	0	0.00
2~5	0	0.00
5~10	0	0.00
10~20	0	0.00
20~35	0	0.00
35~50	1	100.00
≥50	0	0.00

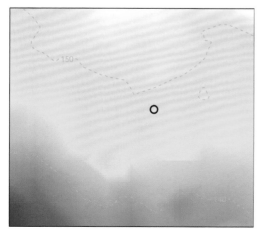

● 1~5 cm DBH　　+ 5~20 cm DBH　　○ ≥20 cm DBH
个体分布图 Distribution of individuals

17 黄毛榕 | huáng máo róng

Ficus esquiroliana H. Lév.

桑科 Moraceae

样地名称（Plot name）= SW
个体数（Individual number/1 hm²）= 4
最大胸径（Max DBH）= 2.8 cm
重要值排序（Importance value rank）= 86

乔木或灌木，高约 4~10 m。树皮灰褐色，具纵棱。叶互生，纸质，广卵形，长（11~）17~27 cm，宽（8~）12~20 cm，急渐尖，具长约 1 cm 尖尾，基部浅心形，表面疏生糙伏状长毛，背面被长约 3~5 mm 褐黄色波状长毛，边缘有细锯齿，齿端被长毛。榕果腋生，圆锥状椭圆形。花期 5~7 月，果期 7 月。

Trees or shrubs, 4–10 m tall. Bark grayish brown to grayish green, with longitudinal ridges. Leaves alternate. Leaf blade broadly obovate, (11–) 17–27 × (8–) 12–20 cm, thickly papery, abaxially with white or yellow soft felted hairs and hairs 3–5 mm, base shallowly cordate, margin sparsely serrate with long hairs at apex of teeth, apex acute to caudate with a ca. 1 cm cauda. Figs axillary, ovoid. Fl. May–Jul., fr. Jul..

果　　Fruits
摄影：椰子　　Photo by: Ye zi

叶　　Leaves
摄影：丁涛　　Photo by: Ding Tao

叶背　　Leaf backs
摄影：丁涛　　Photo by: Ding Tao

径级分布表 DBH class

胸径区间 Diameter class (cm)	个体数 No. of individuals in the plot	比例 Proportion (%)
1~2	0	0.00
2~5	4	100.00
5~10	0	0.00
10~20	0	0.00
20~35	0	0.00
35~50	0	0.00
≥50	0	0.00

● 1~5 cm DBH　　+ 5~20 cm DBH　　○ ≥20 cm DBH
个体分布图 Distribution of individuals

18 水同木 | shuǐ tóng mù 榕属

Ficus fistulosa Reinw. ex Blume
桑科 Moraceae

样地名称（Plot name）= SW
个体数（Individual number/1 hm²）= 1
最大胸径（Max DBH）= 1.9 cm
重要值排序（Importance value rank）= 135

常绿小乔木。树皮黑褐色，枝粗糙，叶互生，纸质，倒卵形至长圆形，长 10~20 cm，宽 4~8 cm，先端具短尖，基部斜楔形或圆形，全缘或微波状，表面无毛，背面微被柔毛或黄色小突体。榕果簇生于老干发出的瘤状枝上，近球形，成熟橘红色，不开裂。花期 5~7 月，果期全年。

Evergreen small trees. Bark dark brown. Leaf blade obovate to oblong, 10–20 × 4–8 cm, papery, abaxially sparsely pubescent or yellow tuberculate, adaxially glabrous, base obliquely cuneate to rounded, margin entire or undulate, apex mucronate. Figs on short ± conic branchlets on main branches, reddish orange when mature, apical pore not open. Fl. May–Jul., fr. throughout year.

果枝　　　　　　　　　Fruiting branch
摄影：杨平　　　　　　Photo by: Yang Ping

叶　　　　　　　　　Leaves
摄影：丁涛　　　　　　Photo by: Ding Tao

果　　　　　　　　　Fruits
摄影：丁涛　　　　　　Photo by: Ding Tao

径级分布表　DBH class

胸径区间 Diameter class (cm)	个体数 No. of individuals in the plot	比例 Proportion (%)
1~2	1	100.00
2~5	0	0.00
5~10	0	0.00
10~20	0	0.00
20~35	0	0.00
35~50	0	0.00
≥50	0	0.00

● 1~5 cm DBH　　＋ 5~20 cm DBH　　○ ≥20 cm DBH
个体分布图　Distribution of individuals

19 褐叶榕 | hè yè róng 榕属

Ficus pubigera (Wall. ex Miq.) Kurz
桑科 Moraceae

样地名称（Plot name）= SW
个体数（Individual number/1 hm²）= 1
最大胸径（Max DBH）= 6.0 cm
重要值排序（Importance value rank）= 121

藤状灌木。叶两列，薄革质，全缘，长椭圆形，长7~11 cm，宽2.5~4 cm，先端短渐尖，基部楔形，稀圆形，干后褐色，表面无毛或沿中脉或小脉疏被柔毛，背面幼时被柔毛，后脱落。榕果生于落叶小枝叶腋，球形，表面疏生瘤状小凸体。花期4~8月，果期6~8月。

Shrubs, scandent. Leaves distichous. Leaf blade brown when dry, oblong, 7–11×2.5–4 cm, ± leathery, abaxially pubescent and glabrescent, adaxially glabrous or pubescent among veins. Figs axillary on leafy or on leafless branchlets, globose, surface sparsely tuberculate. Fl. Apr.–Aug., fr. Jun.–Aug..

枝叶　Branch and leaves
摄影：朱鑫鑫　Photo by: Zhu Xinxin

叶背　Leaf backs
摄影：丁涛　Photo by: Ding Tao

果　Fruits
摄影：朱鑫鑫　Photo by: Zhu Xinxin

径级分布表　DBH class

胸径区间 Diameter class (cm)	个体数 No. of individuals in the plot	比例 Proportion (%)
1~2	0	0.00
2~5	0	0.00
5~10	1	100.00
10~20	0	0.00
20~35	0	0.00
35~50	0	0.00
≥50	0	0.00

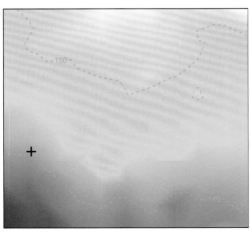

● 1~5 cm DBH　＋ 5~20 cm DBH　○ ≥20 cm DBH
个体分布图　Distribution of individuals

20 斜叶榕 | xié yè róng

榕属

***Ficus tinctoria* subsp. *gibbosa* (Blume) Corner**

桑科 Moraceae

样地名称（Plot name）= SW
个体数（Individual number/1 hm^2）= 2
最大胸径（Max DBH）= 9.0 cm
重要值排序（Importance value rank）= 99

乔木或灌木，幼时多附生。树皮微粗糙，小枝褐色。叶薄革质，排为两列，椭圆形至卵状椭圆形，长8~13 cm，宽4~6 cm，顶端钝或急尖，基部宽楔形，全缘，一侧稍宽，两面无毛，背面略粗糙，网脉明显，干后网眼深褐色。榕果球形，直径1~8 mm。花果期6~7月。

Trees or shrubs, epiphytic. Bark scabrous. Branchlets brown. Leaf blade of various shapes and sizes, ovate-elliptic, strongly asymmetric, 8–13×4–6 cm, ± leathery, abaxially puberulent and not brown when dry, adaxially rough but becoming smooth with age, margin entire or toothed even on same tree. Figs globose, 1–8 mm in diam.. Fl. and fr. Jun.–Jul..

枝叶　Branch and leaves
摄影：杨平　Photo by：Yang Ping

叶　Leaves
摄影：王斌　Photo by：Wang Bin

果　Fruits
摄影：杨平　Photo by：Yang Ping

径级分布表　DBH class

胸径区间 Diameter class (cm)	个体数 No. of individuals in the plot	比例 Proportion (%)
1~2	0	0.00
2~5	1	50.00
5~10	1	50.00
10~20	0	0.00
20~35	0	0.00
35~50	0	0.00
≥50	0	0.00

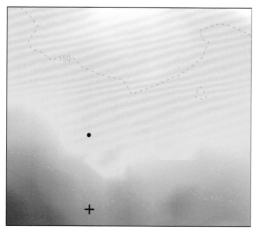

● 1~5 cm DBH　+ 5~20 cm DBH　○ ≥20 cm DBH
个体分布图　Distribution of individuals

21 构棘 | gòu jí （蓑芝）

Maclura cochinchinensis (Lour.) Corner
桑科 Moraceae

样地名称（Plot name）＝ SW
个体数（Individual number/1 hm²）＝ 1
最大胸径（Max DBH）＝ 10.3 cm
重要值排序（Importance value rank）＝ 114

直立或攀援状灌木。枝无毛，具粗壮弯曲无叶的腋生刺，刺长约 2 cm。叶革质，椭圆状披针形或长圆形，长 3~8 cm，宽 2~2.5 cm，全缘，先端钝或短渐尖，基部楔形，两面无毛，侧脉 7~10 对。聚合果肉质，表面微被毛，成熟时橙红色，核果卵圆形，成熟时褐色，光滑。花期 4~5 月，果期 6~7 月。

Shrubs, erect or scandent. Branches glabrous; spines curved or straight, to ca. 2 cm. Leaf blade elliptic-lanceolate to oblong, 3–8 × 2–2.5 cm, papery to leathery, glabrous, base cuneate, margin entire, apex rounded to shortly acuminate; secondary veins 7–10 on each side of midvein, tertiary veins reticulate. Fruiting syncarp reddish orange when mature, pubescent. Drupes brown when mature, ovoid, smooth. Fl. Apr.–May, fr. Jun.–Jul..

果枝　　Fruiting branches
摄影：杨平　　Photo by：Yang Ping

叶背　　Leaf backs
摄影：蒋裕良　　Photo by：Jiang Yuliang

果　　Fruits
摄影：杨平　　Photo by：Yang Ping

径级分布表　DBH class

胸径区间 Diameter class (cm)	个体数 No. of individuals in the plot	比例 Proportion (%)
1~2	0	0.00
2~5	0	0.00
5~10	0	0.00
10~20	1	100.00
20~35	0	0.00
35~50	0	0.00
≥50	0	0.00

● 1~5 cm DBH　　+ 5~20 cm DBH　　○ ≥20 cm DBH
个体分布图　Distribution of individuals

22 刺桑 | cì sāng

***Streblus ilicifolius* (Vidal) Corner**
桑科 Moraceae

鹊肾树属

样地名称（Plot name）= SW
个体数（Individual number/1 hm²）= 48
最大胸径（Max DBH）= 7.0 cm
重要值排序（Importance value rank）= 42

乔木或灌木。树皮灰白色，平滑；小枝具棱，刺长 1~1.5 cm，或更长达 4.5 cm，直。叶厚革质，菱状至圆状倒卵形，长 1~4.5（~9）cm，宽 0.6~2.5（~5）cm，先端急尖至圆钝或内凹，尖端常具 2 小刺齿，基部楔形下延。花序腋生，穗状。小核果扁球形，为宿存花被片半包围。花期 4 月，果期 5~6 月。

Trees or shrubs. Bark grayish white, smooth. Branchlets angular; spines straight, 1–1.5 (–4.5) cm. Leaf blade rhombic to oblong-obovate, 1–4.5 (–9) × 0.6–2.5 (–5) cm, thickly leathery, base cuneate to decurrent, apex acute, blunt, or retuse, with two spiny teeth; inflorescences axillary, spicate. Drupes oblate, half enclosed by persistent calyx lobes. Fl. Apr., fr. May–Jun..

树干　　Trunk
摄影：李健星　Photo by：Li Jianxing

叶背　　Leaf backs
摄影：丁涛　Photo by：Ding Tao

枝叶　　Branch and leaves
摄影：丁涛　Photo by：Ding Tao

径级分布表 DBH class

胸径区间 Diameter class (cm)	个体数 No. of individuals in the plot	比例 Proportion (%)
1~2	26	54.17
2~5	20	41.67
5~10	2	4.16
10~20	0	0.00
20~35	0	0.00
35~50	0	0.00
≥50	0	0.00

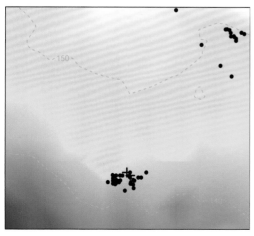

● 1~5 cm DBH　　+ 5~20 cm DBH　　○ ≥20 cm DBH
个体分布图　Distribution of individuals

23 绢毛杜英 | juàn máo dù yīng

Elaeocarpus nitentifolius Merr. Chun
杜英科 Elaeocarpaceae

样地名称（Plot name）= SW
个体数（Individual number/1 hm²）= 1
最大胸径（Max DBH）= 2.0 cm
重要值排序（Importance value rank）= 133

乔木，高 20 m。嫩枝被银灰色绢毛。叶革质，椭圆形，长 8～15 cm，宽 3.5～7.5 cm，先端急尖，尖头长 1～1.5 cm，基部阔楔形，初时两面有绢毛，不久上面变秃净，干后深绿色，发亮，下面有银灰色绢毛。总状花序，长 2～4.5 cm。核果小，椭圆形，内果皮厚 1 mm。花期 4～5 月，果期秋后。

Trees, to 20 m tall. Branchlets rust-colored sericeous. Leaf blade oblong-lanceolate to oblanceolate or elliptic, 8–15 × 3.5–7.5 cm, leathery, abaxially silvery-white sericeous, or rarely glabrescent, not black punctate, apex long acuminate or nearly caudate, acumen 1–1.5 cm. Raceme 2–4.5 cm. Drupe shiny, ellipsoid, small; endocarp ca. 1 mm thick. Fl. Apr.–May, fr. after autumn.

枝叶　Branch and leaves
摄影：从睿　Photo by: Cong Rui

叶背　Leaf backs
摄影：丁涛　Photo by: Ding Tao

果　Fruits
摄影：从睿　Photo by: Cong Rui

径级分布表　DBH class

胸径区间 Diameter class (cm)	个体数 No. of individuals in the plot	比例 Proportion (%)
1～2	0	0.00
2～5	1	100.00
5～10	0	0.00
10～20	0	0.00
20～35	0	0.00
35～50	0	0.00
≥50	0	0.00

● 1～5 cm DBH　＋ 5～20 cm DBH　○ ≥20 cm DBH
个体分布图　Distribution of individuals

24 小盘木 | xiǎo pán mù

Microdesmis caseariifolia Planch. ex Hook. f.
小盘木科 Pandaceae

样地名称（Plot name）= SW
个体数（Individual number/1 hm²）= 2
最大胸径（Max DBH）= 3.9 cm
重要值排序（Importance value rank）= 107

乔木或灌木，高 3～8 m。树皮粗糙。叶片纸质至薄革质，披针形、长圆状披针形至长圆形，长 6～16 cm，宽 2.5～5 cm，顶端渐尖或尾状渐尖，基部楔形或阔楔形，两侧稍不等，边缘具细锯齿或近全缘。花小，黄色，簇生于叶腋。核果圆球状，外面粗糙，成熟时红色，干后呈黑色，外果皮肉质。花期 3～9 月，果期 7～11 月。

Trees or shrubs, 3–8 m tall. Bark scabrous. Leaf blade lanceolate, oblong-lanceolate, or oblong, 6–16 × 2.5–5 cm, papery to thinly leathery, base cuneate or broadly so, inequilateral, margin crenulate or subentire, apex acuminate, sometimes caudate. Flowers yellow, small, in axillary fascicles. Drupe red when mature, globose, scabrous. Fl. Mar.–Sep., fr. Jul.–Nov..

枝叶　Branch and leaves
摄影：丁涛　Photo by：Ding Tao

叶背　Leaf backs
摄影：丁涛　Photo by：Ding Tao

花序　Inflorescence
摄影：丁涛　Photo by：Ding Tao

径级分布表　DBH class

胸径区间 Diameter class (cm)	个体数 No. of individuals in the plot	比例 Proportion (%)
1～2	1	50.00
2～5	1	50.00
5～10	0	0.00
10～20	0	0.00
20～35	0	0.00
35～50	0	0.00
≥50	0	0.00

● 1～5 cm DBH　＋ 5～20 cm DBH　○ ≥20 cm DBH
个体分布图　Distribution of individuals

25 余甘子 | yú gān zǐ （牛甘果）

Phyllanthus emblica L.
叶下珠科 Phyllanthaceae

样地名称（Plot name）＝ SW
个体数（Individual number/1 hm^2）＝ 1
最大胸径（Max DBH）＝ 8.2 cm
重要值排序（Importance value rank）＝ 116

乔木，高达 3~8（~23）m，胸径 50 cm。树皮浅褐色。叶片纸质至革质，二列，线状长圆形，长 8~23 mm，宽 1.5~6 mm，基部浅心形而稍偏斜，顶端截平或钝圆，有锐尖头或微凹。多朵雄花和 1 朵雌花或全为雄花组成腋生的聚伞花序。蒴果呈核果状，圆球形，直径 1~1.3 cm，外果皮肉质，内果皮硬壳质。花期 4~6 月，果期 7~9 月。

Trees, to 3–8 (–23) m tall, to 50 cm d.b.h.. Bark brownish. Leaves papery to leathery, distichous. Leaf blade linear-oblong, 8–23 × 1.5–6 mm, base shallowly cordate and slightly oblique, apex truncate, rounded or obtuse, mucronate or retuse at tip. Fascicles with many male flowers and sometimes 1 or 2 larger female flowers. Fruit a drupe, globose, 1–1.3 cm in diam, exocarp fleshy, endocarp crustaceous. Fl. Apr.–Jun., fr. Jul.–Sep..

花枝　　Flowering branch
摄影：丁涛　　Photo by: Ding Tao

叶　　Leaves
摄影：李健星　　Photo by: Li Jianxing

果　　Fruit
摄影：李健星　　Photo by: Li Jianxing

径级分布表　DBH class

胸径区间 Diameter class (cm)	个体数 No. of individuals in the plot	比例 Proportion (%)
1~2	0	0.00
2~5	0	0.00
5~10	1	100.00
10~20	0	0.00
20~35	0	0.00
35~50	0	0.00
≥50	0	0.00

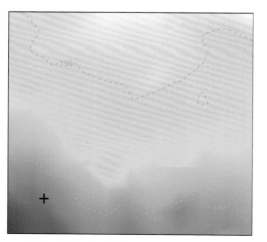

● 1~5 cm DBH　　+ 5~20 cm DBH　　O ≥20 cm DBH
个体分布图　Distribution of individuals

26 青枣核果木 | qīng zǎo hé guǒ mù （青枣柯）

Drypetes cumingii (Baill.) Pax K. Hoffm. in Engler
核果木科 Putranjivaceae

样地名称（Plot name）＝ SW
个体数（Individual number/1 hm²）＝ 2
最大胸径（Max DBH）＝ 5.2 cm
重要值排序（Importance value rank）＝ 104

乔木，高 9～20 m。叶片革质，卵形，长圆形至卵状披针形，长 6～17 cm，宽 2.5～6.5 cm，顶端急尖至长渐尖，基部楔形至钝，稍偏斜，边缘具不规则的波状齿或不明显的钝齿，两面均无毛，叶面有光泽；侧脉每边 7～9 条。花簇生于叶腋。蒴果长圆形至椭圆形，被短柔毛。花期 5～7 月，果期 8～12 月。

Trees, 9–20 m tall. Leaf blade ovate or oblong to ovate-lanceolate, 6–17 × 2.5–6.5 cm, leathery, both surfaces glabrous, adaxially glossy, base cuneate to obtuse, slightly oblique, margin irregularly undulate-serrate or obscurely obtusely so, apex acute to long acuminate; lateral veins 7–9 pairs, reticulate veins prominent. Male flowers clustered. Drupes oblong to ellipsoid, pubescent. Fl. May–Jun., fr. Aug.–Dec..

树干　Trunk
摄影：徐晔春　Photo by: Xu Yechun

叶　Leaves
摄影：蒋裕良　Photo by: Jiang Yuliang

叶背　Leaf backs
摄影：蒋裕良　Photo by: Jiang Yuliang

径级分布表　DBH class

胸径区间 Diameter class (cm)	个体数 No. of individuals in the plot	比例 Proportion (%)
1～2	0	0.00
2～5	1	50.00
5～10	1	50.00
10～20	0	0.00
20～35	0	0.00
35～50	0	0.00
≥50	0	0.00

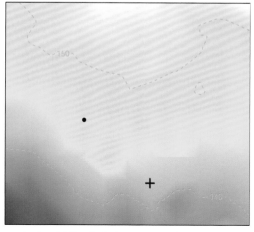

● 1～5 cm DBH　+ 5～20 cm DBH　○ ≥20 cm DBH
个体分布图　Distribution of individuals

27 子楝树 | zǐ liàn shù （桑枝米碎叶）

子楝树属

Decaspermum gracilentum (Hance) Merr. L. M. Perry
桃金娘科 **Myrtaceae**

样地名称（Plot name）= SW
个体数（Individual number/1 hm^2）= 4
最大胸径（Max DBH）= 10.8 cm
重要值排序（Importance value rank）= 73

灌木至乔木，高 4m。嫩枝被灰褐色或灰色柔毛，纤细，有钝棱。叶片纸质或薄革质，椭圆形，有时为长圆形或披针形，长 4～9 cm，宽 2～3.5 cm，先端急锐尖或渐尖，基部楔形，上面干后变黑色，有光泽。聚伞花序腋生。浆果直径约 4 mm，有柔毛。花期 3～5 月，果期 9～10 月。

Shrubs to trees, to 4 m tall. Branchlets often 4-angled or narrowly 4-winged, sparsely sericeous. Leaf blade elliptic to ovate, rarely lanceolate or obovate, 4–9 × 2–3.5 cm adaxially glossy green turning blackish when dry, base cuneate to obtuse. Inflorescences axillary or lateral below leaves. Fruit black, globular, 4 mm in diam., sparsely pubescent. Fl. Mar.–May, fr. Sep.–Oct..

枝叶　Branch and leaves
摄影：丁涛　Photo by: Ding Tao

叶　Leaves
摄影：丁涛　Photo by: Ding Tao

果　Fruits
摄影：丁涛　Photo by: Ding Tao

径级分布表　DBH class

胸径区间 Diameter class (cm)	个体数 No. of individuals in the plot	比例 Proportion (%)
1～2	0	0.00
2～5	1	25.00
5～10	2	50.00
10～20	1	25.00
20～35	0	0.00
35～50	0	0.00
≥50	0	0.00

● 1～5 cm DBH　　+ 5～20 cm DBH　　○ ≥20 cm DBH
个体分布图　Distribution of individuals

28 子凌蒲桃 | zǐ líng pú táo （子凌木）　　蒲桃属

Syzygium championii (Benth.) Merr. L. M. Perry

桃金娘科 Myrtaceae

样地名称（Plot name）= SW
个体数（Individual number/1 hm²）= 26
最大胸径（Max DBH）= 8.5 cm
重要值排序（Importance value rank）= 39

灌木至乔木。嫩枝有4棱，干后灰白色。叶片革质，狭长圆形至椭圆形，长3～6（～9）cm，宽1～2（～3）cm，先端急尖，常有长不及1cm的尖头，基部阔楔形，上面干后灰绿色，下面同色，侧脉多而密，近于水平斜出，脉间相隔1 mm，边脉贴近边缘。聚伞花序顶生。果实长椭圆形，红色。花期8～11月，果期10～12月。

Shrubs to trees. Branchlets grayish white when dry, 4-angled. Leaf blade narrowly oblong to elliptic, 3–6 (–9)×1–2 (–3) cm, leathery, both surfaces grayish green when dry, adaxially not glossy, secondary veins numerous, ca. 1 mm apart, and nearly level with surface, intramarginal veins nearly at margin, base broadly cuneate, apex acute. Inflorescences terminal. Fruit red, long ellipsoid. Fl. Aug.–Nov., fr. Oct.–Dec..

果枝　Fruiting branch
摄影：武丽琼　Photo by：Wu Liqiong

叶背　Leaf backs
摄影：孙观灵　Photo by：Sun Guanling

枝叶　Branch and leaves
摄影：孙观灵　Photo by：Sun Guanling

径级分布表　DBH class

胸径区间 Diameter class (cm)	个体数 No. of individuals in the plot	比例 Proportion (%)
1～2	17	65.38
2～5	6	23.08
5～10	3	11.54
10～20	0	0.00
20～35	0	0.00
35～50	0	0.00
≥50	0	0.00

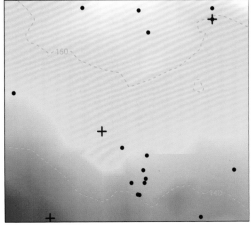

● 1～5 cm DBH　　+ 5～20 cm DBH　　○ ≥20 cm DBH
个体分布图　Distribution of individuals

29 谷木 | gǔ mù

Memecylon ligustrifolium Champ. ex Benth.
野牡丹科 Melastomataceae

样地名称（Plot name）= SW
个体数（Individual number/1 hm²）= 2
最大胸径（Max DBH）= 1.2 cm
重要值排序（Importance value rank）= 119

灌木或乔木，高 1.5~5 (~7) m。小枝圆柱形或不明显的四棱形，分枝多。叶片革质，椭圆形至卵形，或卵状披针形，顶端渐尖，钝头，基部楔形，长 5.5~8 cm，宽 2.5~3.5 cm，全缘，叶面中脉下凹，侧脉不明显，背面中脉隆起。聚伞花序，腋生或生于落叶的叶腋。浆果状核果球形。花期 5~8 月，果期 12 月至翌年 2 月。

Shrubs or trees, 1.5–5 (–7) m tall. Branches terete or sometimes 4-sided, many-branched. Leaf blade elliptic, ovate, or ovate-lanceolate, 5.5–8 × 2.5–3.5 cm, leathery, both surfaces scabrous, base cuneate, margin entire, apex acuminate with an obtuse tip. Inflorescences in axils of leaves or at leaf scars on older branches. Fruit a baccate drupe, globular. Fl. May–Aug., fr. Dec.–Feb. of next year.

枝叶 Branches and leaves
摄影：杨平 Photo by: Yang Ping

花序 Inflorescences
摄影：杨平 Photo by: Yang Ping

花枝 Flowering branches
摄影：杨平 Photo by: Yang Ping

径级分布表 DBH class

胸径区间 Diameter class (cm)	个体数 No. of individuals in the plot	比例 Proportion (%)
1~2	2	100.00
2~5	0	0.00
5~10	0	0.00
10~20	0	0.00
20~35	0	0.00
35~50	0	0.00
≥50	0	0.00

● 1~5 cm DBH ＋ 5~20 cm DBH ○ ≥20 cm DBH
个体分布图 Distribution of individuals

30 山油柑 | shān yóu gān

Acronychia pedunculata (L.) Miq.

芸香科 Rutaceae

样地名称（Plot name）= SW
个体数（Individual number/1 hm²）= 5
最大胸径（Max DBH）= 11.7 cm
重要值排序（Importance value rank）= 66

灌木或小、大乔木，树高达 28 m。叶片椭圆形至长圆形，或倒卵形至倒卵状椭圆形，长 7~18 cm，宽 3.5~7 cm，或有较小的，全缘，基本楔形有时圆形到渐狭，先端渐尖；花序 2~25 cm。果近球形或椭圆形，梨形。花期 4~8 月，果期 8~12 月。

Shrubs or small to large trees, to 28 m tall. Leaflet blades usually elliptic to elliptic-oblong but grading to obovate, oblanceolate, or nearly oblong, 7–18×3.5–7cm, base cuneate or sometimes rounded or attenuate, apex obtusely acuminate. Inflorescences 2–25 cm. Fruit subglobose or sometimes grading to ellipsoid, pyriform. Fl. Apr.–Aug., fr. Aug.–Dec..

树干　Trunk
摄影：李健星　Photo by：Li Jianxing

枝叶　Branch and leaves
摄影：丁涛　Photo by：Ding Tao

果　Fruits
摄影：丁涛　Photo by：Ding Tao

径级分布表　DBH class

胸径区间 Diameter class (cm)	个体数 No. of individuals in the plot	比例 Proportion (%)
1~2	0	0.00
2~5	2	40.00
5~10	2	40.00
10~20	1	20.00
20~35	0	0.00
35~50	0	0.00
≥50	0	0.00

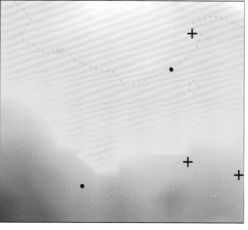

● 1~5 cm DBH　＋ 5~20 cm DBH　○ ≥20 cm DBH
个体分布图　Distribution of individuals

31 云南黄皮 | yún nán huáng pí

黄皮属

Clausena yunnanensis C. C. Huang
芸香科 **Rutaceae**

样地名称（Plot name）＝ SW
个体数（Individual number/1 hm^2）＝ 11
最大胸径（Max DBH）＝ 10.5 cm
重要值排序（Importance value rank）＝ 64

落叶小乔木，高 2~5 m。叶有小叶 5~15 片；小叶卵形至披针形，长 4~10 cm，宽 2~5 cm，稀更大，顶部急尖或渐尖，常钝头，有时微凹，基部两侧不对称，叶边缘有圆或钝裂齿，稀波浪状，两面无毛，或嫩叶的脉上有疏短毛。花序顶生。果近圆球形，直径 10~15 mm，透熟时蓝黑色。花期 6~7 月，果期 10~11 月。

Deciduous small trees, 2–5 m tall. Leaves 5–15-foliolate; petiolules 4–8 mm. Leaflet blades ovate to lanceolate, 4–10 × 2–5 cm, glabrous or villous, base asymmetric, margin serrate or rarely repand, apex acute to acuminate. Inflorescences terminal. Fruit bluish black when ripe, globose, 1–1.5 cm in diam.. Fl. Jun.–Jul., fr. Oct.–Nov..

果枝　Fruiting branches
摄影：丁涛　Photo by：Ding Tao

叶　Leaves
摄影：丁涛　Photo by：Ding Tao

果序　Infructescences
摄影：丁涛　Photo by：Ding Tao

径级分布表　DBH class

胸径区间 Diameter class (cm)	个体数 No. of individuals in the plot	比例 Proportion (%)
1~2	1	9.09
2~5	5	45.46
5~10	4	36.36
10~20	1	9.09
20~35	0	0.00
35~50	0	0.00
≥50	0	0.00

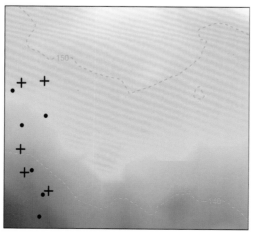

● 1~5 cm DBH　＋ 5~20 cm DBH　○ ≥20 cm DBH
个体分布图　Distribution of individuals

32 楝叶吴萸 | liàn yè wú yú （假茶辣）

Tetradium glabrifolium (Champ. ex Benth.) T. G. Hartley

芸香科 Rutaceae

样地名称（Plot name）= SW
个体数（Individual number/1 hm^2）= 4
最大胸径（Max DBH）= 23.5 cm
重要值排序（Importance value rank）= 62

灌木或乔木，高达 20 m。叶有小叶（3～）5～19 片，小叶斜卵状披针形，通常长 4～15 cm，宽 1.7～6 cm，两则明显不对称叶背灰绿色，干后略呈苍灰色，叶缘有细钝齿或全缘，无毛。花序 9～19 cm；果通常 5 心皮，分果瓣三角形。花期 6～9 月，果期 9～12 月。

Shrubs or trees, to 20 m tall. Leaves (3 or) 5–19 foliolate. Leaflet blades broadly ovate to lanceolate or less often elliptic or elliptic-oblong, 4–15 × 1.7–6 cm, abaxially usually glaucous and not papillate, base in lateral leaflets narrowly cuneate to subrounded to subtruncate, margin entire or ± crenulate, apex acuminate. Inflorescences 9–19 cm. Fruit usually 5-carpelled; follicles trigonous. Fl. Jun.–Sep., fr. Sep.–Dec..

复叶　　Compound Leaves
摄影：丁涛　Photo by: Ding Tao

小叶背部　　Leaflet backs
摄影：丁涛　Photo by: Ding Tao

果　　Fruits
摄影：丁涛　Photo by: Ding Tao

径级分布表　DBH class

胸径区间 Diameter class (cm)	个体数 No. of individuals in the plot	比例 Proportion (%)
1～2	0	0.00
2～5	0	0.00
5～10	1	25.00
10～20	2	50.00
20～35	1	25.00
35～50	0	0.00
≥50	0	0.00

● 1～5 cm DBH　+ 5～20 cm DBH　○ ≥20 cm DBH
个体分布图　Distribution of individuals

33 两面针 | liǎng miàn zhēn 花椒属

Zanthoxylum nitidum (Roxb.) DC.
芸香科 Rutaceae

样地名称（Plot name）= SW
个体数（Individual number/1 hm²）= 1
最大胸径（Max DBH）= 1.3 cm
重要值排序（Importance value rank）= 140

灌木，直立或攀援，或有时木质攀援。茎枝及叶轴均有弯钩锐刺。叶有小叶（3~）5~11片，小叶对生，阔卵形或近圆形，或狭长椭圆形，长3~12 cm，宽1.5~6 cm，顶部长或短尾状。花序腋生。花4基数；果皮红褐色，单个分果瓣径5.5~7 mm，顶端有短芒尖。花期3~5月，果期9~11月。

Shrubs, erect or scrambling, or sometimes woody climbers. Trunk winged. Stems, branchlets, and leaf rachises usually with prickles. Leaves (3 or) 5–11-foliolate. Leaflet blades opposite, broadly ovate, subcordate, elliptic, narrowly elliptic, or rarely ovate, 3–12 × 1.5–6 cm, leathery. Inflorescences axillary. Flowers 4-merous. Follicles reddish brown, 5.5–7 mm in diam., apex beaked. Fl. Mar.–May, fr. Sep.–Nov..

枝叶　Branch and leaves
摄影：丁涛　Photo by：Ding Tao

叶背　Leaflet backs
摄影：丁涛　Photo by：Ding Tao

花序　Inflorescence
摄影：丁涛　Photo by：Ding Tao

径级分布表　DBH class

胸径区间 Diameter class (cm)	个体数 No. of individuals in the plot	比例 Proportion (%)
1~2	1	100.00
2~5	0	0.00
5~10	0	0.00
10~20	0	0.00
20~35	0	0.00
35~50	0	0.00
≥50	0	0.00

● 1~5 cm DBH　+ 5~20 cm DBH　○ ≥20 cm DBH
个体分布图　Distribution of individuals

34 岭南柿 | lǐng nán shì

***Diospyros tutcheri* Dunn**

柿科 Ebenaceae

样地名称（Plot name）= SW
个体数（Individual number/1 hm²）= 3
最大胸径（Max DBH）= 2.4 cm
重要值排序（Importance value rank）= 102

乔木，高约 6 m。树皮粗糙。叶薄革质，椭圆形，长 8~12 cm，宽 2.4~4.5 cm，先端渐尖，基部钝或近圆形，边缘微背卷，上面深绿色，有光泽，下面淡绿色，叶脉在两面均明显，侧脉每边约 5~6 条，纤细。雄聚伞花序由 4 花组成。果球形，直径约 2.5 cm。花期 4~5 月，果期 8~10 月。

Trees, to ca. 6 m tall. Bark asperous. Leaf blade elliptic, elliptic-lanceolate, or oblong-lanceolate, 8–12×2.4–4.5 cm, thinly leathery, both surfaces glabrous, base obtuse to subrounded, margin slightly revolute, apex acuminate, lateral veins 5 or 6 per side, reticulate veinlets dense and conspicuously raised on both surfaces. Male flowers with calyx lobes 4. Berries globose, ca. 2.5 cm in diam.. Fl. Apr.–May, fr. Aug.–Oct..

枝叶　Branch and leaves
摄影：陆昭岑　Photo by: Lu Zhaochen

叶　Leaves
摄影：陆昭岑　Photo by: Lu Zhaochen

果　Fruit
摄影：陆昭岑　Photo by: Lu Zhaochen

径级分布表　DBH class

胸径区间 Diameter class (cm)	个体数 No. of individuals in the plot	比例 Proportion (%)
1~2	2	66.67
2~5	1	33.33
5~10	0	0.00
10~20	0	0.00
20~35	0	0.00
35~50	0	0.00
≥50	0	0.00

● 1~5 cm DBH　＋ 5~20 cm DBH　○ ≥20 cm DBH
个体分布图　Distribution of individuals

35 西藏山茉莉 | xī zàng shān mò lì （脱皮树）

山茉莉属

Huodendron tibeticum (J. Anthony) Rehder
安息香科 **Styracaceae**

样地名称（Plot name）= SW
个体数（Individual number/1 hm²）= 10
最大胸径（Max DBH）= 12.1 cm
重要值排序（Importance value rank）= 60

乔木或灌木，高 6~25 m。叶纸质，披针形或椭圆状披针形，稀卵状披针形，长 6~11 cm，宽 2.5~4 cm，顶端长渐尖或具短尖头，基部宽楔形，边全缘，侧脉每边 5~9 条，和中脉、网脉在两面均明显隆起。伞房状圆锥花序生于小枝顶端。蒴果卵形，长约 3 mm；种子棕色，长约 1 mm，具网纹。花期 3~5 月，果期 8~9 月。

Trees or shrubs, 6–25 m tall. Leaf blade lanceolate to elliptic-lanceolate, rarely ovate-lanceolate, 6–11×2.5–4 cm, papery, base broadly cuneate, margin entire or slightly serrate, apex acuminate to acute, secondary veins 5–9 pairs, tertiary veins reticulate and distinct on both surfaces when dry. Inflorescences terminal, subcorymbose. Fruit ovoid, ca. 3 mm. Seeds brown, ca. 1 mm, reticulately striate. Fl. Mar.–May, fr. Aug.–Sep..

树干　Trunk
摄影：李健星　Photo by: Li Jianxing

叶背　Leaf backs
摄影：李健星　Photo by: Li Jianxing

枝叶　Branch and leaves
摄影：李健星　Photo by: Li Jianxing

径级分布表　DBH class

胸径区间 Diameter class (cm)	个体数 No. of individuals in the plot	比例 Proportion (%)
1~2	3	30.00
2~5	0	0.00
5~10	6	60.00
10~20	1	10.00
20~35	0	0.00
35~50	0	0.00
≥50	0	0.00

● 1~5 cm DBH　+ 5~20 cm DBH　○ ≥20 cm DBH
个体分布图　Distribution of individuals

36 水东哥 | shuǐ dōng gē

***Saurauia tristyla* DC.**

猕猴桃科 **Actinidiaceae**

样地名称（Plot name）＝ SW
个体数（Individual number/1 hm²）＝ 4
最大胸径（Max DBH）＝ 10.8 cm
重要值排序（Importance value rank）＝ 95

灌木或小乔木，高 3～6 m，稀达 12 m。小枝淡红色，粗壮，被爪甲状鳞片。叶倒卵状椭圆形，稀阔椭圆形，长 10～28 cm，宽 4～11 cm，顶端短渐尖，偶有尖头，基部阔楔形，稀钝，叶缘具刺状锯齿，侧脉 8～20 对。花序被绒毛和钻状刺毛。果绿色到白色到浅黄色，球状，直径 6～10 mm。花期 3～6 月，果期 8～12 月。

Shrubs or small trees, 3–6 (–12) m tall. Branchlets tomentose to glabrescent, with unguiculate hairs or subulate scales. Leaf blade obovate to broadly elliptic-obovate, 10–28 × 4–11 cm, papery, lateral veins 8–20 pairs, base cuneate to broadly so, margin setose-serrate, apex shortly acuminate to caudate. Inflorescences, hairy and scaly. Fruit green to white to pale yellow, globose, 6–10 mm in diam.. Fl. Mar.–Jul., fr. Aug.–Dec..

花枝　　Flowering branch
摄影：丁涛　　Photo by：Ding Tao

叶　　Leaves
摄影：丁涛　　Photo by：Ding Tao

花　　Flower
摄影：丁涛　　Photo by：Ding Tao

径级分布表　DBH class

胸径区间 Diameter class (cm)	个体数 No. of individuals in the plot	比例 Proportion (%)
1～2	0	0.00
2～5	3	75.00
5～10	0	0.00
10～20	1	25.00
20～35	0	0.00
35～50	0	0.00
≥50	0	0.00

● 1～5 cm DBH　　＋ 5～20 cm DBH　　○ ≥20 cm DBH
个体分布图　Distribution of individuals

37 水团花 | shuǐ tuán huā

Adina pilulifera (Lam.) Franch. ex Drake
茜草科 Rubiaceae

样地名称（Plot name）= SW
个体数（Individual number/1 hm^2）= 12
最大胸径（Max DBH）= 6.5 cm
重要值排序（Importance value rank）= 52

常绿灌木至小乔木，高 1~5（~10）m。叶对生，厚纸质，椭圆形至椭圆状披针形，或有时倒卵状长圆形至倒卵状披针形，长 4~12 cm，宽 1.5~3 cm，顶端短尖至渐尖而钝头，基部钝或楔形，有时渐狭窄。头状花序明显腋生。果序直径 7~11 mm；小蒴果楔形，长 2~5 mm。花期 6~9 月，果期 7~12 月。

Evergreen shrubs to small trees, 1–5 (–10) m tall. Leaves decussate; blade drying papery to stiffly papery, narrowly elliptic, elliptic-lanceolate, obovate-oblong, oblanceolate, or obovate-oblanceolate, 4–12 × 1.5–3 cm, base acute to cuneate or obtuse, apex acute to acuminate with tip usually ultimately blunt. Inflorescences puberulent to glabrous; Fruiting heads 7–11 mm in diam.. Capsules obcuneate, 2–5 mm. Fl. Jun.–Sep., fr. Jul.–Dec..

果枝　Fruiting branches
摄影：丁涛　Photo by: Ding Tao

叶　Leaf
摄影：丁涛　Photo by: Ding Tao

果序　Infructescence
摄影：丁涛　Photo by: Ding Tao

径级分布表　DBH class

胸径区间 Diameter class (cm)	个体数 No. of individuals in the plot	比例 Proportion (%)
1~2	5	41.67
2~5	4	33.33
5~10	3	25.00
10~20	0	0.00
20~35	0	0.00
35~50	0	0.00
≥50	0	0.00

● 1~5 cm DBH　+ 5~20 cm DBH　○ ≥20 cm DBH
个体分布图　Distribution of individuals

38 裂果金花 | liè guǒ jīn huā

Schizomussaenda henryi (Hutch.) X. F. Deng D. X. Zhang

茜草科 Rubiaceae

样地名称（Plot name）= SW
个体数（Individual number/1 hm^2）= 1
最大胸径（Max DBH）= 9.3 cm
重要值排序（Importance value rank）= 115

灌木至小乔木，高 7～8 m。叶薄纸质，倒披针形，长圆状倒披针形或卵状披针形，顶端渐尖或短尖，基部楔形，长 10～17 cm，宽 2.5～6 cm，上面被疏散硬毛，下面细脉疏被糙伏毛；侧脉 7～10 对。穗形蝎尾状聚伞花序顶生。蒴果倒卵圆形或椭圆状倒卵形，顶部室间开裂。花期 5～10 月，果期 7～12 月。

Shrubs to small trees, 7–8 m tall. Leaf blade drying thinly papery, lanceolate, lanceolate-elliptic, or ovate-lanceolate, 10–17 × 2.5–6 cm, abaxially strigose to strigillose along principal lateral veins, sparsely strigose along higher order veins, base cuneate to rounded, apex acuminate or acute; secondary veins 7–10 pairs. Inflorescences densely hirtellous to strigillose. Capsule. Fl. May–Oct., fr. Jul.–Dec..

花枝　Flowering branches
摄影：杨平　Photo by: Yang Ping

叶　Leaves
摄影：杨平　Photo by: Yang Ping

花序　Inflorescence
摄影：杨平　Photo by: Yang Ping

径级分布表 DBH class

胸径区间 Diameter class (cm)	个体数 No. of individuals in the plot	比例 Proportion (%)
1～2	0	0.00
2～5	0	0.00
5～10	1	100.00
10～20	0	0.00
20～35	0	0.00
35～50	0	0.00
≥50	0	0.00

● 1～5 cm DBH　　＋ 5～20 cm DBH　　○ ≥20 cm DBH
个体分布图　Distribution of individuals

39 蓝树 | lán shù （羊角汁）

Wrightia laevis Hook. f.
夹竹桃科 Apocynaceae

样地名称（Plot name）= SW
个体数（Individual number/1 hm²）= 1
最大胸径（Max DBH）= 2.0 cm
重要值排序（Importance value rank）= 132

乔木，高 40 m，除花外，均无毛，具乳汁。叶膜质，长圆状披针形或狭椭圆形至椭圆形，稀卵圆形，顶端渐尖至尾状渐尖，基部楔形，长 7~18 cm，宽 2.5~8 cm，无毛；叶脉在叶面扁平，在叶背略凸起，侧脉每边 5~11 条。顶生聚伞花序。蓇葖 2 个离生，圆柱状，顶部渐尖；种子线状披针形，长 1.5~2 cm，顶端具白色绢质种毛。花期 4~8 月，果期 7~12 月。

Trees, to 40 m tall, glabrous except for flowers, with latex. Leaf blade oblong or narrowly elliptic, rarely ovate, 7–18×2.5–8 cm, apex acuminate to caudate-acuminate; lateral veins 5–11 pairs. Cymes ca. 6 cm; peduncle ca. 1 cm. Ovaries 2, distinct. Follicles cylindric, distinct, lenticellate. Seeds sublinear, 1.5–2 cm. Fl. Apr.–Aug., fr. Jul.–Dec..

树干 Trunk
摄影：李健星 Photo by: Li Jianxing

叶背 Leaf backs
摄影：李健星 Photo by: Li Jianxing

枝叶 Branch and leaves
摄影：李健星 Photo by: Li Jianxing

径级分布表 DBH class

胸径区间 Diameter class (cm)	个体数 No. of individuals in the plot	比例 Proportion (%)
1~2	0	0.00
2~5	1	100.00
5~10	0	0.00
10~20	0	0.00
20~35	0	0.00
35~50	0	0.00
≥50	0	0.00

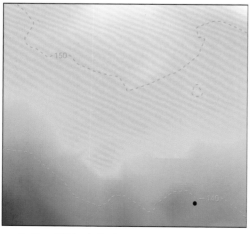

● 1~5 cm DBH　+ 5~20 cm DBH　○ ≥20 cm DBH
个体分布图 Distribution of individuals

40 厚壳树 | hòu qiào shù

Ehretia acuminata R. Br.
紫草科 Boraginaceae

样地名称（Plot name）＝ SW
个体数（Individual number/1 hm^2）＝ 1
最大胸径（Max DBH）＝ 18.5 cm
重要值排序（Importance value rank）＝ 101

乔木，高达 15 m，具条裂的黑灰色树皮。叶椭圆形、倒卵形或长圆状倒卵形，长 5~13 cm，宽 4~6 cm，先端尖，基部宽楔形，稀圆形，边缘有整齐的锯齿，齿端向上而内弯，无毛或被稀疏柔毛。聚伞花序圆锥状。核果黄色或橘黄色，直径 3~4 mm；核具皱折，成熟时分裂为 2 个具 2 粒种子的分核，花果期 4~6 月。

Trees, to 15 m tall. Bark black-gray, laciniate. Leaf blade elliptic to obovate or oblong-obovate, 5–13 × 4–6 cm, glabrous or sparsely puberulent, base broadly cuneate, margin regularly serrate with teeth curved upward, apex acute, apiculate. Cymes paniculate. Drupes yellow or orange, 3–4 mm in diam.; endocarp wrinkled, divided at maturity into 2, 2-seeded pyrenes. Fl. and fr. Apr.–Jun..

花枝　　　Flowering branches
摄影：杨平　　Photo by: Yang Ping

叶　　　Leaves
摄影：杨平　　Photo by: Yang Ping

花序　　　Inflorescence
摄影：杨平　　Photo by: Yang Ping

径级分布表 DBH class

胸径区间 Diameter class (cm)	个体数 No. of individuals in the plot	比例 Proportion (%)
1~2	0	0.00
2~5	0	0.00
5~10	0	0.00
10~20	1	100.00
20~35	0	0.00
35~50	0	0.00
≥50	0	0.00

● 1~5 cm DBH　　＋ 5~20 cm DBH　　○ ≥20 cm DBH
个体分布图　Distribution of individuals

41 海南荚蒾 | hǎi nán jiá mí （油炸木）

Viburnum hainanense Merr. Chun
五福花科 Adoxaceae

样地名称（Plot name）= SW
个体数（Individual number/1 hm²）= 2
最大胸径（Max DBH）= 1.4 cm
重要值排序（Importance value rank）= 118

常绿灌木，高达 3 m。叶亚革质，矩圆形、宽矩圆状披针形或椭圆形，长 3.5~7（~10）cm，宽 1.5~4 cm，顶端短渐尖或尖，基部宽楔形或有时圆形，全缘，两面无毛或下面中脉及侧脉被疏或密的簇状毛。复伞形式聚伞花序顶生。果实红色，扁，卵圆形。花期 4~7 月，果期 8~12 月。

Evergreen shrubs, to 3 m tall. Leaf blade yellowish green when young, oblong, broadly oblong-lanceolate, or elliptic, 3.5–7 (–10) × 1.5–4 cm, subleathery, adaxially slightly shiny, both surfaces glabrous or stellate-pubescent on midvein and lateral veins, base broadly cuneate or sometimes rounded, without glands. Inflorescence a compound umbel-like cyme. Fruit maturing red, compressed ovoid. Fl. Apr.–Jul., fr. Aug.–Dec..

果枝　　Fruiting branches
摄影：林春蕊　　Photo by: Lin Chunrui

枝叶　　Branch and leaves
摄影：林春蕊　　Photo by: Lin Chunrui

花序　　Inflorescences
摄影：林春蕊　　Photo by: Lin Chunrui

径级分布表 DBH class

胸径区间 Diameter class (cm)	个体数 No. of individuals in the plot	比例 Proportion (%)
1~2	2	100.00
2~5	0	0.00
5~10	0	0.00
10~20	0	0.00
20~35	0	0.00
35~50	0	0.00
≥50	0	0.00

● 1~5 cm DBH　　+ 5~20 cm DBH　　○ ≥20 cm DBH
个体分布图 Distribution of individuals

42 淡黄荚蒾 | dàn huáng jiá mí （黄荚蒾）

Viburnum lutescens Blume

五福花科 Adoxaceae

样地名称（Plot name）= SW
个体数（Individual number/1 hm²）= 2
最大胸径（Max DBH）= 4.2 cm
重要值排序（Importance value rank）= 106

常绿灌木，高可达 8（~11）m。叶亚革质，宽椭圆形至矩圆形或矩圆状倒卵形，长 7~15 cm，宽 3~4.5 cm，顶端常短渐尖，基部狭窄而多少下延，边缘，基部除外有粗大钝锯齿，齿端微凸，侧脉 5~6 对，弧形。复伞形聚伞花序。果实先红色后变黑色，宽椭圆形。花期 2~4 月，果期 8~12 月。

Evergreen shrubs, to 8 (–11) m tall. Leaf blade broadly elliptic to oblong or oblong-obovate, 7–15×3–4.5 cm, subleathery, midvein raised abaxially, lateral veins 5–6 jugate, pinnate, arched, base narrowed, without glands, margin serrate except at base, apex shortly acuminate. Inflorescence a compound umbel-like or panicle-like cyme. Fruit initially turning red, maturing black, broadly ellipsoid. Fl. Feb.–Apr., fr. Aug.–Dec..

果枝　Fruiting branches
摄影：杨平　Photo by：Yang Ping

叶背　Leaf backs
摄影：杨平　Photo by：Yang Ping

果序　Infructescence
摄影：杨平　Photo by：Yang Ping

径级分布表　DBH class

胸径区间 Diameter class (cm)	个体数 No. of individuals in the plot	比例 Proportion (%)
1~2	0	0.00
2~5	2	100.00
5~10	0	0.00
10~20	0	0.00
20~35	0	0.00
35~50	0	0.00
≥50	0	0.00

● 1~5 cm DBH　+ 5~20 cm DBH　○ ≥20 cm DBH
个体分布图　Distribution of individuals

43 海桐 | hǎi tóng

海桐属

***Pittosporum tobira* (Thunb.) W. T. Aiton**
海桐科 Pittosporaceae

样地名称（Plot name）= SW
个体数（Individual number/1 hm²）= 1
最大胸径（Max DBH）= 1.6 cm
重要值排序（Importance value rank）= 136

常绿灌木或小乔木，高达 6 m。叶聚生于枝顶，二年生，革质，倒卵形或倒卵状披针形，长 4~9 cm，宽 1.5~4 cm，上面深绿色，发亮、干后暗晦无光，先端圆形或钝，常微凹入或为微心形，基部窄楔形。伞形花序顶生。蒴果圆球形，有棱或呈三角形，直径 12 mm。花期 3~5 月，果期 5~10 月。

Evergreen shrubs or small trees, to 6 m tall. Leaves clustered at branchlet apex, biennial. Leaf blade dark green and shiny adaxially, dull after drying, obovate or obovate-lanceolate, 4–9 × 1.5–4 cm, leathery, base narrowly cuneate, margin entire, revolute, apex rounded, usually emarginate. Inflorescences terminal, umbellate. Capsule globose, angular, dehiscing by 3 valves, 12 mm in diam.. Fl. Mar.–May, fr. May–Oct..

花枝　Flowering branches
摄影：杨平　Photo by：Yang Ping

叶　Leaves
摄影：杨平　Photo by：Yang Ping

花序　Inflorescence
摄影：杨平　Photo by：Yang Ping

径级分布表　DBH class

胸径区间 Diameter class (cm)	个体数 No. of individuals in the plot	比例 Proportion (%)
1~2	1	100.00
2~5	0	0.00
5~10	0	0.00
10~20	0	0.00
20~35	0	0.00
35~50	0	0.00
≥50	0	0.00

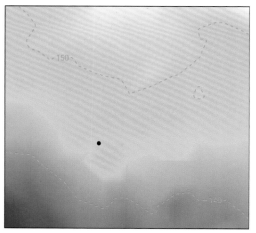
● 1~5 cm DBH　+ 5~20 cm DBH　○ ≥20 cm DBH
个体分布图　Distribution of individuals

44　马尾松　| mǎ wěi sōng　　　　松属

***Pinus massoniana* Lamb.**

松科 Pinaceae

（物种照片见第14页。See page 14 for photos of species.）

样地名称（Plot name）＝ SW
个体数（Individual number/1 hm²）＝ 1
最大胸径（Max DBH）＝ 26.6 cm
重要值排序（Importance value rank）＝ 87

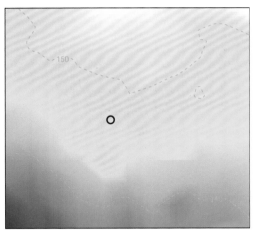

● 1～5 cm DBH　＋ 5～20 cm DBH　○ ≥20 cm DBH
个体分布图 Distribution of individuals

径级分布表　DBH class

胸径区间 Diameter class (cm)	个体数 No. of individuals in the plot	比例 Proportion (%)
1～2	0	0.00
2～5	0	0.00
5～10	0	0.00
10～20	0	0.00
20～35	1	100.00
35～50	0	0.00
≥50	0	0.00

45　假鹰爪　| jiǎ yīng zhuǎ　（鸡爪枫）　　假鹰爪属

***Desmos chinensis* Lour.**

番荔枝科 Annonaceae

（物种照片见第17页。See page 17 for photos of species.）

样地名称（Plot name）＝ SW
个体数（Individual number/1 hm²）＝ 18
最大胸径（Max DBH）＝ 2.4 cm
重要值排序（Importance value rank）＝ 37

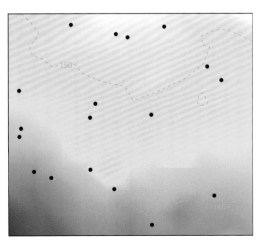

● 1～5 cm DBH　＋ 5～20 cm DBH　○ ≥20 cm DBH
个体分布图 Distribution of individuals

径级分布表　DBH class

胸径区间 Diameter class (cm)	个体数 No. of individuals in the plot	比例 Proportion (%)
1～2	16	88.89
2～5	2	11.11
5～10	0	0.00
10～20	0	0.00
20～35	0	0.00
35～50	0	0.00
≥50	0	0.00

46 紫玉盘 | zǐ yù pán （那大柒玉盘） 紫玉盘属

Uvaria macrophylla Roxb.
番荔枝科 Annonaceae

（物种照片见第18页。See page 18 for photos of species.）

样地名称（Plot name）= SW
个体数（Individual number/1 hm²）= 17
最大胸径（Max DBH）= 2.6 cm
重要值排序（Importance value rank）= 45

径级分布表 DBH class

胸径区间 Diameter class (cm)	个体数 No. of individuals in the plot	比例 Proportion (%)
1～2	16	94.12
2～5	1	5.88
5～10	0	0.00
10～20	0	0.00
20～35	0	0.00
35～50	0	0.00
≥50	0	0.00

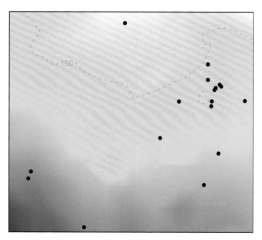

● 1～5 cm DBH ＋ 5～20 cm DBH ○ ≥20 cm DBH
个体分布图 Distribution of individuals

47 黄樟 | huáng zhāng 樟属

Cinnamomum parthenoxylon (Jack) Meisner
樟科 Lauraceae

（物种照片见第19页。See page 19 for photos of species.）

样地名称（Plot name）= SW
个体数（Individual number/1 hm²）= 44
最大胸径（Max DBH）= 30.2 cm
重要值排序（Importance value rank）= 9

径级分布表 DBH class

胸径区间 Diameter class (cm)	个体数 No. of individuals in the plot	比例 Proportion (%)
1～2	4	9.09
2～5	2	4.55
5～10	5	11.36
10～20	23	52.27
20～35	10	22.73
35～50	0	0.00
≥50	0	0.00

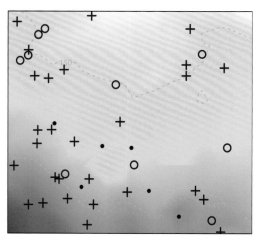

● 1～5 cm DBH ＋ 5～20 cm DBH ○ ≥20 cm DBH
个体分布图 Distribution of individuals

48 鼎湖钓樟 | dǐng hú diào zhāng （白胶木） 山胡椒属

Lindera chunii Merr.

樟科 Lauraceae

（物种照片见第20页。See page 20 for photos of species.）

样地名称（Plot name）= SW
个体数（Individual number/1 hm²）= 1
最大胸径（Max DBH）= 6.1 cm
重要值排序（Importance value rank）= 120

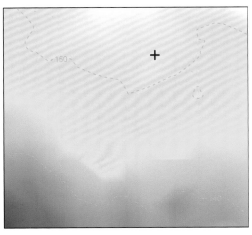

● 1～5 cm DBH　＋ 5～20 cm DBH　○ ≥20 cm DBH
个体分布图 Distribution of individuals

径级分布表 DBH class

胸径区间 Diameter class (cm)	个体数 No. of individuals in the plot	比例 Proportion (%)
1～2	0	0.00
2～5	0	0.00
5～10	1	100.00
10～20	0	0.00
20～35	0	0.00
35～50	0	0.00
≥50	0	0.00

49 尖脉木姜子 | jiān mài mù jiāng zǐ 木姜子属

Litsea acutivena Hayata

樟科 Lauraceae

（物种照片见第23页。See page 23 for photos of species.）

样地名称（Plot name）= SW
个体数（Individual number/1 hm²）= 18
最大胸径（Max DBH）= 17.2 cm
重要值排序（Importance value rank）= 34

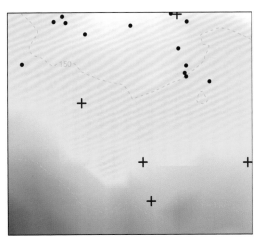

● 1～5 cm DBH　＋ 5～20 cm DBH　○ ≥20 cm DBH
个体分布图 Distribution of individuals

径级分布表 DBH class

胸径区间 Diameter class (cm)	个体数 No. of individuals in the plot	比例 Proportion (%)
1～2	3	16.67
2～5	10	55.55
5～10	3	16.67
10～20	2	11.11
20～35	0	0.00
35～50	0	0.00
≥50	0	0.00

50 豹皮樟 | chái pí zhāng （圆叶豹皮樟） 木姜子属

Litsea rotundifolia var. *oblongifolia* (Nees) C. K. Allen

樟科 Lauraceae

（物种照片见第 25 页。See page 25 for photos of species.）

样地名称（Plot name）= SW
个体数（Individual number/1 hm^2）= 2
最大胸径（Max DBH）= 20.0 cm
重要值排序（Importance value rank）= 85

径级分布表 DBH class

胸径区间 Diameter class (cm)	个体数 No. of individuals in the plot	比例 Proportion (%)
1～2	1	50.00
2～5	0	0.00
5～10	0	0.00
10～20	0	0.00
20～35	1	50.00
35～50	0	0.00
≥50	0	0.00

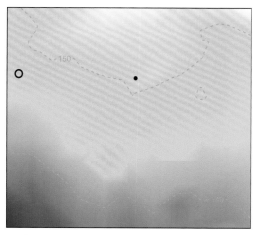

● 1～5 cm DBH　+ 5～20 cm DBH　○ ≥20 cm DBH
个体分布图 Distribution of individuals

51 黄椿木姜子 | huáng chūn mù jiāng zǐ （雄鸡树） 木姜子属

Litsea variabilis Hemsl.

樟科 Lauraceae

（物种照片见第 26 页。See page 26 for photos of species.）

样地名称（Plot name）= SW
个体数（Individual number/1 hm^2）= 198
最大胸径（Max DBH）= 26.4 cm
重要值排序（Importance value rank）= 11

径级分布表 DBH class

胸径区间 Diameter class (cm)	个体数 No. of individuals in the plot	比例 Proportion (%)
1～2	89	44.95
2～5	92	46.46
5～10	14	7.07
10～20	2	1.01
20～35	1	0.51
35～50	0	0.00
≥50	0	0.00

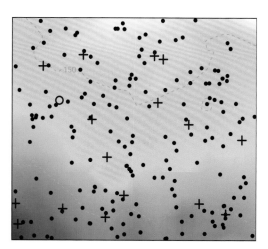

● 1～5 cm DBH　+ 5～20 cm DBH　○ ≥20 cm DBH
个体分布图 Distribution of individuals

52 华润楠 | huá rùn nán （桢楠） 润楠属

Machilus chinensis (Champ. ex Benth.) Hemsl.

樟科 Lauraceae

（物种照片见第 28 页。See page 28 for photos of species.）

样地名称（Plot name）＝ SW
个体数（Individual number/1 hm²）＝ 194
最大胸径（Max DBH）＝ 28.6 cm
重要值排序（Importance value rank）＝ 3

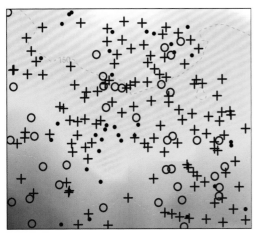

• 1～5 cm DBH + 5～20 cm DBH ○ ≥20 cm DBH
个体分布图 Distribution of individuals

径级分布表 DBH class

胸径区间 Diameter class (cm)	个体数 No. of individuals in the plot	比例 Proportion (%)
1～2	8	4.12
2～5	27	13.92
5～10	48	24.74
10～20	73	37.63
20～35	38	19.59
35～50	0	0.00
≥50	0	0.00

53 赛短花润楠 | sài duǎn huā rùn nán 润楠属

Machilus parabreviflora Hung T. Chang

樟科 Lauraceae

（物种照片见第 29 页。See page 29 for photos of species.）

样地名称（Plot name）＝ SW
个体数（Individual number/1 hm²）＝ 13
最大胸径（Max DBH）＝ 30.1 cm
重要值排序（Importance value rank）＝ 31

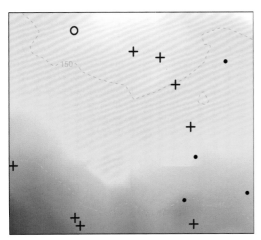

• 1～5 cm DBH + 5～20 cm DBH ○ ≥20 cm DBH
个体分布图 Distribution of individuals

径级分布表 DBH class

胸径区间 Diameter class (cm)	个体数 No. of individuals in the plot	比例 Proportion (%)
1～2	1	7.69
2～5	3	23.08
5～10	5	38.46
10～20	3	23.08
20～35	1	7.69
35～50	0	0.00
≥50	0	0.00

54 细枝龙血树 | xì zhī lóng xuè shù 龙血树属

Dracaena elliptica Thunb.
天门冬科 Asparagaceae

（物种照片见第31页。See page 31 for photos of species.）

样地名称（Plot name）= SW
个体数（Individual number/1 hm²）= 1
最大胸径（Max DBH）= 1.4 cm
重要值排序（Importance value rank）= 139

径级分布表 DBH class

胸径区间 Diameter class (cm)	个体数 No. of individuals in the plot	比例 Proportion (%)
1~2	1	100.00
2~5	0	0.00
5~10	0	0.00
10~20	0	0.00
20~35	0	0.00
35~50	0	0.00
≥50	0	0.00

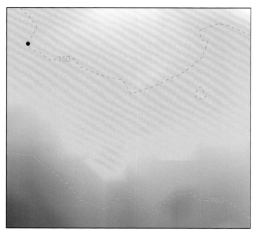

● 1~5 cm DBH ＋ 5~20 cm DBH ○ ≥20 cm DBH
个体分布图 Distribution of individuals

55 狭叶泡花树 | xiá yè pāo huā shù 泡花树属

Meliosma angustifolia Merr.
清风藤科 Sabiaceae

（物种照片见第32页。See page 32 for photos of species.）

样地名称（Plot name）= SW
个体数（Individual number/1 hm²）= 11
最大胸径（Max DBH）= 19.1 cm
重要值排序（Importance value rank）= 36

径级分布表 DBH class

胸径区间 Diameter class (cm)	个体数 No. of individuals in the plot	比例 Proportion (%)
1~2	1	9.09
2~5	2	18.18
5~10	4	36.37
10~20	4	36.36
20~35	0	0.00
35~50	0	0.00
≥50	0	0.00

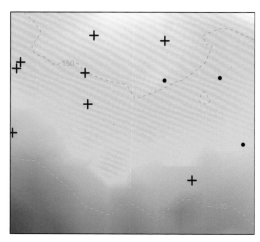

● 1~5 cm DBH ＋ 5~20 cm DBH ○ ≥20 cm DBH
个体分布图 Distribution of individuals

56 小果山龙眼 | xiǎo guǒ shān lóng yǎn （红叶树） 山龙眼属

Helicia cochinchinensis Lour.
山龙眼科 Proteaceae

（物种照片见第33页。See page 33 for photos of species.）

样地名称（Plot name）= SW
个体数（Individual number/1 hm²）= 2
最大胸径（Max DBH）= 33.2 cm
重要值排序（Importance value rank）= 68

径级分布表 DBH class

胸径区间 Diameter class (cm)	个体数 No. of individuals in the plot	比例 Proportion (%)
1~2	0	0.00
2~5	0	0.00
5~10	0	0.00
10~20	1	50.00
20~35	1	50.00
35~50	0	0.00
≥50	0	0.00

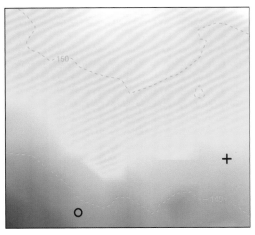

● 1~5 cm DBH + 5~20 cm DBH ○ ≥20 cm DBH
个体分布图 Distribution of individuals

57 假山龙眼 | jiǎ shān lóng yǎn 假山龙眼属

Heliciopsis henryi (Diels) W. T. Wang
山龙眼科 Proteaceae

（物种照片见第34页。See page 34 for photos of species.）

样地名称（Plot name）= SW
个体数（Individual number/1 hm²）= 8
最大胸径（Max DBH）= 20.8 cm
重要值排序（Importance value rank）= 56

径级分布表 DBH class

胸径区间 Diameter class (cm)	个体数 No. of individuals in the plot	比例 Proportion (%)
1~2	5	62.50
2~5	1	12.50
5~10	1	12.50
10~20	0	0.00
20~35	1	12.50
35~50	0	0.00
≥50	0	0.00

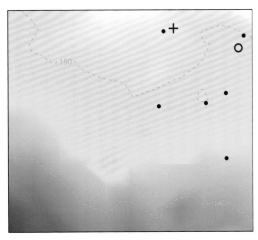

● 1~5 cm DBH + 5~20 cm DBH ○ ≥20 cm DBH
个体分布图 Distribution of individuals

58 枫香树 | fēng xiāng shù （山枫香树） 枫香树属

Liquidambar formosana Hance
蕈树科 Altingiaceae

（物种照片见第35页。See page 35 for photos of species.）

样地名称（Plot name）＝ SW
个体数（Individual number/1 hm²）＝ 64
最大胸径（Max DBH）＝ 43.8 cm
重要值排序（Importance value rank）＝ 5

径级分布表 DBH class

胸径区间 Diameter class (cm)	个体数 No. of individuals in the plot	比例 Proportion (%)
1～2	1	1.56
2～5	0	0.00
5～10	8	12.50
10～20	24	37.50
20～35	28	43.75
35～50	3	4.69
≥50	0	0.00

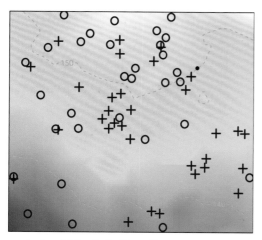

● 1～5 cm DBH　＋ 5～20 cm DBH　○ ≥20 cm DBH
个体分布图 Distribution of individuals

59 碟腺棋子豆 | dié xiàn qí zǐ dòu 猴耳环属

Archidendron kerrii (Gagnep.) I. C. Nielsen
豆科 Fabaceae

（物种照片见第37页。See page 37 for photos of species.）

样地名称（Plot name）＝ SW
个体数（Individual number/1 hm²）＝ 21
最大胸径（Max DBH）＝ 10.4 cm
重要值排序（Importance value rank）＝ 33

径级分布表 DBH class

胸径区间 Diameter class (cm)	个体数 No. of individuals in the plot	比例 Proportion (%)
1～2	5	23.81
2～5	11	52.38
5～10	4	19.05
10～20	1	4.76
20～35	0	0.00
35～50	0	0.00
≥50	0	0.00

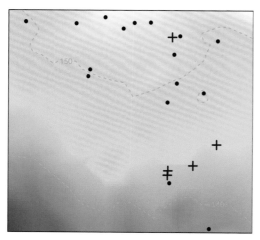

● 1～5 cm DBH　＋ 5～20 cm DBH　○ ≥20 cm DBH
个体分布图 Distribution of individuals

60 亮叶猴耳环 | liàng yè hóu ěr huán 猴耳环属

***Archidendron lucidum* (Benth.) I. C. Nielsen**

豆科 Fabaceae

（物种照片见第 38 页。See page 38 for photos of species.）

样地名称（Plot name）= SW
个体数（Individual number/1 hm^2）= 28
最大胸径（Max DBH）= 15.8 cm
重要值排序（Importance value rank）= 29

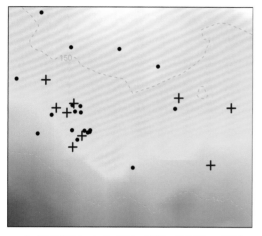

- 1～5 cm DBH   + 5～20 cm DBH   ○ ≥20 cm DBH
个体分布图 Distribution of individuals

径级分布表 DBH class

胸径区间 Diameter class (cm)	个体数 No. of individuals in the plot	比例 Proportion (%)
1～2	11	39.29
2～5	8	28.57
5～10	3	10.71
10～20	6	21.43
20～35	0	0.00
35～50	0	0.00
≥50	0	0.00

61 臀果木 | tún guǒ mù　（臀形果） 臀果木属

***Pygeum topengii* Merr.**

蔷薇科 Rosaceae

（物种照片见第 40 页。See page 40 for photos of species.）

样地名称（Plot name）= SW
个体数（Individual number/1 hm^2）= 62
最大胸径（Max DBH）= 25.4 cm
重要值排序（Importance value rank）= 22

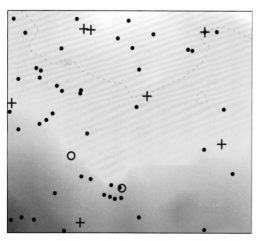

- 1～5 cm DBH   + 5～20 cm DBH   ○ ≥20 cm DBH
个体分布图 Distribution of individuals

径级分布表 DBH class

胸径区间 Diameter class (cm)	个体数 No. of individuals in the plot	比例 Proportion (%)
1～2	25	40.32
2～5	28	45.16
5～10	4	6.45
10～20	3	4.84
20～35	2	3.23
35～50	0	0.00
≥50	0	0.00

62 石斑木 | shí bān mù （春花木）

石斑木属

Rhaphiolepis indica (L.) Lindl.
蔷薇科 Rosaceae

（物种照片见第41页。See page 41 for photos of species.）

样地名称（Plot name）= SW
个体数（Individual number/1 hm²）= 3
最大胸径（Max DBH）= 18.0 cm
重要值排序（Importance value rank）= 77

径级分布表 DBH class

胸径区间 Diameter class (cm)	个体数 No. of individuals in the plot	比例 Proportion (%)
1~2	0	0.00
2~5	0	0.00
5~10	1	33.33
10~20	2	66.67
20~35	0	0.00
35~50	0	0.00
≥50	0	0.00

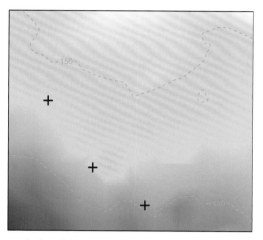

● 1~5 cm DBH ✚ 5~20 cm DBH ○ ≥20 cm DBH
个体分布图 Distribution of individuals

63 假玉桂 | jiǎ yù guì （米吃）

朴属

Celtis timorensis Span.
大麻科 Cannabaceae

（物种照片见第42页。See page 42 for photos of species.）

样地名称（Plot name）= SW
个体数（Individual number/1 hm²）= 24
最大胸径（Max DBH）= 49.2 cm
重要值排序（Importance value rank）= 24

径级分布表 DBH class

胸径区间 Diameter class (cm)	个体数 No. of individuals in the plot	比例 Proportion (%)
1~2	5	20.83
2~5	5	20.83
5~10	4	16.67
10~20	8	33.33
20~35	1	4.17
35~50	1	4.17
≥50	0	0.00

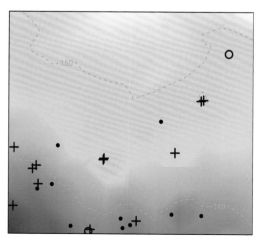

● 1~5 cm DBH ✚ 5~20 cm DBH ○ ≥20 cm DBH
个体分布图 Distribution of individuals

64 大果榕 | dà guǒ róng

榕属

Ficus auriculata Lour.
桑科 Moraceae

（物种照片见第44页。See page 44 for photos of species.）

样地名称（Plot name）= SW
个体数（Individual number/1 hm²）= 4
最大胸径（Max DBH）= 7.2 cm
重要值排序（Importance value rank）= 94

径级分布表 DBH class

胸径区间 Diameter class (cm)	个体数 No. of individuals in the plot	比例 Proportion (%)
1～2	0	0.00
2～5	1	25.00
5～10	3	75.00
10～20	0	0.00
20～35	0	0.00
35～50	0	0.00
≥50	0	0.00

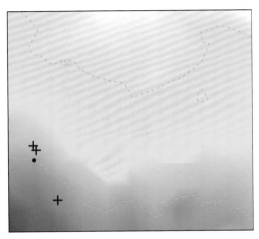

• 1～5 cm DBH   + 5～20 cm DBH   ○ ≥20 cm DBH
个体分布图 Distribution of individuals

65 粗叶榕 | cū yè róng

榕属

Ficus hirta Vahl
桑科 Moraceae

（物种照片见第46页。See page 46 for photos of species.）

样地名称（Plot name）= SW
个体数（Individual number/1 hm²）= 2
最大胸径（Max DBH）= 1.0 cm
重要值排序（Importance value rank）= 103

径级分布表 DBH class

胸径区间 Diameter class (cm)	个体数 No. of individuals in the plot	比例 Proportion (%)
1～2	2	100.00
2～5	0	0.00
5～10	0	0.00
10～20	0	0.00
20～35	0	0.00
35～50	0	0.00
≥50	0	0.00

• 1～5 cm DBH   + 5～20 cm DBH   ○ ≥20 cm DBH
个体分布图 Distribution of individuals

66 青藤公 | qīng téng gōng （山榕） 榕属

***Ficus langkokensis* Drake**

桑科 **Moraceae**

（物种照片见第47页。See page 47 for photos of species.）

样地名称（Plot name）= SW
个体数（Individual number/1 hm²）= 19
最大胸径（Max DBH）= 14.1 cm
重要值排序（Importance value rank）= 40

径级分布表 DBH class

胸径区间 Diameter class (cm)	个体数 No. of individuals in the plot	比例 Proportion (%)
1~2	4	21.05
2~5	7	36.84
5~10	5	26.32
10~20	3	15.79
20~35	0	0.00
35~50	0	0.00
≥50	0	0.00

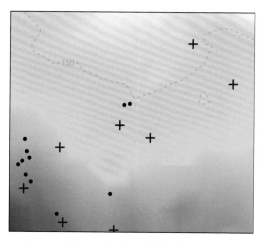

● 1~5 cm DBH ＋ 5~20 cm DBH ○ ≥20 cm DBH
个体分布图 Distribution of individuals

67 杂色榕 | zá sè róng 榕属

***Ficus variegata* Blume**

桑科 **Moraceae**

（物种照片见第48页。See page 48 for photos of species.）

样地名称（Plot name）= SW
个体数（Individual number/1 hm²）= 254
最大胸径（Max DBH）= 16.2 cm
重要值排序（Importance value rank）= 8

径级分布表 DBH class

胸径区间 Diameter class (cm)	个体数 No. of individuals in the plot	比例 Proportion (%)
1~2	60	23.62
2~5	135	53.15
5~10	51	20.08
10~20	8	3.15
20~35	0	0.00
35~50	0	0.00
≥50	0	0.00

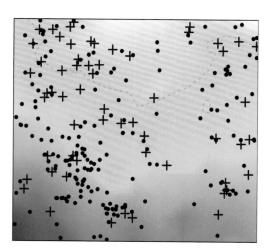

● 1~5 cm DBH ＋ 5~20 cm DBH ○ ≥20 cm DBH
个体分布图 Distribution of individuals

| 68 | 变叶榕 | biàn yè róng | （细叶牛乳树） | 榕属 |

Ficus variolosa Lindl. ex Benth.

桑科 Moraceae

（物种照片见第 49 页。See page 49 for photos of species.）

样地名称（Plot name）= SW
个体数（Individual number/1 hm^2）= 2
最大胸径（Max DBH）= 21.8 cm
重要值排序（Importance value rank）= 83

径级分布表 DBH class

胸径区间 Diameter class (cm)	个体数 No. of individuals in the plot	比例 Proportion (%)
1～2	0	0.00
2～5	1	50.00
5～10	0	0.00
10～20	0	0.00
20～35	1	50.00
35～50	0	0.00
≥50	0	0.00

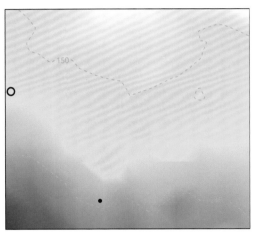

● 1～5 cm DBH　＋ 5～20 cm DBH　○ ≥20 cm DBH
个体分布图 Distribution of individuals

| 69 | 白肉榕 | bái ròu róng | （突脉榕） | 榕属 |

Ficus vasculosa Wall. ex Miq.

桑科 Moraceae

（物种照片见第 50 页。See page 50 for photos of species.）

样地名称（Plot name）= SW
个体数（Individual number/1 hm^2）= 6
最大胸径（Max DBH）= 13.2 cm
重要值排序（Importance value rank）= 63

径级分布表 DBH class

胸径区间 Diameter class (cm)	个体数 No. of individuals in the plot	比例 Proportion (%)
1～2	2	33.33
2～5	1	16.67
5～10	1	16.67
10～20	2	33.33
20～35	0	0.00
35～50	0	0.00
≥50	0	0.00

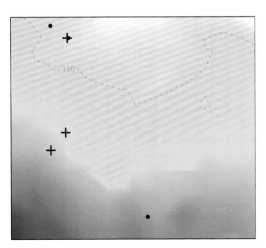

● 1～5 cm DBH　＋ 5～20 cm DBH　○ ≥20 cm DBH
个体分布图 Distribution of individuals

70 黄葛树 | huáng gé shù （绿黄葛树） 榕属

Ficus virens Aiton

桑科 Moraceae

（物种照片见第51页。See page 51 for photos of species.）

样地名称（Plot name）＝ SW
个体数（Individual number/1 hm²）＝ 19
最大胸径（Max DBH）＝ 24.0 cm
重要值排序（Importance value rank）＝ 28

径级分布表 DBH class

胸径区间 Diameter class (cm)	个体数 No. of individuals in the plot	比例 Proportion (%)
1～2	1	5.26
2～5	3	15.79
5～10	2	10.53
10～20	9	47.37
20～35	4	21.05
35～50	0	0.00
≥50	0	0.00

● 1～5 cm DBH   ＋ 5～20 cm DBH   ○ ≥20 cm DBH
个体分布图 Distribution of individuals

71 上思青冈 | shàng sī qīng gāng 栎属

Quercus delicatula Chun Tsiang

壳斗科 Fagaceae

（物种照片见第52页。See page 52 for photos of species.）

样地名称（Plot name）＝ SW
个体数（Individual number/1 hm²）＝ 1
最大胸径（Max DBH）＝ 3.0 cm
重要值排序（Importance value rank）＝ 127

径级分布表 DBH class

胸径区间 Diameter class (cm)	个体数 No. of individuals in the plot	比例 Proportion (%)
1～2	0	0.00
2～5	1	100.00
5～10	0	0.00
10～20	0	0.00
20～35	0	0.00
35～50	0	0.00
≥50	0	0.00

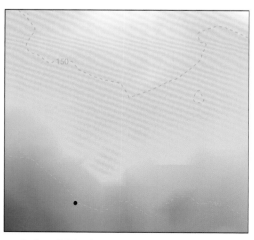

● 1～5 cm DBH   ＋ 5～20 cm DBH   ○ ≥20 cm DBH
个体分布图 Distribution of individuals

72 疏花卫矛 | shū huā wèi máo

卫矛属

Euonymus laxiflorus Champ. Benth.
卫矛科 **Celastraceae**

（物种照片见第 53 页。See page 53 for photos of species.）

样地名称（Plot name）= SW
个体数（Individual number/1 hm²）= 1
最大胸径（Max DBH）= 2.8 cm
重要值排序（Importance value rank）= 128

● 1~5 cm DBH   ＋ 5~20 cm DBH   ○ ≥20 cm DBH
个体分布图 Distribution of individuals

径级分布表 DBH class

胸径区间 Diameter class (cm)	个体数 No. of individuals in the plot	比例 Proportion (%)
1~2	0	0.00
2~5	1	100.00
5~10	0	0.00
10~20	0	0.00
20~35	0	0.00
35~50	0	0.00
≥50	0	0.00

73 中华杜英 | zhōng huá dù yīng （华杜英）

杜英属

Elaeocarpus chinensis (Gardner Champ.) Hook. f. ex Benth.
杜英科 **Elaeocarpaceae**

（物种照片见第 54 页。See page 54 for photos of species.）

样地名称（Plot name）= SW
个体数（Individual number/1 hm²）= 62
最大胸径（Max DBH）= 24.1 cm
重要值排序（Importance value rank）= 20

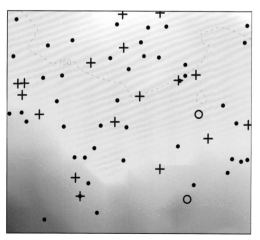

● 1~5 cm DBH   ＋ 5~20 cm DBH   ○ ≥20 cm DBH
个体分布图 Distribution of individuals

径级分布表 DBH class

胸径区间 Diameter class (cm)	个体数 No. of individuals in the plot	比例 Proportion (%)
1~2	14	22.58
2~5	29	46.77
5~10	11	17.74
10~20	6	9.68
20~35	2	3.23
35~50	0	0.00
≥50	0	0.00

74 山杜英 | shān dù yīng （羊屎树） 杜英属

Elaeocarpus sylvestris **(Lour.) Poir. in Lamarck**

杜英科 Elaeocarpaceae

（物种照片见第 55 页。See page 55 for photos of species.）

样地名称（Plot name）= SW
个体数（Individual number/1 hm^2）= 3
最大胸径（Max DBH）= 2.3 cm
重要值排序（Importance value rank）= 92

径级分布表 DBH class

胸径区间 Diameter class (cm)	个体数 No. of individuals in the plot	比例 Proportion (%)
1~2	2	66.67
2~5	1	33.33
5~10	0	0.00
10~20	0	0.00
20~35	0	0.00
35~50	0	0.00
≥50	0	0.00

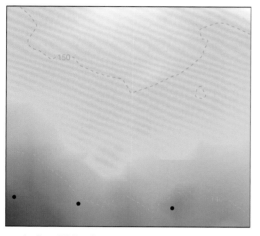

● 1~5 cm DBH ＋ 5~20 cm DBH ○ ≥20 cm DBH
个体分布图 Distribution of individuals

75 旁杞木 | páng qǐ mù （百六齿） 竹节树属

Carallia pectinifolia **W. C. Ko**

红树科 Rhizophoraceae

（物种照片见第 56 页。See page 56 for photos of species.）

样地名称（Plot name）= SW
个体数（Individual number/1 hm^2）= 149
最大胸径（Max DBH）= 6.2 cm
重要值排序（Importance value rank）= 16

径级分布表 DBH class

胸径区间 Diameter class (cm)	个体数 No. of individuals in the plot	比例 Proportion (%)
1~2	55	36.91
2~5	90	60.40
5~10	4	2.69
10~20	0	0.00
20~35	0	0.00
35~50	0	0.00
≥50	0	0.00

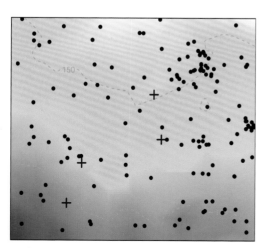

● 1~5 cm DBH ＋ 5~20 cm DBH ○ ≥20 cm DBH
个体分布图 Distribution of individuals

76 巴豆 | bā dòu （小巴豆） 巴豆属

***Croton tiglium* L.**

大戟科 **Euphorbiaceae**

（物种照片见第 57 页。See page 57 for photos of species.）

样地名称（Plot name）= SW
个体数（Individual number/1 hm²）= 5
最大胸径（Max DBH）= 27.3 cm
重要值排序（Importance value rank）= 51

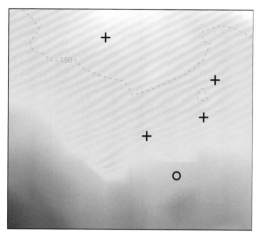

● 1~5 cm DBH ＋ 5~20 cm DBH ○ ≥20 cm DBH
个体分布图 Distribution of individuals

径级分布表 DBH class

胸径区间 Diameter class (cm)	个体数 No. of individuals in the plot	比例 Proportion (%)
1~2	0	0.00
2~5	0	0.00
5~10	1	20.00
10~20	3	60.00
20~35	1	20.00
35~50	0	0.00
≥50	0	0.00

77 印度血桐 | yìn dù xuè tóng 血桐属

***Macaranga indica* Wight**

大戟科 **Euphorbiaceae**

（物种照片见第 60 页。See page 60 for photos of species.）

样地名称（Plot name）= SW
个体数（Individual number/1 hm²）= 1
最大胸径（Max DBH）= 1.2 cm
重要值排序（Importance value rank）= 142

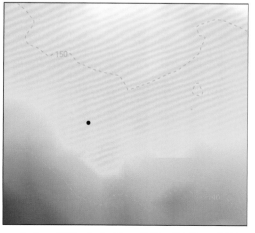

● 1~5 cm DBH ＋ 5~20 cm DBH ○ ≥20 cm DBH
个体分布图 Distribution of individuals

径级分布表 DBH class

胸径区间 Diameter class (cm)	个体数 No. of individuals in the plot	比例 Proportion (%)
1~2	1	100.00
2~5	0	0.00
5~10	0	0.00
10~20	0	0.00
20~35	0	0.00
35~50	0	0.00
≥50	0	0.00

78 木油桐 | mù yóu tóng （皱桐） 油桐属

***Vernicia montana* Lour.**
大戟科 **Euphorbiaceae**

（物种照片见第 62 页。See page 62 for photos of species.）

样地名称（Plot name）= SW
个体数（Individual number/1 hm^2）= 1
最大胸径（Max DBH）= 18.0 cm
重要值排序（Importance value rank）= 111

径级分布表 DBH class

胸径区间 Diameter class (cm)	个体数 No. of individuals in the plot	比例 Proportion (%)
1～2	0	0.00
2～5	0	0.00
5～10	0	0.00
10～20	1	100.00
20～35	0	0.00
35～50	0	0.00
≥50	0	0.00

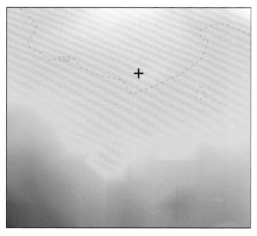

● 1～5 cm DBH　+ 5～20 cm DBH　○ ≥20 cm DBH
个体分布图 Distribution of individuals

79 黄毛五月茶 | huáng máo wǔ yuè chá （旱禾子树） 五月茶属

***Antidesma fordii* Hemsl.**
叶下珠科 **Phyllanthaceae**

（物种照片见第 63 页。See page 63 for photos of species.）

样地名称（Plot name）= SW
个体数（Individual number/1 hm^2）= 109
最大胸径（Max DBH）= 7.3 cm
重要值排序（Importance value rank）= 18

径级分布表 DBH class

胸径区间 Diameter class (cm)	个体数 No. of individuals in the plot	比例 Proportion (%)
1～2	37	33.95
2～5	69	63.30
5～10	3	2.75
10～20	0	0.00
20～35	0	0.00
35～50	0	0.00
≥50	0	0.00

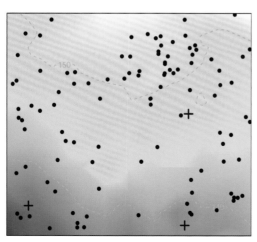

● 1～5 cm DBH　+ 5～20 cm DBH　○ ≥20 cm DBH
个体分布图 Distribution of individuals

80 银柴 | yín chái （大沙叶） 银柴属

Aporosa dioica **(Roxb.) Müll. Arg.**
叶下珠科 **Phyllanthaceae**

（物种照片见第 64 页。See page 64 for photos of species.）

样地名称（Plot name）= SW
个体数（Individual number/1 hm^2）= 414
最大胸径（Max DBH）= 26.0 cm
重要值排序（Importance value rank）= 4

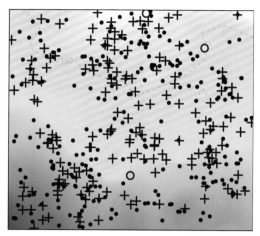

● 1~5 cm DBH   + 5~20 cm DBH   ○ ≥20 cm DBH
个体分布图 Distribution of individuals

径级分布表 DBH class

胸径区间 Diameter class (cm)	个体数 No. of individuals in the plot	比例 Proportion (%)
1~2	102	24.64
2~5	141	34.06
5~10	122	29.47
10~20	46	11.11
20~35	3	0.72
35~50	0	0.00
≥50	0	0.00

81 重阳木 | chóng yáng mù （水蚬木） 秋枫属

Bischofia polycarpa **(H. Lév.) Airy Shaw**
叶下珠科 **Phyllanthaceae**

（物种照片见第 65 页。See page 65 for photos of species.）

样地名称（Plot name）= SW
个体数（Individual number/1 hm^2）= 2
最大胸径（Max DBH）= 42.8 cm
重要值排序（Importance value rank）= 54

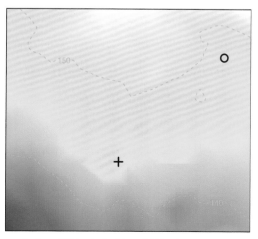

● 1~5 cm DBH   + 5~20 cm DBH   ○ ≥20 cm DBH
个体分布图 Distribution of individuals

径级分布表 DBH class

胸径区间 Diameter class (cm)	个体数 No. of individuals in the plot	比例 Proportion (%)
1~2	0	0.00
2~5	0	0.00
5~10	0	0.00
10~20	1	50.00
20~35	0	0.00
35~50	1	50.00
≥50	0	0.00

82 黑面神 | hēi miàn shén （鬼划符）

黑面神属

Breynia fruticosa (L.) Hook. f.
叶下珠科 Phyllanthaceae

（物种照片见第 66 页。See page 66 for photos of species.）

样地名称（Plot name）= SW
个体数（Individual number/1 hm^2）= 12
最大胸径（Max DBH）= 21.3 cm
重要值排序（Importance value rank）= 43

径级分布表 DBH class

胸径区间 Diameter class (cm)	个体数 No. of individuals in the plot	比例 Proportion (%)
1～2	3	25.00
2～5	6	50.00
5～10	2	16.67
10～20	0	0.00
20～35	1	8.33
35～50	0	0.00
≥50	0	0.00

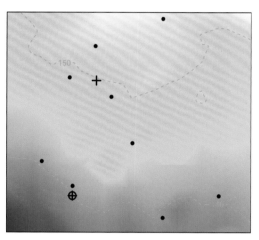

● 1～5 cm DBH ＋ 5～20 cm DBH ○ ≥20 cm DBH
个体分布图 Distribution of individuals

83 禾串树 | hé chuàn shù （禾串土蜜树）

土蜜树属

Bridelia balansae Tutcher
叶下珠科 Phyllanthaceae

（物种照片见第 67 页。See page 67 for photos of species.）

样地名称（Plot name）= SW
个体数（Individual number/1 hm^2）= 59
最大胸径（Max DBH）= 23.6 cm
重要值排序（Importance value rank）= 17

径级分布表 DBH class

胸径区间 Diameter class (cm)	个体数 No. of individuals in the plot	比例 Proportion (%)
1～2	16	27.12
2～5	19	32.20
5～10	11	18.64
10～20	9	15.26
20～35	4	6.78
35～50	0	0.00
≥50	0	0.00

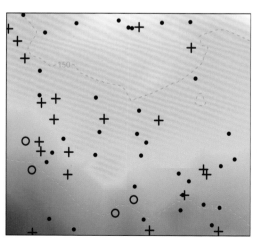

● 1～5 cm DBH ＋ 5～20 cm DBH ○ ≥20 cm DBH
个体分布图 Distribution of individuals

84 膜叶土蜜树 | mó yè tǔ mì shù

Bridelia glauca Blume

叶下珠科 Phyllanthaceae

（物种照片见第 68 页。See page 68 for photos of species.）

土蜜树属

样地名称（Plot name）= SW
个体数（Individual number/1 hm²）= 5
最大胸径（Max DBH）= 16.1 cm
重要值排序（Importance value rank）= 67

径级分布表 DBH class

胸径区间 Diameter class (cm)	个体数 No. of individuals in the plot	比例 Proportion (%)
1~2	3	60.00
2~5	1	20.00
5~10	0	0.00
10~20	1	20.00
20~35	0	0.00
35~50	0	0.00
≥50	0	0.00

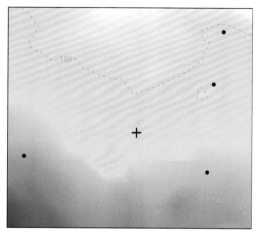

● 1~5 cm DBH ＋ 5~20 cm DBH ○ ≥20 cm DBH
个体分布图 Distribution of individuals

85 毛果算盘子 | máo guǒ suàn pán zǐ （漆大姑）

Glochidion puberum (L.) Hutch.

叶下珠科 Phyllanthaceae

（物种照片见第 69 页。See page 69 for photos of species.）

算盘子属

样地名称（Plot name）= SW
个体数（Individual number/1 hm²）= 4
最大胸径（Max DBH）= 3.7 cm
重要值排序（Importance value rank）= 88

径级分布表 DBH class

胸径区间 Diameter class (cm)	个体数 No. of individuals in the plot	比例 Proportion (%)
1~2	2	50.00
2~5	2	50.00
5~10	0	0.00
10~20	0	0.00
20~35	0	0.00
35~50	0	0.00
≥50	0	0.00

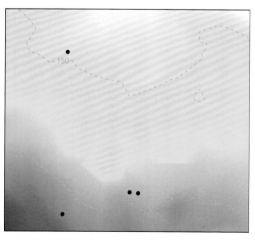

● 1~5 cm DBH ＋ 5~20 cm DBH ○ ≥20 cm DBH
个体分布图 Distribution of individuals

86 山桂花 | shān guì huā （大叶山桂花） 山桂花属

Bennettiodendron leprosipes (Clos) Merr.
杨柳科 Salicaceae

（物种照片见第 70 页。See page 70 for photos of species.）

样地名称（Plot name）= SW
个体数（Individual number/1 hm²）= 2
最大胸径（Max DBH）= 3.0 cm
重要值排序（Importance value rank）= 108

径级分布表 DBH class

胸径区间 Diameter class (cm)	个体数 No. of individuals in the plot	比例 Proportion (%)
1~2	1	50.00
2~5	1	50.00
5~10	0	0.00
10~20	0	0.00
20~35	0	0.00
35~50	0	0.00
≥50	0	0.00

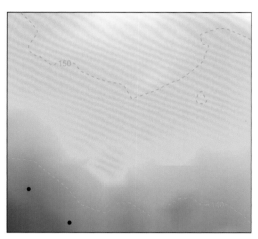

● 1~5 cm DBH　＋ 5~20 cm DBH　○ ≥20 cm DBH
个体分布图 Distribution of individuals

87 爪哇脚骨脆 | zhuǎ wājiǎo gǔ cuì （毛叶脚骨脆） 脚骨脆属

Casearia velutina Blume
杨柳科 Salicaceae

（物种照片见第 72 页。See page 72 for photos of species.）

样地名称（Plot name）= SW
个体数（Individual number/1 hm²）= 3
最大胸径（Max DBH）= 2.6 cm
重要值排序（Importance value rank）= 93

径级分布表 DBH class

胸径区间 Diameter class (cm)	个体数 No. of individuals in the plot	比例 Proportion (%)
1~2	2	66.67
2~5	1	33.33
5~10	0	0.00
10~20	0	0.00
20~35	0	0.00
35~50	0	0.00
≥50	0	0.00

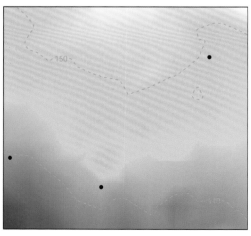

● 1~5 cm DBH　＋ 5~20 cm DBH　○ ≥20 cm DBH
个体分布图 Distribution of individuals

88 箣柊 | cè zhōng （土乌药） 箣柊属

Scolopia chinensis (Lour.) Clos
杨柳科 Salicaceae

（物种照片见第 73 页。See page 73 for photos of species.）

样地名称（Plot name）= SW
个体数（Individual number/1 hm^2）= 9
最大胸径（Max DBH）= 13.2 cm
重要值排序（Importance value rank）= 53

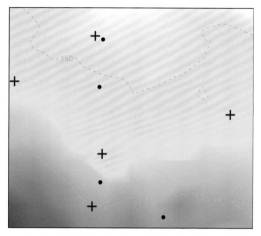

● 1~5 cm DBH + 5~20 cm DBH ○ ≥20 cm DBH
个体分布图 Distribution of individuals

径级分布表 DBH class

胸径区间 Diameter class (cm)	个体数 No. of individuals in the plot	比例 Proportion (%)
1~2	1	11.11
2~5	3	33.34
5~10	3	33.33
10~20	2	22.22
20~35	0	0.00
35~50	0	0.00
≥50	0	0.00

89 南岭柞木 | nán lǐng zhà mù （光叶柞木） 柞木属

Xylosma controversa Clos
杨柳科 Salicaceae

（物种照片见第 74 页。See page 74 for photos of species.）

样地名称（Plot name）= SW
个体数（Individual number/1 hm^2）= 29
最大胸径（Max DBH）= 17.1 cm
重要值排序（Importance value rank）= 26

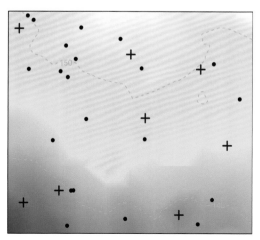

● 1~5 cm DBH + 5~20 cm DBH ○ ≥20 cm DBH
个体分布图 Distribution of individuals

径级分布表 DBH class

胸径区间 Diameter class (cm)	个体数 No. of individuals in the plot	比例 Proportion (%)
1~2	8	27.58
2~5	13	44.83
5~10	6	20.69
10~20	2	6.90
20~35	0	0.00
35~50	0	0.00
≥50	0	0.00

90 岭南山竹子 | lǐng nán shān zhú zǐ   藤黄属

Garcinia oblongifolia Champ. ex Benth.
藤黄科 Clusiaceae

（物种照片见第 75 页。See page 75 for photos of species.）

样地名称（Plot name）＝ SW
个体数（Individual number/1 hm²）＝ 2
最大胸径（Max DBH）＝ 10.5 cm
重要值排序（Importance value rank）＝ 98

径级分布表 DBH class

胸径区间 Diameter class (cm)	个体数 No. of individuals in the plot	比例 Proportion (%)
1～2	0	0.00
2～5	0	0.00
5～10	1	50.00
10～20	1	50.00
20～35	0	0.00
35～50	0	0.00
≥50	0	0.00

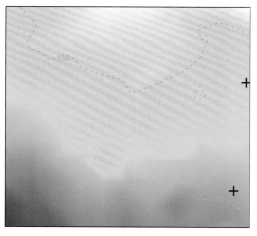

● 1～5 cm DBH　＋ 5～20 cm DBH　○ ≥20 cm DBH
个体分布图 Distribution of individuals

91 黄牛木 | huáng niú mù （黄芽木）   黄牛木属

Cratoxylum cochinchinense (Lour.) Blume
金丝桃科 Hypericaceae

（物种照片见第 76 页。See page 76 for photos of species.）

样地名称（Plot name）＝ SW
个体数（Individual number/1 hm²）＝ 60
最大胸径（Max DBH）＝ 15.2 cm
重要值排序（Importance value rank）＝ 21

径级分布表 DBH class

胸径区间 Diameter class (cm)	个体数 No. of individuals in the plot	比例 Proportion (%)
1～2	10	16.67
2～5	23	38.33
5～10	22	36.67
10～20	5	8.33
20～35	0	0.00
35～50	0	0.00
≥50	0	0.00

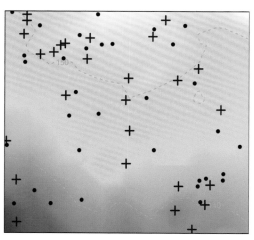

● 1～5 cm DBH　＋ 5～20 cm DBH　○ ≥20 cm DBH
个体分布图 Distribution of individuals

92 黑嘴蒲桃 | hēi zuǐ pú táo　　蒲桃属

Syzygium bullockii (Hance) Merr. L. M. Perry
桃金娘科 **Myrtaceae**

（物种照片见第 78 页。See page 78 for photos of species.）

样地名称（Plot name）= SW
个体数（Individual number/1 hm²）= 113
最大胸径（Max DBH）= 22.9 cm
重要值排序（Importance value rank）= 15

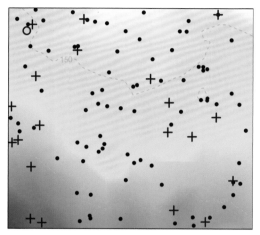

● 1～5 cm DBH　　＋ 5～20 cm DBH　　○ ≥20 cm DBH
个体分布图 Distribution of individuals

径级分布表 DBH class

胸径区间 Diameter class (cm)	个体数 No. of individuals in the plot	比例 Proportion (%)
1～2	39	34.51
2～5	52	46.02
5～10	15	13.27
10～20	6	5.31
20～35	1	0.89
35～50	0	0.00
≥50	0	0.00

93 红鳞蒲桃 | hóng lín pú táo　　蒲桃属

Syzygium hancei Merr. L. M. Perry
桃金娘科 **Myrtaceae**

（物种照片见第 79 页。See page 79 for photos of species.）

样地名称（Plot name）= SW
个体数（Individual number/1 hm²）= 123
最大胸径（Max DBH）= 22.8 cm
重要值排序（Importance value rank）= 10

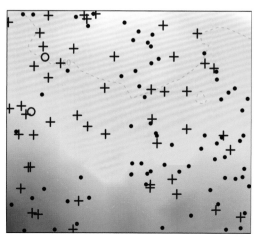

● 1～5 cm DBH　　＋ 5～20 cm DBH　　○ ≥20 cm DBH
个体分布图 Distribution of individuals

径级分布表 DBH class

胸径区间 Diameter class (cm)	个体数 No. of individuals in the plot	比例 Proportion (%)
1～2	23	18.70
2～5	49	39.84
5～10	34	27.64
10～20	15	12.19
20～35	2	1.63
35～50	0	0.00
≥50	0	0.00

94 狭叶蒲桃 | xiá yè pú táo 蒲桃属

***Syzygium tsoongii* (Merr.) Merr. L. M. Perry**
桃金娘科 **Myrtaceae**

（物种照片见第80页。See page 80 for photos of species.）

样地名称（Plot name）= SW
个体数（Individual number/1 hm²）= 10
最大胸径（Max DBH）= 18.4 cm
重要值排序（Importance value rank）= 48

径级分布表 DBH class

胸径区间 Diameter class (cm)	个体数 No. of individuals in the plot	比例 Proportion (%)
1~2	4	40.00
2~5	4	40.00
5~10	0	0.00
10~20	2	20.00
20~35	0	0.00
35~50	0	0.00
≥50	0	0.00

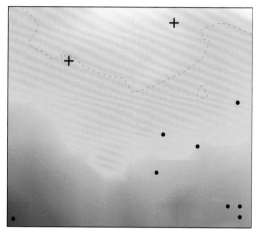

● 1~5 cm DBH + 5~20 cm DBH ○ ≥20 cm DBH
个体分布图 Distribution of individuals

95 野漆 | yě qī （漆木） 漆树属

***Toxicodendron succedaneum* (L.) Kuntze**
漆树科 **Anacardiaceae**

（物种照片见第84页。See page 84 for photos of species.）

样地名称（Plot name）= SW
个体数（Individual number/1 hm²）= 6
最大胸径（Max DBH）= 9.7 cm
重要值排序（Importance value rank）= 58

径级分布表 DBH class

胸径区间 Diameter class (cm)	个体数 No. of individuals in the plot	比例 Proportion (%)
1~2	1	16.67
2~5	0	0.00
5~10	5	83.33
10~20	0	0.00
20~35	0	0.00
35~50	0	0.00
≥50	0	0.00

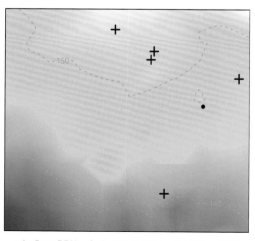

● 1~5 cm DBH + 5~20 cm DBH ○ ≥20 cm DBH
个体分布图 Distribution of individuals

96 赤才 | chì cái

鳞花木属

Lepisanthes rubiginosa **(Roxb.) Leenh.**
无患子科 **Sapindaceae**

（物种照片见第 85 页。See page 85 for photos of species.）

样地名称（Plot name）= SW
个体数（Individual number/1 hm^2）= 8
最大胸径（Max DBH）= 14.8 cm
重要值排序（Importance value rank）= 71

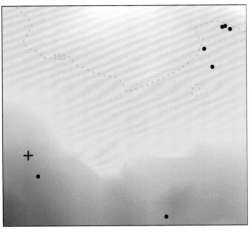

• 1~5 cm DBH   + 5~20 cm DBH   ○ ≥20 cm DBH
个体分布图 Distribution of individuals

径级分布表 DBH class

胸径区间 Diameter class (cm)	个体数 No. of individuals in the plot	比例 Proportion (%)
1~2	7	87.50
2~5	0	0.00
5~10	0	0.00
10~20	1	12.50
20~35	0	0.00
35~50	0	0.00
≥50	0	0.00

97 韶子 | sháo zǐ

韶子属

Nephelium chryseum **Blume**
无患子科 **Sapindaceae**

（物种照片见第 86 页。See page 86 for photos of species.）

样地名称（Plot name）= SW
个体数（Individual number/1 hm^2）= 11
最大胸径（Max DBH）= 6.3 cm
重要值排序（Importance value rank）= 72

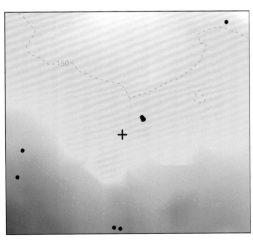

• 1~5 cm DBH   + 5~20 cm DBH   ○ ≥20 cm DBH
个体分布图 Distribution of individuals

径级分布表 DBH class

胸径区间 Diameter class (cm)	个体数 No. of individuals in the plot	比例 Proportion (%)
1~2	7	63.64
2~5	3	27.27
5~10	1	9.09
10~20	0	0.00
20~35	0	0.00
35~50	0	0.00
≥50	0	0.00

98 假黄皮 | jiǎ huáng pí — 黄皮属

Clausena excavata N. L. Burman
芸香科 Rutaceae

（物种照片见第 87 页。See page 87 for photos of species.）

样地名称（Plot name）= SW
个体数（Individual number/1 hm²）= 2
最大胸径（Max DBH）= 1.5 cm
重要值排序（Importance value rank）= 110

径级分布表 DBH class

胸径区间 Diameter class (cm)	个体数 No. of individuals in the plot	比例 Proportion (%)
1~2	2	100.00
2~5	0	0.00
5~10	0	0.00
10~20	0	0.00
20~35	0	0.00
35~50	0	0.00
≥50	0	0.00

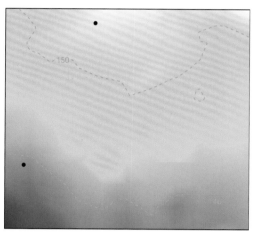

● 1~5 cm DBH   + 5~20 cm DBH   ○ ≥20 cm DBH
个体分布图 Distribution of individuals

99 少花山小橘 | shǎo huā shān xiǎo jú — 山小橘属

Glycosmis oligantha C. C. Huang
芸香科 Rutaceae

（物种照片见第 88 页。See page 88 for photos of species.）

样地名称（Plot name）= SW
个体数（Individual number/1 hm²）= 15
最大胸径（Max DBH）= 2.6 cm
重要值排序（Importance value rank）= 47

径级分布表 DBH class

胸径区间 Diameter class (cm)	个体数 No. of individuals in the plot	比例 Proportion (%)
1~2	12	80.00
2~5	3	20.00
5~10	0	0.00
10~20	0	0.00
20~35	0	0.00
35~50	0	0.00
≥50	0	0.00

● 1~5 cm DBH   + 5~20 cm DBH   ○ ≥20 cm DBH
个体分布图 Distribution of individuals

100 三桠苦 | sān yā kǔ （密茱萸） 蜜茱萸属

Melicope pteleifolia (Champ. ex Benth.) T. G. Hartley

芸香科 Rutaceae

（物种照片见第89页。See page 89 for photos of species.）

样地名称（Plot name）＝ SW
个体数（Individual number/1 hm²）＝ 15
最大胸径（Max DBH）＝ 8.1 cm
重要值排序（Importance value rank）＝ 46

径级分布表 DBH class

胸径区间 Diameter class (cm)	个体数 No. of individuals in the plot	比例 Proportion (%)
1~2	7	46.67
2~5	7	46.67
5~10	1	6.66
10~20	0	0.00
20~35	0	0.00
35~50	0	0.00
≥50	0	0.00

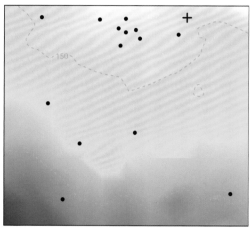

● 1~5 cm DBH　＋ 5~20 cm DBH　○ ≥20 cm DBH
个体分布图 Distribution of individuals

101 九里香 | jiǔ lǐ xiāng （千里香） 九里香属

Murraya exotica L.

芸香科 Rutaceae

（物种照片见第90页。See page 90 for photos of species.）

样地名称（Plot name）＝ SW
个体数（Individual number/1 hm²）＝ 36
最大胸径（Max DBH）＝ 11.3 cm
重要值排序（Importance value rank）＝ 27

径级分布表 DBH class

胸径区间 Diameter class (cm)	个体数 No. of individuals in the plot	比例 Proportion (%)
1~2	12	33.33
2~5	14	38.89
5~10	8	22.22
10~20	2	5.56
20~35	0	0.00
35~50	0	0.00
≥50	0	0.00

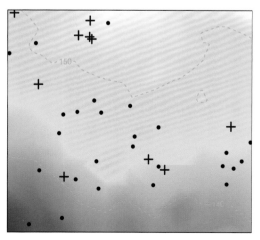

● 1~5 cm DBH　＋ 5~20 cm DBH　○ ≥20 cm DBH
个体分布图 Distribution of individuals

102 簕欓花椒 | lè dǎng huā jiāo （簕档花椒） 花椒属

Zanthoxylum avicennae (Lam.) DC.
芸香科 Rutaceae

（物种照片见第91页。See page 91 for photos of species.）

样地名称（Plot name）= SW
个体数（Individual number/1 hm^2）= 4
最大胸径（Max DBH）= 11.0 cm
重要值排序（Importance value rank）= 80

径级分布表 DBH class

胸径区间 Diameter class (cm)	个体数 No. of individuals in the plot	比例 Proportion (%)
1~2	0	0.00
2~5	0	0.00
5~10	3	75.00
10~20	1	25.00
20~35	0	0.00
35~50	0	0.00
≥50	0	0.00

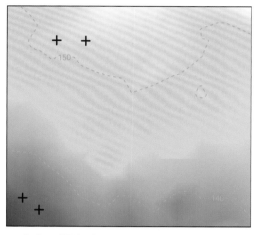

● 1~5 cm DBH ＋ 5~20 cm DBH ○ ≥20 cm DBH
个体分布图 Distribution of individuals

103 米仔兰 | mǐ zǎi lán （小叶米仔兰） 米仔兰属

Aglaia odorata Lour.
楝科 Meliaceae

（物种照片见第92页。See page 92 for photos of species.）

样地名称（Plot name）= SW
个体数（Individual number/1 hm^2）= 10
最大胸径（Max DBH）= 6.7 cm
重要值排序（Importance value rank）= 49

径级分布表 DBH class

胸径区间 Diameter class (cm)	个体数 No. of individuals in the plot	比例 Proportion (%)
1~2	3	30.00
2~5	6	60.00
5~10	1	10.00
10~20	0	0.00
20~35	0	0.00
35~50	0	0.00
≥50	0	0.00

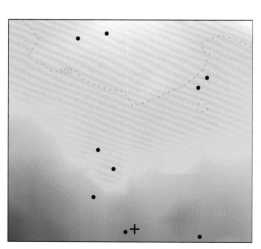

● 1~5 cm DBH ＋ 5~20 cm DBH ○ ≥20 cm DBH
个体分布图 Distribution of individuals

104 破布叶 | pò bù yè （布渣叶） 破布叶属

Microcos paniculata L.

锦葵科 Malvaceae

（物种照片见第93页。See page 93 for photos of species.）

样地名称（Plot name）＝ SW
个体数（Individual number/1 hm²）＝ 6
最大胸径（Max DBH）＝ 16.4 cm
重要值排序（Importance value rank）＝ 61

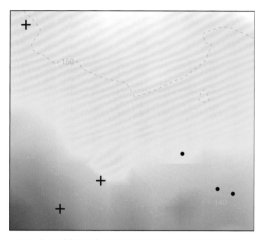

● 1～5 cm DBH   ＋ 5～20 cm DBH   ○ ≥20 cm DBH
个体分布图 Distribution of individuals

径级分布表 DBH class

胸径区间 Diameter class (cm)	个体数 No. of individuals in the plot	比例 Proportion (%)
1～2	0	0.00
2～5	3	50.00
5～10	2	33.33
10～20	1	16.67
20～35	0	0.00
35～50	0	0.00
≥50	0	0.00

105 翻白叶树 | fān bái yè shù （异叶翅子木） 翅子树属

Pterospermum heterophyllum Hance

锦葵科 Malvaceae

（物种照片见第94页。See page 94 for photos of species.）

样地名称（Plot name）＝ SW
个体数（Individual number/1 hm²）＝ 3
最大胸径（Max DBH）＝ 22.6 cm
重要值排序（Importance value rank）＝ 75

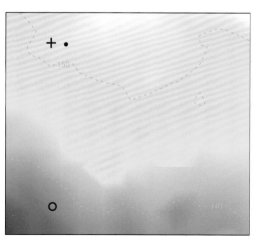

● 1～5 cm DBH   ＋ 5～20 cm DBH   ○ ≥20 cm DBH
个体分布图 Distribution of individuals

径级分布表 DBH class

胸径区间 Diameter class (cm)	个体数 No. of individuals in the plot	比例 Proportion (%)
1～2	1	33.34
2～5	0	0.00
5～10	1	33.33
10～20	0	0.00
20～35	1	33.33
35～50	0	0.00
≥50	0	0.00

106 梭罗树 | suō luó shù （两广梭罗） 梭罗树属

Reevesia pubescens Mast.
锦葵科 Malvaceae

（物种照片见第 95 页。See page 95 for photos of species.）

样地名称（Plot name）= SW
个体数（Individual number/1 hm^2）= 3
最大胸径（Max DBH）= 3.5 cm
重要值排序（Importance value rank）= 113

径级分布表 DBH class

胸径区间 Diameter class (cm)	个体数 No. of individuals in the plot	比例 Proportion (%)
1～2	1	33.33
2～5	2	66.67
5～10	0	0.00
10～20	0	0.00
20～35	0	0.00
35～50	0	0.00
≥50	0	0.00

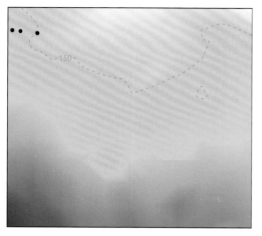

● 1～5 cm DBH ＋ 5～20 cm DBH ○ ≥20 cm DBH
个体分布图 Distribution of individuals

107 假苹婆 | jiǎ píng pó 苹婆属

Sterculia lanceolata Cav.
锦葵科 Malvaceae

（物种照片见第 96 页。See page 96 for photos of species.）

样地名称（Plot name）= SW
个体数（Individual number/1 hm^2）= 434
最大胸径（Max DBH）= 28.4 cm
重要值排序（Importance value rank）= 6

径级分布表 DBH class

胸径区间 Diameter class (cm)	个体数 No. of individuals in the plot	比例 Proportion (%)
1～2	180	41.48
2～5	154	35.48
5～10	71	16.36
10～20	27	6.22
20～35	2	0.46
35～50	0	0.00
≥50	0	0.00

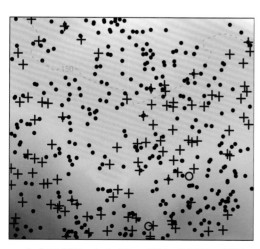

● 1～5 cm DBH ＋ 5～20 cm DBH ○ ≥20 cm DBH
个体分布图 Distribution of individuals

108 八角枫 | bā jiǎo fēng

Alangium chinense (Lour.) Harms
山茱萸科 Cornaceae

（物种照片见第97页。See page 97 for photos of species.）

样地名称（Plot name）= SW
个体数（Individual number/1 hm²）= 3
最大胸径（Max DBH）= 3.7 cm
重要值排序（Importance value rank）= 90

径级分布表 DBH class

胸径区间 Diameter class (cm)	个体数 No. of individuals in the plot	比例 Proportion (%)
1～2	2	66.67
2～5	1	33.33
5～10	0	0.00
10～20	0	0.00
20～35	0	0.00
35～50	0	0.00
≥50	0	0.00

● 1～5 cm DBH   ＋ 5～20 cm DBH   ○ ≥20 cm DBH
个体分布图 Distribution of individuals

109 黑柃 | hēi líng

Eurya macartneyi Champion
五列木科 Pentaphylacaceae

（物种照片见第98页。See page 98 for photos of species.）

样地名称（Plot name）= SW
个体数（Individual number/1 hm²）= 4
最大胸径（Max DBH）= 15.5 cm
重要值排序（Importance value rank）= 89

径级分布表 DBH class

胸径区间 Diameter class (cm)	个体数 No. of individuals in the plot	比例 Proportion (%)
1～2	1	25.00
2～5	2	50.00
5～10	0	0.00
10～20	1	25.00
20～35	0	0.00
35～50	0	0.00
≥50	0	0.00

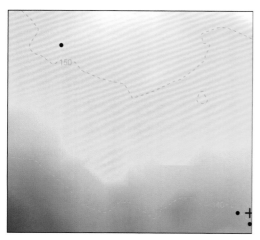

● 1～5 cm DBH   ＋ 5～20 cm DBH   ○ ≥20 cm DBH
个体分布图 Distribution of individuals

110 大果毛柃 | dà guǒ máo líng （大毛果柃） 柃属

Eurya megatrichocarpa Hung T. Chang
五列木科 **Pentaphylacaceae**

（物种照片见第99页。See page 99 for photos of species.）

样地名称（Plot name）= SW
个体数（Individual number/1 hm²）= 4
最大胸径（Max DBH）= 9.8 cm
重要值排序（Importance value rank）= 81

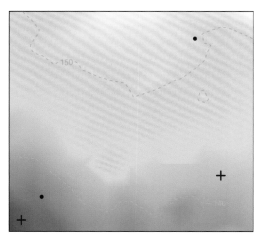

● 1~5 cm DBH   + 5~20 cm DBH   ○ ≥20 cm DBH
个体分布图 Distribution of individuals

径级分布表 DBH class

胸径区间 Diameter class (cm)	个体数 No. of individuals in the plot	比例 Proportion (%)
1~2	0	0.00
2~5	2	50.00
5~10	2	50.00
10~20	0	0.00
20~35	0	0.00
35~50	0	0.00
≥50	0	0.00

111 细齿叶柃 | xì chǐ yè líng 柃属

Eurya nitida?Korth.
五列木科 **Pentaphylacaceae**

（物种照片见第100页。See page 100 for photos of species.）

样地名称（Plot name）= SW
个体数（Individual number/1 hm²）= 2
最大胸径（Max DBH）= 2.8 cm
重要值排序（Importance value rank）= 109

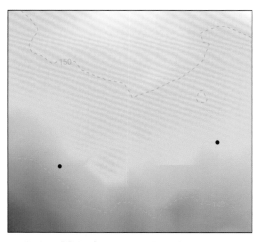

● 1~5 cm DBH   + 5~20 cm DBH   ○ ≥20 cm DBH
个体分布图 Distribution of individuals

径级分布表 DBH class

胸径区间 Diameter class (cm)	个体数 No. of individuals in the plot	比例 Proportion (%)
1~2	1	50.00
2~5	1	50.00
5~10	0	0.00
10~20	0	0.00
20~35	0	0.00
35~50	0	0.00
≥50	0	0.00

112 大叶五室柃 | dà yè wǔ shì líng 柃属

Eurya quinquelocularis Kobuski
五列木科 **Pentaphylacaceae**

（物种照片见第101页。See page 101 for photos of species.）

样地名称（Plot name）= SW
个体数（Individual number/1 hm²）= 5
最大胸径（Max DBH）= 6.5 cm
重要值排序（Importance value rank）= 82

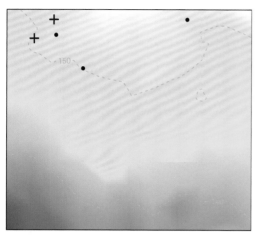

• 1～5 cm DBH   + 5～20 cm DBH   ○ ≥20 cm DBH
个体分布图 Distribution of individuals

径级分布表 DBH class

胸径区间 Diameter class (cm)	个体数 No. of individuals in the plot	比例 Proportion (%)
1～2	1	20.00
2～5	2	40.00
5～10	2	40.00
10～20	0	0.00
20～35	0	0.00
35～50	0	0.00
≥50	0	0.00

113 紫荆木 | zǐ jīng mù （铁色） 紫荆木属

Madhuca pasquieri (Dubard) H. J. Lam
山榄科 **Sapotaceae**

（物种照片见第103页。See page 103 for photos of species.）

样地名称（Plot name）= SW
个体数（Individual number/1 hm²）= 1
最大胸径（Max DBH）= 5.7 cm
重要值排序（Importance value rank）= 122

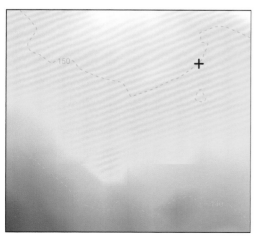

• 1～5 cm DBH   + 5～20 cm DBH   ○ ≥20 cm DBH
个体分布图 Distribution of individuals

径级分布表 DBH class

胸径区间 Diameter class (cm)	个体数 No. of individuals in the plot	比例 Proportion (%)
1～2	0	0.00
2～5	0	0.00
5～10	1	100.00
10～20	0	0.00
20～35	0	0.00
35～50	0	0.00
≥50	0	0.00

114 肉实树 | ròu shí shù （水石梓） 肉实树属

Sarcosperma laurinum (Benth.) Hook. f.
山榄科 Sapotaceae

（物种照片见第 104 页。See page 104 for photos of species.）

样地名称（Plot name）＝ SW
个体数（Individual number/1 hm^2）＝ 90
最大胸径（Max DBH）＝ 33.9 cm
重要值排序（Importance value rank）＝ 12

径级分布表 DBH class

胸径区间 Diameter class (cm)	个体数 No. of individuals in the plot	比例 Proportion (%)
1～2	19	21.11
2～5	26	28.89
5～10	21	23.33
10～20	19	21.11
20～35	5	5.56
35～50	0	0.00
≥50	0	0.00

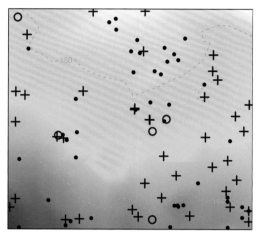

● 1～5 cm DBH ＋ 5～20 cm DBH ○ ≥20 cm DBH
个体分布图 Distribution of individuals

115 罗浮柿 | luó fú shì 柿属

Diospyros morrisiana Hance
柿科 Ebenaceae

（物种照片见第 105 页。See page 105 for photos of species.）

样地名称（Plot name）＝ SW
个体数（Individual number/1 hm^2）＝ 28
最大胸径（Max DBH）＝ 14.6 cm
重要值排序（Importance value rank）＝ 25

径级分布表 DBH class

胸径区间 Diameter class (cm)	个体数 No. of individuals in the plot	比例 Proportion (%)
1～2	6	21.43
2～5	9	32.14
5～10	4	14.29
10～20	9	32.14
20～35	0	0.00
35～50	0	0.00
≥50	0	0.00

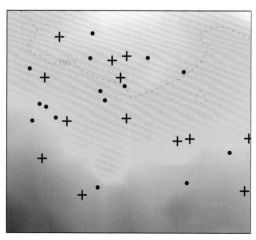

● 1～5 cm DBH ＋ 5～20 cm DBH ○ ≥20 cm DBH
个体分布图 Distribution of individuals

116 大罗伞树 | dà luó sǎn shù （郎伞树） 紫金牛属

Ardisia hanceana Mez
报春花科 Primulaceae

（物种照片见第 106 页。See page 106 for photos of species.）

样地名称（Plot name）＝ SW
个体数（Individual number/1 hm²）＝ 6
最大胸径（Max DBH）＝ 8.8 cm
重要值排序（Importance value rank）＝ 74

径级分布表 DBH class

胸径区间 Diameter class (cm)	个体数 No. of individuals in the plot	比例 Proportion (%)
1～2	2	33.34
2～5	2	33.33
5～10	2	33.33
10～20	0	0.00
20～35	0	0.00
35～50	0	0.00
≥50	0	0.00

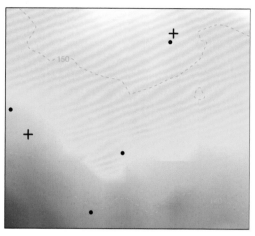

● 1～5 cm DBH　　＋ 5～20 cm DBH　　○ ≥20 cm DBH
个体分布图 Distribution of individuals

117 罗伞树 | luó sǎn shù （海南罗伞树） 紫金牛属

Ardisia quinquegona Blume
报春花科 Primulaceae

（物种照片见第 107 页。See page 107 for photos of species.）

样地名称（Plot name）＝ SW
个体数（Individual number/1 hm²）＝ 1
最大胸径（Max DBH）＝ 2.1 cm
重要值排序（Importance value rank）＝ 131

径级分布表 DBH class

胸径区间 Diameter class (cm)	个体数 No. of individuals in the plot	比例 Proportion (%)
1～2	0	0.00
2～5	1	100.00
5～10	0	0.00
10～20	0	0.00
20～35	0	0.00
35～50	0	0.00
≥50	0	0.00

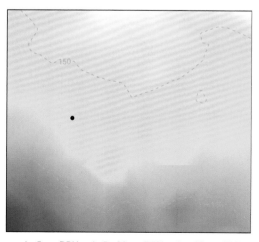

● 1～5 cm DBH　　＋ 5～20 cm DBH　　○ ≥20 cm DBH
个体分布图 Distribution of individuals

118 米珍果 | mǐ zhēn guǒ

杜茎山属

Maesa acuminatissima Merr.
报春花科 **Primulaceae**

（物种照片见第 108 页。See page 108 for photos of species.）

样地名称（Plot name）＝ SW
个体数（Individual number/1 hm²）＝ 73
最大胸径（Max DBH）＝ 18.8 cm
重要值排序（Importance value rank）＝ 23

径级分布表 DBH class

胸径区间 Diameter class (cm)	个体数 No. of individuals in the plot	比例 Proportion (%)
1～2	33	45.21
2～5	39	53.42
5～10	0	0.00
10～20	1	1.37
20～35	0	0.00
35～50	0	0.00
≥50	0	0.00

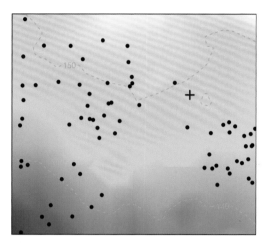

● 1～5 cm DBH   ＋ 5～20 cm DBH   ○ ≥20 cm DBH
个体分布图 Distribution of individuals

119 密花树 | mì huā shù

铁仔属

Myrsine seguinii H. Lév.
报春花科 **Primulaceae**

（物种照片见第 109 页。See page 109 for photos of species.）

样地名称（Plot name）＝ SW
个体数（Individual number/1 hm²）＝ 89
最大胸径（Max DBH）＝ 15.0 cm
重要值排序（Importance value rank）＝ 19

径级分布表 DBH class

胸径区间 Diameter class (cm)	个体数 No. of individuals in the plot	比例 Proportion (%)
1～2	32	35.96
2～5	36	40.45
5～10	17	19.10
10～20	4	4.49
20～35	0	0.00
35～50	0	0.00
≥50	0	0.00

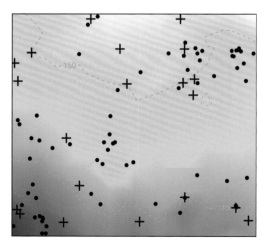

● 1～5 cm DBH   ＋ 5～20 cm DBH   ○ ≥20 cm DBH
个体分布图 Distribution of individuals

120 显脉金花茶 | xiǎn mài jīn huā chá　　　　　　山茶属

Camellia euphlebia Merr. ex Sealy

山茶科 **Theaceae**

（物种照片见第111页。See page 111 for photos of species.）

样地名称（Plot name）= SW
个体数（Individual number/1 hm²）= 7
最大胸径（Max DBH）= 1.4 cm
重要值排序（Importance value rank）= 50

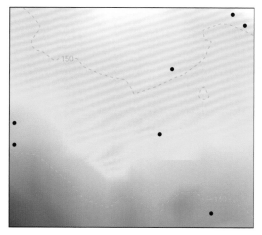

● 1~5 cm DBH　　＋ 5~20 cm DBH　　○ ≥20 cm DBH
个体分布图 Distribution of individuals

径级分布表 DBH class

胸径区间 Diameter class (cm)	个体数 No. of individuals in the plot	比例 Proportion (%)
1~2	7	100.00
2~5	0	0.00
5~10	0	0.00
10~20	0	0.00
20~35	0	0.00
35~50	0	0.00
≥50	0	0.00

121 老鼠屎 | lǎo shǔ shǐ （老鼠矢）　　　　　　山矾属

Symplocos stellaris Brand

山矾科 **Symplocaceae**

（物种照片见第117页。See page 117 for photos of species.）

样地名称（Plot name）= SW
个体数（Individual number/1 hm²）= 1
最大胸径（Max DBH）= 4.0 cm
重要值排序（Importance value rank）= 125

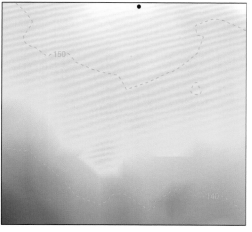

● 1~5 cm DBH　　＋ 5~20 cm DBH　　○ ≥20 cm DBH
个体分布图 Distribution of individuals

径级分布表 DBH class

胸径区间 Diameter class (cm)	个体数 No. of individuals in the plot	比例 Proportion (%)
1~2	0	0.00
2~5	1	100.00
5~10	0	0.00
10~20	0	0.00
20~35	0	0.00
35~50	0	0.00
≥50	0	0.00

122 赤杨叶 | chì yáng yè （水冬瓜） 赤杨叶属

Alniphyllum fortunei (Hemsl.) Makino
安息香科 Styracaceae

（物种照片见第118页。See page 118 for photos of species.）

样地名称（Plot name）= SW
个体数（Individual number/1 hm²）= 3
最大胸径（Max DBH）= 18.7 cm
重要值排序（Importance value rank）= 76

径级分布表 DBH class

胸径区间 Diameter class (cm)	个体数 No. of individuals in the plot	比例 Proportion (%)
1~2	0	0.00
2~5	0	0.00
5~10	0	0.00
10~20	3	100.00
20~35	0	0.00
35~50	0	0.00
≥50	0	0.00

● 1~5 cm DBH　+ 5~20 cm DBH　○ ≥20 cm DBH
个体分布图 Distribution of individuals

123 茜树 | qiàn shù （越南香楠） 茜树属

Aidia cochinchinensis Lour.
茜草科 Rubiaceae

（物种照片见第119页。See page 119 for photos of species.）

样地名称（Plot name）= SW
个体数（Individual number/1 hm²）= 26
最大胸径（Max DBH）= 15.7 cm
重要值排序（Importance value rank）= 30

径级分布表 DBH class

胸径区间 Diameter class (cm)	个体数 No. of individuals in the plot	比例 Proportion (%)
1~2	10	38.46
2~5	14	53.84
5~10	1	3.85
10~20	1	3.85
20~35	0	0.00
35~50	0	0.00
≥50	0	0.00

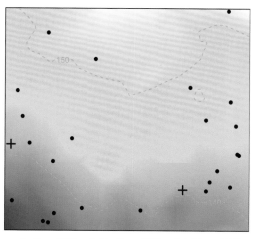

● 1~5 cm DBH　+ 5~20 cm DBH　○ ≥20 cm DBH
个体分布图 Distribution of individuals

124 猪肚木 | zhū dù mù （刺鱼骨木） 猪肚木属

Canthium horridum Blume
茜草科 Rubiaceae

（物种照片见第 120 页。See page 120 for photos of species.）

样地名称（Plot name）＝ SW
个体数（Individual number/1 hm²）＝ 21
最大胸径（Max DBH）＝ 6.0 cm
重要值排序（Importance value rank）＝ 32

径级分布表 DBH class

胸径区间 Diameter class (cm)	个体数 No. of individuals in the plot	比例 Proportion (%)
1～2	6	28.57
2～5	12	57.14
5～10	3	14.29
10～20	0	0.00
20～35	0	0.00
35～50	0	0.00
≥50	0	0.00

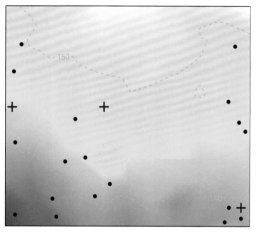

● 1～5 cm DBH　＋ 5～20 cm DBH　○ ≥20 cm DBH
个体分布图 Distribution of individuals

125 山石榴 | shān shí liú 山石榴属

Catunaregam spinosa (Thunb.) Tirveng.
茜草科 Rubiaceae

（物种照片见第 122 页。See page 122 for photos of species.）

样地名称（Plot name）＝ SW
个体数（Individual number/1 hm²）＝ 26
最大胸径（Max DBH）＝ 6.9 cm
重要值排序（Importance value rank）＝ 41

径级分布表 DBH class

胸径区间 Diameter class (cm)	个体数 No. of individuals in the plot	比例 Proportion (%)
1～2	3	11.54
2～5	17	65.38
5～10	6	23.08
10～20	0	0.00
20～35	0	0.00
35～50	0	0.00
≥50	0	0.00

● 1～5 cm DBH　＋ 5～20 cm DBH　○ ≥20 cm DBH
个体分布图 Distribution of individuals

126 斜基粗叶木 | xié jī cū yè mù   粗叶木属

Lasianthus attenuatus Jack

茜草科 Rubiaceae

（物种照片见第 124 页。See page 124 for photos of species.）

样地名称（Plot name）＝ SW
个体数（Individual number/1 hm²）＝ 1
最大胸径（Max DBH）＝ 1.1 cm
重要值排序（Importance value rank）＝ 143

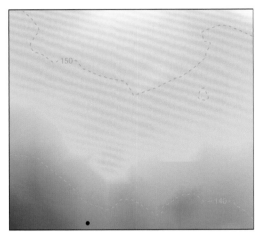

● 1～5 cm DBH   ＋ 5～20 cm DBH   ○ ≥20 cm DBH
个体分布图 Distribution of individuals

径级分布表 DBH class

胸径区间 Diameter class (cm)	个体数 No. of individuals in the plot	比例 Proportion (%)
1～2	1	100.00
2～5	0	0.00
5～10	0	0.00
10～20	0	0.00
20～35	0	0.00
35～50	0	0.00
≥50	0	0.00

127 南山花 | nán shān huā （四蕊三角瓣花）   南山花属

Prismatomeris tetrandra (Roxb.) K. Schum. in Engler Prantl

茜草科 Rubiaceae

（物种照片见第 129 页。See page 129 for photos of species.）

样地名称（Plot name）＝ SW
个体数（Individual number/1 hm²）＝ 377
最大胸径（Max DBH）＝ 13.3 cm
重要值排序（Importance value rank）＝ 7

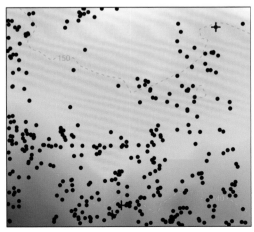

● 1～5 cm DBH   ＋ 5～20 cm DBH   ○ ≥20 cm DBH
个体分布图 Distribution of individuals

径级分布表 DBH class

胸径区间 Diameter class (cm)	个体数 No. of individuals in the plot	比例 Proportion (%)
1～2	241	63.93
2～5	132	35.01
5～10	3	0.80
10～20	1	0.26
20～35	0	0.00
35～50	0	0.00
≥50	0	0.00

128 九节 | jiǔ jié 九节属

***Psychotria asiatica* L.**

茜草科 Rubiaceae

（物种照片见第 130 页。See page 130 for photos of species.）

样地名称（Plot name）= SW
个体数（Individual number/1 hm^2）= 1494
最大胸径（Max DBH）= 18.9 cm
重要值排序（Importance value rank）= 2

径级分布表 DBH class

胸径区间 Diameter class (cm)	个体数 No. of individuals in the plot	比例 Proportion (%)
1～2	786	52.61
2～5	686	45.92
5～10	18	1.20
10～20	4	0.27
20～35	0	0.00
35～50	0	0.00
≥50	0	0.00

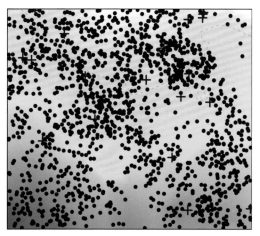

● 1～5 cm DBH ＋ 5～20 cm DBH ○ ≥20 cm DBH
个体分布图 Distribution of individuals

129 鱼骨木 | yú gǔ mù （铁屎米） 鱼骨木属

***Psydrax dicocca* Gaertn.**

茜草科 Rubiaceae

（物种照片见第 132 页。See page 132 for photos of species.）

样地名称（Plot name）= SW
个体数（Individual number/1 hm^2）= 6
最大胸径（Max DBH）= 4.1 cm
重要值排序（Importance value rank）= 70

径级分布表 DBH class

胸径区间 Diameter class (cm)	个体数 No. of individuals in the plot	比例 Proportion (%)
1～2	0	0.00
2～5	6	100.00
5～10	0	0.00
10～20	0	0.00
20～35	0	0.00
35～50	0	0.00
≥50	0	0.00

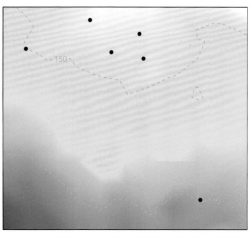

● 1～5 cm DBH ＋ 5～20 cm DBH ○ ≥20 cm DBH
个体分布图 Distribution of individuals

130 短花水金京 | duǎn huā shuǐ jīn jīng 水锦树属

Wendlandia formosana subsp. **breviflora** F. C. How

茜草科 Rubiaceae

（物种照片见第 134 页。See page 134 for photos of species.）

样地名称（Plot name）＝ SW
个体数（Individual number/1 hm²）＝ 7
最大胸径（Max DBH）＝ 33.1 cm
重要值排序（Importance value rank）＝ 44

径级分布表 DBH class

胸径区间 Diameter class (cm)	个体数 No. of individuals in the plot	比例 Proportion (%)
1~2	2	28.57
2~5	3	42.86
5~10	0	0.00
10~20	0	0.00
20~35	2	28.57
35~50	0	0.00
≥50	0	0.00

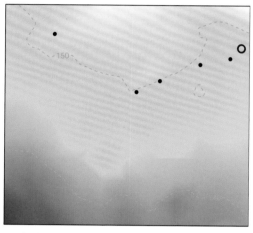

● 1~5 cm DBH   ＋ 5~20 cm DBH   ○ ≥20 cm DBH
个体分布图 Distribution of individuals

131 水锦树 | shuǐ jǐn shù （中华水锦树） 水锦树属

Wendlandia uvariifolia Hance

茜草科 Rubiaceae

（物种照片见第 135 页。See page 135 for photos of species.）

样地名称（Plot name）＝ SW
个体数（Individual number/1 hm²）＝ 91
最大胸径（Max DBH）＝ 18.2 cm
重要值排序（Importance value rank）＝ 14

径级分布表 DBH class

胸径区间 Diameter class (cm)	个体数 No. of individuals in the plot	比例 Proportion (%)
1~2	4	4.39
2~5	22	24.18
5~10	45	49.45
10~20	20	21.98
20~35	0	0.00
35~50	0	0.00
≥50	0	0.00

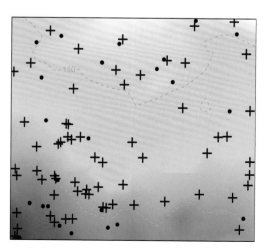

● 1~5 cm DBH   ＋ 5~20 cm DBH   ○ ≥20 cm DBH
个体分布图 Distribution of individuals

132 大青 | dà qīng （鸡屎青） 大青属

Clerodendrum cyrtophyllum Turca.
唇形科 Lamiaceae

（物种照片见第 139 页。See page 139 for photos of species.）

样地名称（Plot name）＝ SW
个体数（Individual number/1 hm^2）＝ 9
最大胸径（Max DBH）＝ 4.0 cm
重要值排序（Importance value rank）＝ 57

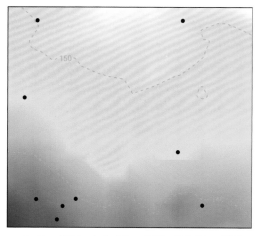

● 1～5 cm DBH ＋ 5～20 cm DBH ○ ≥20 cm DBH
个体分布图 Distribution of individuals

径级分布表 DBH class

胸径区间 Diameter class (cm)	个体数 No. of individuals in the plot	比例 Proportion (%)
1～2	5	55.56
2～5	4	44.44
5～10	0	0.00
10～20	0	0.00
20～35	0	0.00
35～50	0	0.00
≥50	0	0.00

133 垂茉莉 | chuí mò lì （长花龙吐珠） 大青属

Clerodendrum wallichii Merr.
唇形科 Lamiaceae

（物种照片见第 140 页。See page 140 for photos of species.）

样地名称（Plot name）＝ SW
个体数（Individual number/1 hm^2）＝ 1
最大胸径（Max DBH）＝ 2.4 cm
重要值排序（Importance value rank）＝ 130

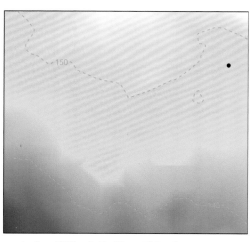

● 1～5 cm DBH ＋ 5～20 cm DBH ○ ≥20 cm DBH
个体分布图 Distribution of individuals

径级分布表 DBH class

胸径区间 Diameter class (cm)	个体数 No. of individuals in the plot	比例 Proportion (%)
1～2	0	0.00
2～5	1	100.00
5～10	0	0.00
10～20	0	0.00
20～35	0	0.00
35～50	0	0.00
≥50	0	0.00

134 山牡荆 | shān mǔ jīng 牡荆属

Vitex quinata (Lour.) F. W. Williams
唇形科 Lamiaceae

（物种照片见第 141 页。See page 141 for photos of species.）

样地名称（Plot name）= SW
个体数（Individual number/1 hm²）= 5
最大胸径（Max DBH）= 3.6 cm
重要值排序（Importance value rank）= 96

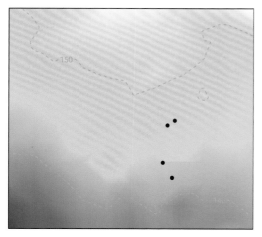

● 1～5 cm DBH   ＋ 5～20 cm DBH   ○ ≥20 cm DBH
个体分布图 Distribution of individuals

径级分布表 DBH class

胸径区间 Diameter class (cm)	个体数 No. of individuals in the plot	比例 Proportion (%)
1～2	3	60.00
2～5	2	40.00
5～10	0	0.00
10～20	0	0.00
20～35	0	0.00
35～50	0	0.00
≥50	0	0.00

135 微毛布惊 | wēi máo bù jīng （微毛布荆） 牡荆属

Vitex quinata var. puberula (H. J. Lam) Moldenke
唇形科 Lamiaceae

（物种照片见第 142 页。See page 142 for photos of species.）

样地名称（Plot name）= SW
个体数（Individual number/1 hm²）= 7
最大胸径（Max DBH）= 15.2 cm
重要值排序（Importance value rank）= 78

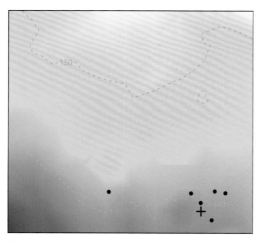

● 1～5 cm DBH   ＋ 5～20 cm DBH   ○ ≥20 cm DBH
个体分布图 Distribution of individuals

径级分布表 DBH class

胸径区间 Diameter class (cm)	个体数 No. of individuals in the plot	比例 Proportion (%)
1～2	1	14.29
2～5	5	71.43
5～10	0	0.00
10～20	1	14.28
20～35	0	0.00
35～50	0	0.00
≥50	0	0.00

136 粗丝木 | cū sī mù （海南粗丝木）

粗丝木属

Gomphandra tetrandra (Wall.) Sleumer

粗丝木科 Stemonuraceae

（物种照片见第 143 页。See page 143 for photos of species.）

样地名称（Plot name）= SW
个体数（Individual number/1 hm²）= 20
最大胸径（Max DBH）= 3.6 cm
重要值排序（Importance value rank）= 38

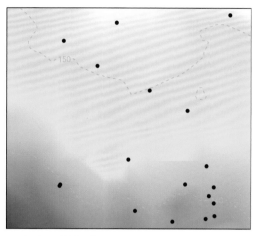

● 1～5 cm DBH   ✚ 5～20 cm DBH   ○ ≥20 cm DBH
个体分布图 Distribution of individuals

径级分布表 DBH class

胸径区间 Diameter class (cm)	个体数 No. of individuals in the plot	比例 Proportion (%)
1～2	10	50.00
2～5	10	50.00
5～10	0	0.00
10～20	0	0.00
20～35	0	0.00
35～50	0	0.00
≥50	0	0.00

137 棱枝冬青 | léng zhī dōng qīng

冬青属

Ilex angulata Merr. Chun

冬青科 Aquifoliaceae

（物种照片见第 144 页。See page 144 for photos of species.）

样地名称（Plot name）= SW
个体数（Individual number/1 hm²）= 155
最大胸径（Max DBH）= 31.2 cm
重要值排序（Importance value rank）= 13

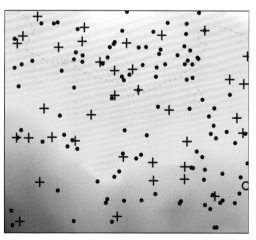

● 1～5 cm DBH   ✚ 5～20 cm DBH   ○ ≥20 cm DBH
个体分布图 Distribution of individuals

径级分布表 DBH class

胸径区间 Diameter class (cm)	个体数 No. of individuals in the plot	比例 Proportion (%)
1～2	41	26.45
2～5	76	49.03
5～10	33	21.29
10～20	4	2.58
20～35	1	0.65
35～50	0	0.00
≥50	0	0.00

138 铁冬青 | tiě dōng qīng （小果铁冬青） 冬青属

Ilex rotunda Thunb.
冬青科 Aquifoliaceae

（物种照片见第 146 页。See page 146 for photos of species.）

样地名称（Plot name）= SW
个体数（Individual number/1 hm²）= 7
最大胸径（Max DBH）= 15.9 cm
重要值排序（Importance value rank）= 55

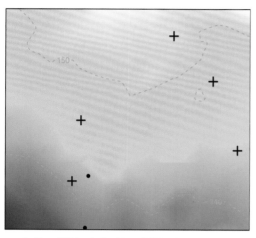

● 1～5 cm DBH　＋ 5～20 cm DBH　○ ≥20 cm DBH
个体分布图 Distribution of individuals

径级分布表 DBH class

胸径区间 Diameter class (cm)	个体数 No. of individuals in the plot	比例 Proportion (%)
1～2	0	0.00
2～5	2	28.57
5～10	3	42.86
10～20	2	28.57
20～35	0	0.00
35～50	0	0.00
≥50	0	0.00

139 常绿荚蒾 | cháng lǜ jiá mí 荚蒾属

Viburnum sempervirens K. Koch
五福花科 Adoxaceae

（物种照片见第 149 页。See page 149 for photos of species.）

样地名称（Plot name）= SW
个体数（Individual number/1 hm²）= 3
最大胸径（Max DBH）= 2.3 cm
重要值排序（Importance value rank）= 91

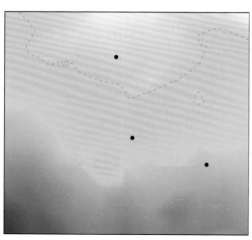

● 1～5 cm DBH　＋ 5～20 cm DBH　○ ≥20 cm DBH
个体分布图 Distribution of individuals

径级分布表 DBH class

胸径区间 Diameter class (cm)	个体数 No. of individuals in the plot	比例 Proportion (%)
1～2	2	66.67
2～5	1	33.33
5～10	0	0.00
10～20	0	0.00
20～35	0	0.00
35～50	0	0.00
≥50	0	0.00

140 广西海桐 | guǎng xī hǎi tóng 海桐属

Pittosporum kwangsiense H. T. Chang S. Z. Yan

海桐科 Pittosporaceae

（物种照片见第 150 页。See page 150 for photos of species.）

样地名称（Plot name）= SW
个体数（Individual number/1 hm^2）= 21
最大胸径（Max DBH）= 14.3 cm
重要值排序（Importance value rank）= 35

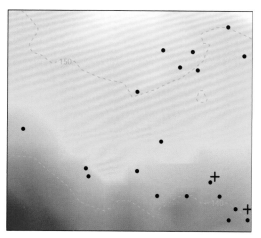

● 1～5 cm DBH　＋ 5～20 cm DBH　○ ≥20 cm DBH
个体分布图 Distribution of individuals

径级分布表 DBH class

胸径区间 Diameter class (cm)	个体数 No. of individuals in the plot	比例 Proportion (%)
1～2	12	57.14
2～5	7	33.34
5～10	1	4.76
10～20	1	4.76
20～35	0	0.00
35～50	0	0.00
≥50	0	0.00

141 罗伞 | luó sǎn （短梗罗伞） 罗伞属

Brassaiopsis glomerulata (Blume) Regel

五加科 Araliaceae

（物种照片见第 151 页。See page 151 for photos of species.）

样地名称（Plot name）= SW
个体数（Individual number/1 hm^2）= 10
最大胸径（Max DBH）= 2.8 cm
重要值排序（Importance value rank）= 59

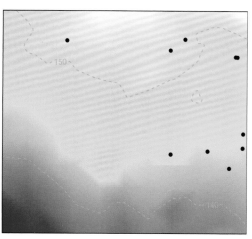

● 1～5 cm DBH　＋ 5～20 cm DBH　○ ≥20 cm DBH
个体分布图 Distribution of individuals

径级分布表 DBH class

胸径区间 Diameter class (cm)	个体数 No. of individuals in the plot	比例 Proportion (%)
1～2	4	40.00
2～5	6	60.00
5～10	0	0.00
10～20	0	0.00
20～35	0	0.00
35～50	0	0.00
≥50	0	0.00

142 鹅掌柴 | é zhǎng chái

鹅掌柴属

Schefflera heptaphylla (L.) Frodin
五加科 Araliaceae

（物种照片见第 152 页。See page 152 for photos of species.）

样地名称（Plot name）＝ SW
个体数（Individual number/1 hm²）＝ 543
最大胸径（Max DBH）＝ 52.8 cm
重要值排序（Importance value rank）＝ 1

径级分布表 DBH class

胸径区间 Diameter class (cm)	个体数 No. of individuals in the plot	比例 Proportion (%)
1~2	116	21.36
2~5	170	31.31
5~10	81	14.92
10~20	130	23.94
20~35	45	8.29
35~50	0	0.00
≥50	1	0.18

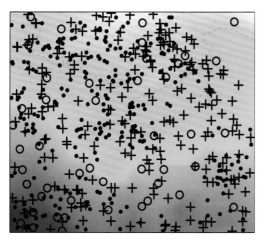

• 1~5 cm DBH   + 5~20 cm DBH   ○ ≥20 cm DBH
个体分布图 Distribution of individuals

广西防城季节性雨林物种及其分布格局
GUANGXI FANGCHENG SEASONAL RAIN FORESTS: SPECIES AND THEIR DISTRIBUTION PATTERNS

5 防城样地的草本层植物和藤本植物
Plants in the Herb Layer and Lianas in the two Guangxi Fangcheng Plots

1 铁芒萁 | tiě máng qí

芒萁属

Dicranopteris linearis **(Burm. f.) Underw.**　　里白科 **Gleicheniaceae**

植株高达 0.5~2（~3）m。根状茎横走，粗 2~3 mm，深棕色，被锈毛。叶远生；柄长约 60 cm，深棕色，无毛；叶轴 1~3（~5）回两叉分枝，一回叶轴长 10~15 cm，被棕色毛，二回以上的羽轴较短，末回叶轴长 3~5 cm，上面具 1 纵沟；叶坚纸质，上面绿色，下面灰白色，无毛。孢子囊群圆形，细小，一列，由 5~8 个孢子囊组成。

Plants 0.5–2 (–3) m tall. Rhizomes creeping, 2–3 mm in diam, covered with dense dark brown or brown hairs. Stipe stramineous, 60cm, glabrous; rachis 1–3 (–5) times dichotomously branched, basal internode 10–15 cm, covered with dark brown hairs, glabrescent, second internode 3–5 cm; lamina papery, glaucous abaxially, yellowish green or green adaxially. Sori in 1 line on each side of costule; sporangia 5–8.

蕨叶　　Fronds　　摄影：丁涛　　Photo by: Ding Tao

羽片　　Pinna　　摄影：丁涛　　Photo by: Ding Tao

2 曲轴海金沙 | qū zhóu hǎi jīn shā

海金沙属

Lygodium flexuosum **(L.) Sw.**　　海金沙科 **Lygodiaceae**

根状茎匍匐，密被根，紧密地贴在一起。叶缘有细锯齿，先端近尖。叶草质，干后暗绿褐色，小羽轴两侧有狭翅和棕色短毛，叶面沿中脉及小脉略被刚毛。孢子囊穗长 3~5 mm，线形，棕褐色，无毛，小羽片顶部通常不育。

Rhizome shortly creeping and densely covered with roots, stipes very close togethe. Margin serrate, apex subacute; costae usually with scattered long hairs, less often with dense short hairs, veins often with scattered short hairs on abaxial surface, lamina herbaceous to papery, dark green brown when dry. Sorophores 3–5 mm, at apices of small triangular lobes; indusia glabrous or with a few hairs like those of abaxial surface of lamina; spores finely evenly verrucose.

蕨叶　　Fronds　　摄影：丁涛　　Photo by: Ding Tao

羽片　　Pinnae　　摄影：丁涛　　Photo by: Ding Tao

3 团叶鳞始蕨 | tuán yè lín shǐ jué （团叶陵齿蕨）　　鳞始蕨属

Lindsaea orbiculata (Lam.) Mett. ex Kuhn　　鳞始蕨科 Lindsaeaceae

根状茎短而横走，先端密被红棕色的狭小鳞片。叶近生；叶柄长 4~35 cm；叶片线状披针形，长 9~25 cm，宽 1.5~15 cm，一回羽状，下部往往二回羽状；羽片 10~22 对；在二回羽状植株上，有 1~5 回侧羽片。叶草质。孢子囊群连续不断成长线形，或偶为缺刻所中断；囊群盖线形，有细齿牙，几达叶缘。

Rhizomes shortly creeping, sparsely scaly; scales appressed or spreading. Fronds approximate; stipe castaneous, 4–35 cm, lamina 9–25 × 1.5–15 cm, herbaceous to papery, 1- or 2-pinnate; if 1-pinnate then lamina linear, pinnae 10–22 pairs; if 2-pinnate then lamina with 1–5 pairs lateral pinnae. Sori marginal or submarginal, terminal on all veins; indusia linear, continuous, or rarely interrupted by incisions.

植株　　Whole plant
摄影：丁涛　　Photo by：Ding Tao

羽片　　Pinnae
摄影：丁涛　　Photo by：Ding Tao

4 边缘鳞盖蕨 | biān yuán lín gài jué　　鳞盖蕨属

Microlepia marginata (Panzer) C. Chr.　　碗蕨科 Dennstaedtiaceae

陆生植株高约 60~120 cm。根状茎长而横走，密被锈色长柔毛。叶柄长 50~70 cm，深禾杆色；叶片长圆三角形，先端渐尖，羽状深裂，基部不变狭，长与叶柄略等，一回羽状。叶纸质，叶轴密被锈色开展的硬毛。孢子囊群圆形，向边缘着生；囊群盖杯形，长宽几相等，上边截形，多少被短硬毛。

Plants terrestrial, 0.6–1.2 m tall. Rhizome with dense, red-brown, subulate hairs. Stipe straw-colored, 50–70 cm, thick and strong; lamina brown-green when dried, 1- or 2-pinnate, oblong in outline, papery, glabrous or hairy, costa densely pubescent, apex long caudate. Sori orbicular, near margin; indusium hemitelioid, wider, truncate above, glabrous or hairy.

孢子囊群　　Sori
摄影：丁涛　　Photo by：Ding Tao

羽片　　Pinnae
摄影：丁涛　　Photo by：Ding Tao

5 扇叶铁线蕨 | shàn yè tiě xiàn jué　　铁线蕨属

***Adiantum flabellulatum* L.**　　凤尾蕨科 **Pteridaceae**

植株高 20~45 cm。根状茎短而直立，密被淡黄色至棕色、有光泽的钻状披针形鳞片。叶簇生；柄长 10~30 cm，粗 2.5 mm，紫黑色；叶片扇形，长 10~25 cm，二至三回不对称的二叉分枝；小羽片 8~15 对，互生，平展。孢子囊群每羽片 2~5 枚；囊群盖半圆形或长圆形，上缘平直，革质，褐黑色，全缘，宿存。

Plants terrestrial, 20–45 cm tall. Rhizomes erect, short, scales dense, yellowish to brown, glossy, linear-lanceolate. Fronds clustered; stipe black-purple, glossy, 10–30 cm, 2.5 mm in diam.; lamina pedately 2-3-dichotomously branched, flabellate in outline, 10–25 cm, pinnules 8–15 pairs per pinna, alternate, horizontally spreading. Sori 2–5 per pinnule; false indusia dark brown, semi-orbicular or oblong, upper margins flat and straight, entire, persistent.

蕨叶　　Fronds
摄影：丁涛　　Photo by: Ding Tao

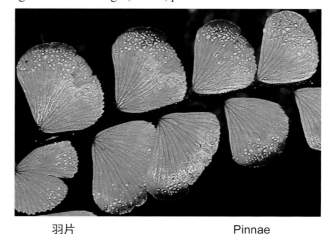

羽片　　Pinnae
摄影：丁涛　　Photo by: Ding Tao

6 剑叶凤尾蕨 | jiàn yè fèng wěi jué　　凤尾蕨属

***Pteris ensiformis* Burm. f.**　　凤尾蕨科 **Pteridaceae**

植株高 30~50 cm。根状茎细长，斜升或横卧，粗 4~5 mm，被黑褐色鳞片。叶密生，二型；柄长 10~30 cm（不育叶的柄较短），与叶轴同为禾秆色，稍光泽，光滑；叶片长圆状卵形，长 10~25 cm，宽 5~15 cm 羽状，羽片 2~6 对，对生，稍斜向上，上部的无柄，下部的有短柄。叶干后草质，灰绿色至褐绿色。

Plants 30–50 cm tall. Rhizome ascending or prostrate, slender, 4–5 mm in diam, apex with black-brown scales. Fronds dense, dimorphic; stipe and rachis straw-colored, slightly lustrous, stipe 10–30 cm (stipes of sterile fronds shorter), glabrescent; lamina oblong-ovate, 10–25 × 5–15 cm, pinnate to bipinnate; pinnae 2–6 pairs, opposite, slightly decumbent, upper ones sessile, lower pairs shortly stalked; herbaceous when dried.

植株　　Whole plant
摄影：丁涛　　Photo by: Ding Tao

羽片　　Pinnae
摄影：丁涛　　Photo by: Ding Tao

7 百越凤尾蕨 | bǎi yuè fèng wěi jué

凤尾蕨属

Pteris fauriei var. *chinensis* Ching S. H. Wu

凤尾蕨科 **Pteridaceae**

植株高50~90 cm。根状茎短，斜升，先端密被鳞片。叶簇生；柄长30~50 cm，下部粗2~4 mm，暗褐色并被鳞片，向上与叶轴均为禾秆色；叶片卵形至卵状三角形，长25~45 cm，宽17~24（~30）cm，二回深羽裂（或基部三回深羽裂）。孢子囊群线形，沿裂片边缘延伸；囊群盖线形，灰棕色，膜质，宿存。

Plants 50–90 cm tall. Rhizome ascending, short, apex densely scaly. Fronds clustered; stipe straw-colored, 30–50 cm, 2–4 mm in diam., with dark brownish scales; rachis similar; lamina 2 (or 3)-pinnatipartite, ovate to ovate-triangular in outline, 25–45 × 17–24 (–30) cm. Sori linear, along segment margins, absent at apex and sinuses; indusia gray-brown or brown, linear, membranous, persistent.

蕨叶　　　Frond
摄影：丁涛　　Photo by: Ding Tao

羽片　　　Pinnae
摄影：丁涛　　Photo by: Ding Tao

8 林下凤尾蕨 | lín xià fèng wěi jué

凤尾蕨属

Pteris grevilleana Wall. ex J. Agardh

凤尾蕨科 **Pteridaceae**

植株高20~45 cm。根状茎短而直立，先端被黑褐色鳞片。叶簇生（10~15片），同型；能育叶的柄比不育叶的柄长2倍以上，长20~30 cm，粗1~1.5 m；叶片阔卵状三角形，长10~15 cm，宽8~12 cm，二回深羽裂。叶干后坚草质，暗绿色，两面均有密接的细斜条纹。

Plants 20–45 cm tall. Rhizome erect, short, apex with black-brown scales. Fronds clustered (10–15 per plant), monomorphic; stipes of fertile fronds ca. 2 × as long as those of sterile fronds, castaneous-brown, shiny, 20–30 cm × 1–1.5 mm; lamina 2-pinnatipartite, broadly ovate-triangular, 10–15 × 8–12 cm. Lamina pale green, firmly herbaceous when dried, both surfaces with short raised false veins between veins.

植株　　　Whole plant
摄影：丁涛　　Photo by: Ding Tao

羽片　　　Pinna
摄影：丁涛　　Photo by: Ding Tao

9 半边旗 | bàn biān qí 凤尾蕨属

Pteris semipinnata L. 凤尾蕨科 Pteridaceae

植株高 35～80（～120）cm。根状茎长而横走，先端及叶柄基部被褐色鳞片。叶簇生，近一型；叶柄长 15～55 cm，粗 1.5～3 mm，连同叶轴均为栗红有光泽，光滑；叶片长圆披针形，长 15～40（～60）cm，宽 6～15（～18）cm，二回半边深裂。侧脉明显，斜上，二叉或回二叉。叶干后草质，灰绿色，无毛。

Plants 35–80 (–120) cm tall. Rhizome long creeping, apex with blackish brown scales; scales also at base of stipes. Fronds clustered, submonomorphic; stipe 15–55× 0.15–0.3 cm, stipe and rachis castaneous-reddish, shiny, glabrous; lamina pinnate, oblong-lanceolate in outline, 15–40 (–60) × 6–15 (–18) cm, at one side deeply bipinnate-lobed. Veins conspicuous, decumbent, 2-forked or bipinnate-forked. Lamina gray-green, herbaceous when dried, glabrous.

蕨叶 Fronds 摄影：丁涛 Photo by: Ding Tao

羽片 Pinnae 摄影：丁涛 Photo by: Ding Tao

10 长叶铁角蕨 | cháng yè tiě jiǎo jué 铁角蕨属

Asplenium prolongatum Hook. 铁角蕨科 Aspleniaceae

植株高 20～40 cm。根状茎短而直立。叶簇生；叶柄长 8～18 cm，淡绿色，上面有纵沟，干后压扁；叶片线状披针形，长 10～25 cm，宽 3～4.5 cm，尾头，二回羽状；羽片 20～24 对。孢子囊群狭线形，深棕色，每小羽片或裂片 1 枚，位于小羽片的中部上侧边；囊群盖狭线形，灰绿色，膜质，开向叶边，宿存。

Plants 20–40 cm tall. Rhizome erect, short. Fronds caespitose; stipe green, 8–18 cm, sulcate adaxially; lamina linear-ovate, 10–25 × 3–4.5 cm, bipinnate, apex caudate; pinnae 20–24 pairs. Sori 1 per pinnule or segment, median on acroscopic side of subtending vein, linear; indusia grayish green, linear, membranous, opening toward costa and margin, persistent.

植株 Whole plant 摄影：丁涛 Photo by: Ding Tao

羽片 Pinnae 摄影：丁涛 Photo by: Ding Tao

11 华南毛蕨 | huá nán máo jué 毛蕨属

Cyclosorus parasiticus **(L.) Farw.** 金星蕨科 **Thelypteridaceae**

植株高达（30～）50～70（～100）cm。根状茎横走，连同叶柄基部有深棕色披针形鳞片。叶近生；叶柄长达（10～）20～30（～40）cm，深禾秆色；叶片长圆披针形，先端羽裂，尾状渐尖头，基部不变狭，二回羽裂。孢子囊群圆形，生侧脉中部以上；囊群盖小，膜质，棕色，上面密生柔毛，宿存。

Plants (30–) 50–70 (–100) cm tall. Rhizomes shortly to long creeping, including stipe bases with dark brown lanceolate scales. Fronds approximate to distant; stipes (10–) 20–30 (–40) cm, stramineous; bases not narrowed (sometimes slightly narrowed), apices caudate-acuminate. Sori orbicular, medial; indusia densely hairy. Sporangia bearing reddish orange glands on stalks.

植株　Whole plant
摄影：丁涛　Photo by: Ding Tao

孢子囊群　Sori
摄影：丁涛　Photo by: Ding Tao

12 乌毛蕨 | wū máo jué 乌毛蕨属

Blechnum orientale **L.** 乌毛蕨科 **Blechnaceae**

根茎深褐色，直立，短，密被鳞片，狭线形，约 1 cm，全缘。叶簇生于根状茎顶端；柄长 10～60 cm，粗 3～10 mm，基部往往为黑褐色，被鳞片覆盖。叶片卵状披针形，近革质，长达 55～100 cm，宽 20～60 cm，一回羽状，羽片多数。孢子囊群线形，连续，紧靠主脉两侧，与主脉平行。

Rhizome dark brown, erect, short, densely scaly; narrowly linear, ca. 1 cm, entire. Stipe 10–60 cm, 3–10 mm in diam., base dark brown and covered with scales as rhizome; lamina imparipinnate, monomorphic, ovate-lanceolate, 55–100 × 20–60 cm, subleathery; pinnae numerous. Sori linear, continuous, close to both sides of the main vein, parallel to the main vein.

植株　Whole plant
摄影：丁涛　Photo by: Ding Tao

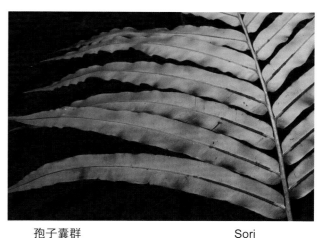

孢子囊群　Sori
摄影：丁涛　Photo by: Ding Tao

13 中华复叶耳蕨 | zhōng huá fù yè ěr jué 复叶耳蕨属

Arachniodes chinensis (Rosenst.) Ching — 鳞毛蕨科 Dryopteridaceae

根状茎短匍匐，硬，密被鳞片。叶柄长 15～55 cm，粗 2.5～5（～6）mm，禾秆色，基部密鳞片。叶片卵状三角形，长 25～65 cm，宽 15～35 cm，顶部略狭缩呈长三角形，渐尖头，基部心形，二回羽状或三回羽状。孢子囊群每小羽片 1～8 对，于中脉与叶边之间；囊群盖棕色，膜质。

Rhizome shortly creeping, stiff, densely scaly. Stipe stramineous, 15–55 cm, 2.5–5 (–6) mm in diam., base densely scaly. Lamina 2- or 3-pinnate, deltoid-ovate or ovate-oblong, 25–65 × 15–35 cm, papery or subleathery, base cordate, apex acuminate, attenuate or abruptly narrowed and acuminate. Sori terminal on veinlets, 1–8 pairs per ultimate segment, medial between midvein and margin or closer to midvein; indusia brown, firmly membranous.

蕨叶　Frond　摄影：丁涛　Photo by: Ding Tao

孢子囊群　Sori　摄影：丁涛　Photo by: Ding Tao

14 肾蕨 | shèn jué 肾蕨属

Nephrolepis cordifolia (L.) C. Presl — 肾蕨科 Nephrolepidaceae

附生或土生。根状茎直立，被蓬松的淡棕色长钻形鳞片。叶簇生，柄长 5～15 cm，密被淡棕色线形鳞片；叶片线状披针形或狭披针形，长 25～75 cm，宽 3～6 cm，先端短尖，叶轴两侧被纤维状鳞片，一回羽状，羽状多数，约 40～120 对，互生，披针形。孢子囊群成 1 行位于主脉两侧，肾形，少有为圆肾形或近圆形。

Plants terrestrial or epiphytic. Rhizome erect, short, covered with yellowish brown, narrowly lanceolate scales. Stipe 5–15 cm, densely covered with same scales as on rhizome; lamina linear-lanceolate or narrowly lanceolate, 25–75 × 3–6 cm, pinnate; pinnae 40–120 pairs, approximate, lanceolate. Sori lunulate or rarely orbicular-reniform; indusia brown, elongate.

植株　Whole plant　摄影：丁涛　Photo by: Ding Tao

羽片　Pinnae　摄影：丁涛　Photo by: Ding Tao

15 三叉蕨 | sān chà jué

Tectaria subtriphylla (Hook. Arnott) Copel.　　三叉蕨科 Tectariaceae　　三叉蕨属

植株高 20~70 cm。根状茎长而横走，粗状，顶部及叶柄基部均密被鳞片。叶近生；叶柄长 10~40 cm，上面有浅沟；叶二型，可育的叶高和窄，长 10~35 cm，基部宽 10~25 cm，一回羽状，能育叶与不育叶形状相似但各部均缩狭。孢子囊群圆形，生于小脉联结处；囊群盖圆肾形，坚膜质，棕色，脱落。

Plants terrestrial, 20–70 cm tall. Rhizome long creeping or ascendant, stout, densely scaly at apex and stipe bases. Fronds widely spaced; stipe dark stramineous, 10–40 cm. Laminae subdimorphic, fertile fronds rather tall but narrowed, pinnate to bipinnatifid at base, 10–35 × 10–25 cm, papery. Sori small, orbicular, located at coupling veinlets; indusia brown, reniform, small, deciduous.

蕨叶　Fronds　摄影：丁涛　Photo by: Ding Tao

孢子囊群　Sori　摄影：丁涛　Photo by: Ding Tao

16 大叶骨碎补 | dà yè gǔ suì bǔ

Davallia divaricata Blume　　骨碎补科 Davalliaceae　　骨碎补属

植株高达 1 m。根状茎粗壮，长而横走，粗 10~15 mm，密被蓬松的鳞片。叶柄长 30~60 cm，与叶轴均为亮棕色或暗褐色；叶片大，三角形或卵状三角形，长 55~100 cm，宽 40~90 cm。孢子囊群多数，每裂片有 1 枚，生于小脉中部稍下的弯弓处或生于小脉分叉处；囊群盖管状，约为宽的 2 倍。

Plants terrestrial, 1m tall. Rhizome 10–15 mm in diam. Scales brown. Stipe pale, adaxially grooved, 30–60 cm, glabrous or with few scales; tripinnate toward base and in middle part, deltoid and broadest toward base, 55–100 × 40–90 cm, glabrous. Sori separate, borne several on a segment, at forking point of veins; indusium also attached along sides, pouch-shaped, oblong, ± as wide as long.

蕨叶　Frond　摄影：丁涛　Photo by: Ding Tao

羽片　Pinnae　摄影：丁涛　Photo by: Ding Tao

17 团叶槲蕨 | tuán yè hú jué 槲蕨属

Drynaria bonii Christ 水龙骨科 Polypodiaceae

根状茎横走，粗 1~2 cm，顶端密被鳞片；鳞片盾形，长 2~12 mm，宽 1.5~3 mm。叶二型，无毛，基生叶邻接或重叠，无梗，近全缘到浅裂，能育叶，长 30~70 cm，宽 20~30 cm，裂片 3~7 对。孢子囊群细小，在中肋两侧不规则地排成 2 行，在相邻两对侧脉间有 2 至 4 行。

Rhizome shortly creeping, 1–2 cm wide; scales peltate, 2–12 × 1.5–3 mm. Fronds dimorphic, glabrous; basal fronds contiguous or overlapping, sessile, subentire to shallowly lobed; foliage fronds stalked, 30–70 × 20–30 cm; pinnae 3–7 pairs. Sori in 2 irregular rows between costa and margin, 2–4 rows between lateral veins. Spores with spines.

植株　Whole plant
摄影：丁涛　Photo by: Ding Tao

蕨叶　Frond
摄影：丁涛　Photo by: Ding Tao

18 披针骨牌蕨 | pī zhēn gǔ pái jué （披针骨自牌蕨） 伏石蕨属

Lemmaphyllum diversum (Rosenst.) Tagawa 水龙骨科 Polypodiaceae

植株高 3~10 cm。根状茎直径 1.5 mm，密被鳞片；鳞片棕色，钻状披针形，边缘有锯齿。叶远生，一型或近二型；叶柄变化大，长 0.5~3 cm，禾秆色，光滑；叶片通常为阔卵状披针形，短尖头，长 3.5~10 cm，宽 0.5~2.5 cm，具短柄。孢子囊群圆形，在主脉两侧各成一行。

Plants 3–10 cm tall. Rhizomes to 1.5 mm in diam., densely scaly. Fronds remote, ± dimorphic; stipe strawcolored, 0.5–3 cm, smooth; fronds lamina oblong or narrowly to broadly lanceolate, 3.5–10 × 0.5–2.5 cm. Sporangia in discrete sori in 1 line on each side of midrib, sori mostly orbicular.

植株　Whole plant
摄影：丁涛　Photo by: Ding Tao

孢子囊群　Sori
摄影：丁涛　Photo by: Ding Tao

19 宽羽线蕨 | kuān yǔ xiàn jué

薄唇蕨属

Leptochilus ellipticus var. *pothifolius* (Buch.-Ham. ex D. Don) X. C. Zhang 水龙骨科 **Polypodiaceae**

根状茎长而横走，密生鳞片。叶大，70～100 cm，叶轴圆柱状到具狭翅，裂片羽状，边缘全缘或有时不明显的稍波状；羽片或裂片（5～）7～14 对，狭长披针形或线形，长 13～24（～31）cm，宽 13～24（～31）cm。孢子囊群线形，斜展，在每对侧脉间各排列成一行，伸达叶边；无囊群盖。

Rhizome slender, long creeping; scales brown. Fronds large, 70–100 cm; rachis terete to narrowly winged; lamina pinnate to pinnatisect, margin entire or sometimes indistinctly slightly undulate; pinnae (5–) 7–14 pairs, largest lobe 13–24 (–31) ×13–24 (–31) cm. Sori linear, 1 regular row between lateral veins, up to margin of lamina, without paraphyses.

| 植株 Whole plant | 蕨叶 Frond |
| 摄影：丁涛 Photo by: Ding Tao | 摄影：丁涛 Photo by: Ding Tao |

20 绿叶线蕨 | lǜ yè xiàn jué

薄唇蕨属

Leptochilus leveillei (Christ) X. C. Zhang Noot. 水龙骨科 **Polypodiaceae**

根状茎长而横走，密生鳞片；鳞片褐棕色，质薄。叶疏生或近生，通常一型，近无柄；叶片线状披针形至线形，长 20～40 cm，宽 0.8～4 cm，顶端长渐尖或呈长尾状，中部以下渐变狭而下延，近达基部，边缘浅波状。孢子囊群线形，在每对侧脉间排列成一行，直达叶边，无囊群盖。

Rhizome slender, long creeping; scales brown, ovate-lanceolate. Fronds monomorphic or slightly dimorphic, distant; lamina linear or linear-lanceolate, 20–40 cm, 0.8–4 cm wide at middle, glabrous, gradually decurrent nearly to base, margin slightly undulate, apex long acuminate or caudate; Sori linear, 1 regular row between lateral veins, up to margin of lamina, without paraphyses.

| 植株 Whole plant | 孢子囊群 Sori |
| 摄影：丁涛 Photo by: Ding Tao | 摄影：丁涛 Photo by: Ding Tao |

21 买麻藤 | mǎi má téng （倪藤）　　　　买麻藤属

***Gnetum montanum* Markgraf**　　　　买麻藤科 Gnetaceae

大藤本，高达 10 m 以上，小枝圆或扁圆，光滑。叶形大小多变，通常呈矩圆形，革质或半革质，长 10～25 cm，宽 4～11 cm，先端具短钝尖头，基部圆或宽楔形，侧脉 8～13 对。雄球花序松散；雌球花序侧生，单生或数序丛生。种子矩圆状卵圆形或矩圆形，熟时黄褐色或红褐色。花期 4～6 月，种子 8～10 月成熟。

Vines to more than 10 m tall; branchlets orbicular or compressed orbicular in cross section, smooth. Leaf blade usually oblong, 10–25 × 4–11 cm, leathery or nearly so, lateral veins 8–13 on each side, base rounded or broadly cuneate, apex obtuse to acute. Male inflorescences lax. Female inflorescences lateral, solitary or fascicled. Seeds yellowish brown or reddish brown, cylindric-ovoid or cylindric. Pollination Apr.–Jun., seed maturity Aug.–Oct..

叶　　Leaves　　　　　　　　　　　花　　Flowers
摄影：丁涛　　Photo by: Ding Tao　　摄影：丁涛　　Photo by: Ding Tao

22 草珊瑚 | cǎo shān hú （接骨金粟兰）　　　　草珊瑚属

***Sarcandra glabra* (Thunb.) Nakai**　　　　金粟兰科 Chloranthaceae

常绿半灌木，高 50～150 cm。茎有膨大的节。叶革质，椭圆形、卵形至卵状披针形，长 6～20 cm，宽 2～8 cm，顶端渐尖，基部尖或楔形，边缘具粗锐锯齿，齿尖有一腺体。穗状花序顶生。核果球形，熟时亮红色。花期 6 月，果期 8～10 月。

Subshrubs, evergreen, 50–150 cm tall. Stems nodes swollen. Leaf blade elliptic or ovate to ovate-lanceolate, or broadly elliptic to oblong, 6–20 × 2–8 cm, leathery or papery, glandular mucronate on marginal teeth, base acute, cuneate, margin sharply coarsely-serrate or dully serrate, apex acute to acuminate. Inflorescences terminal. Drupes shiny red at maturity, globose. Fl. Jun., fr. Aug.–Oct..

叶背　　Leaf backs　　　　　　　　　果序　　Infructescence
摄影：丁涛　　Photo by: Ding Tao　　摄影：丁涛　　Photo by: Ding Tao

23 石南藤 | shí nán téng （毛山蒟）　　胡椒属

***Piper wallichii* (Miq.) Hand.-Mazz.**　　胡椒科 **Piperaceae**

攀援藤本。茎被短硬毛，干时呈黑色，有纵棱。叶硬纸质，干时变淡黄色，无明显腺点，卵状披针形的或狭椭圆形，长 5~14 cm，宽 2~6.5 cm，顶端长渐尖，有小尖头，基部短狭或钝圆，两侧近相等。雄花穗状花序 2 倍于叶片。雌穗 1.5~3 cm。浆果球形，有疣状凸起。花期 2~6 月，果期 4~10 月。

Climbers dioecious. Stems black when dry, ridged, usually hispidulous. Leaf blade ovate-lanceolate or narrowly elliptic, 5–14 × 2–6.5 cm, papery, drying grayish, base rounded to shortly tapered, symmetric to slightly oblique, apex acuminate. Male spikes more than 2 as long as leaf blades. Female spikes 1.5–3 cm. Drupe subglobose, ± tuberculate. Fl. Feb.–Jun., fr. Apr.–Oct..

植株　　Whole plant　　摄影：丁涛　　Photo by: Ding Tao

叶　　Leaves　　摄影：丁涛　　Photo by: Ding Tao

24 白叶瓜馥木 | bái yè guā fù mù （火索藤）　　瓜馥木属

***Fissistigma glaucescens* (Hance) Merr.**　　番荔枝科 **Annonaceae**

攀援灌木，长达 6 m。枝条无毛。叶近革质，长圆形或长圆状椭圆形，有时倒卵状长圆形，长 3~20 cm，宽 1.2~6 cm，顶端通常圆形，少数微凹，基部圆形或钝形，两面无毛，叶背白绿色，干后苍白色。花数朵集成聚伞式的总状花序，花序顶生。果圆球状，无毛。花期 1~9 月，果期 3~12 月。

Climbers to 6 m tall, most parts glabrous except for inflorescences. Leaf blade oblong, oblong-elliptic, or sometimes obovate-oblong, 3–20 × 1.2–6 cm, thinly leathery, abaxially grayish green and glaucous when dry. Inflorescences terminal, thyrsoid. Monocarp spheroidal, glabrous. Fl. Jan.–Sep., fr. Mar.–Dec..

叶背　　Leaf backs　　摄影：丁涛　　Photo by: Ding Tao

枝叶　　Branch and leaves　　摄影：丁涛　　Photo by: Ding Tao

25 心叶青藤 | xīn yè qīng téng 青藤属

***Illigera cordata* Dunn** 莲叶桐科 **Hernandiaceae**

藤本。茎具纵向条纹，初被短柔毛，后变无毛。叶为指状，小叶 3 枚。小叶卵形、椭圆形至长圆状椭圆形，长 8~12 cm，宽 4~8 cm，纸质，全缘，先端短渐尖，基部心形，两侧不对称，上面沿脉被柔毛，下面疏被毛或无毛。聚伞花序较紧密地排列成近伞房状，生于叶腋；厚纸质。花期 5~6 月，果期 8~9 月。

Lianas. Stem striate, pubescent at first, soon glabrescent. Leaves 3-foliolate; blade ovate or elliptic to oblong-elliptic, 8–12 × 4–8 cm, papery, lateral veins 4- or 5-paired, base cordate, asymmetric, apex shortly acuminate. Cymes axillary, compact, subcorymbose. Fruit 4-winged; thickly papery. Fl. May–Jun., fr. Aug.–Sep..

复叶　Compound leaf
摄影：丁涛　Photo by: Ding Tao

枝叶　Branch and leaves
摄影：丁涛　Photo by: Ding Tao

26 魔芋 | mó yù 魔芋属

***Amorphophallus konjac* K. Koch** 天南星科 **Araceae**

块茎扁球形，直径 20~30 cm，顶部中央多少下凹。叶单生；叶柄白色带粉红色或脏奶油色，基部无毛或具散在的斑点状疣；叶片高度裂片化，直径约 200 cm，小枝狭具翅；小叶正面深绿色，椭圆形，长 3~10 cm，宽 2~6 cm，渐尖。佛焰苞漏斗形，浆果球形或扁球形，成熟时黄绿色。花期 4 月，果期 8~9 月。

Tuber depressed globose, to ca. 30 cm in diam.. Leaf solitary; petiole background color dirty whitish pinkish or dirty cream-colored, glabrous or with scattered punctiform warts at base. Leaf blade highly dissected, to ca. 200 cm in diam., rachises narrowly winged. Leaflets dull green adaxially, elliptic, 3–10 × 2–6 cm, acuminate. Spathe funnel shape. Fl. Apr., fr. Aug.–Sep..

叶背　Leaf back
摄影：丁涛　Photo by: Ding Tao

枝叶　Branch and leaves
摄影：丁涛　Photo by: Ding Tao

27 石柑子 | shí gān zǐ （上树葫芦） 石柑属

Pothos chinensis (Raf.) Merr. 天南星科 **Araceae**

附生藤本，长达 10 m。茎亚木质，近圆柱形。叶片纸质，鲜时表面深绿色，背面淡绿色，椭圆形，披针状卵形至披针状长圆形，长 3~20.5 cm，宽 1.5~20.5 cm，先端渐尖至长渐尖；叶柄倒卵状长圆形或楔形或狭三角形，具翅，先端截形。花序腋生。浆果黄绿色至红色。花果期四季。

Lianas, to 10 m, root-climbing. Stem weakly 4-angled or terete in cross section. Leaves paler abaxially, bright to mid-green adaxially; petiole obovate-oblong to linear-oblong or narrowly triangular, broadly winged, apex truncate. Leaf blade ovate to elliptic or lanceolate, 3–20.5 × 1.5–20.5 cm, apex attenuate-mucronate to acute or attenuate. Inflorescences solitary or in pairs. Fruit mid-green, ripening to scarlet. Fl. and fr. throughout year.

植株 Whole plant 摄影：丁涛 Photo by: Ding Tao

叶背 Leaf backs 摄影：丁涛 Photo by: Ding Tao

28 山菅兰 | shān jiān lán （山菅） 山菅兰属

Dianella ensifolia (L.) DC. 阿福花科 **Asphodelaceae**

根状茎圆柱状，横走，粗 5~8 mm。叶狭条状披针形，长 30~80 cm，宽 1~2.5 cm，基部稍收狭成鞘状，边缘和背面中脉具锯齿。顶端圆锥花序长 10~40 cm，分枝疏散；花常多朵生于侧枝上端；花被片绿白色、淡黄色至青紫色。浆果近球形，深蓝色。花果期 3~8 月。

Rhizome creeping, 5–8 mm thick. Leaves sword-shaped, gradually narrowed at both ends, 30–80 × 1–2.5 cm, leathery, midvein abaxially and margin usually scabrous, apex obtuse. Pan-icle laxly branched, 10–40 cm, usually with flowers borne distally. Tepals spreading, white, greenish white, yellowish, or bluish purple. Berries deep blue, subglobose. Fl. and fr. Mar.–Aug..

植株 Whole plant 摄影：丁涛 Photo by: Ding Tao

果序 Infructescences 摄影：丁涛 Photo by: Ding Tao

29 大百部 | dà bǎi bù （对叶百部） 百部属

Stemona tuberosa **Lour.** 百部科 **Stemonaceae**

块根通常纺锤状，长 9~13（~30）cm。茎常具少数分枝，攀援状，下部木质化。叶对生或轮生，极少兼有互生，卵状披针形、卵形或宽卵形，长 6~24 cm，宽（2~）5~17 cm，顶端渐尖，基部心形，边缘稍波状，纸质或薄革质。花单生或 1~3 朵排成总状花序。蒴果光滑，具多数种子。花期 4~7 月，果期 7~8 月。

Vines. Roots 9–13 (–30) cm. Stems often branched, base woody. Leaves opposite or whorled, rarely alternate. Leaf blade ovate to ovate-lanceolate, 6–24 × (2–) 5–17 cm, membranous, base cordate, margin slightly undulate, apex acuminate. Inflorescences racemes, 1–3-flowered. Capsule ovoid-oblong. Seeds several. Fl. Apr.–Jul., fr. Jul.–Aug..

枝叶 — Branch and leaves
摄影：丁涛 — Photo by: Ding Tao

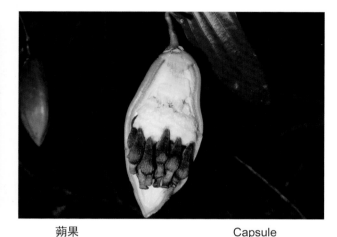

蒴果 — Capsule
摄影：丁涛 — Photo by: Ding Tao

30 露兜树 | lù dōu shù （露兜簕） 露兜树属

Pandanus tectorius **Parkinson** 露兜树科 **Pandanaceae**

灌木或小乔木，高 3~10 m。茎直立或上升，多分枝，具大量的气根。叶绿色，通常背面被白霜，条形，长达 180 cm，宽 10 cm，叶缘和背面中脉均有粗壮的锐刺。雄花序由若干穗状花序组成，长 60 cm；雌花序头状，单生于枝顶。聚花果大，由 40~80 个核果束组成，圆球形或长圆形。花期 1~5 月，果期 10 月。

Trees or shrubs, 3–10 m. Stems erect or ascending, many branched; numerous aerial roots often present. Leaves green, often glaucous abaxially, linear-ensiform, to 180 × 10 cm, spinose-serrate on margin and midvein abaxially. Male inflorescence to 60 cm, pedunculate, female inflorescence capitate, solitary. Syncarp globose or cylindric; phalanges 40–80 per aggregate-head. Fl. Jan.–May, fr. Oct..

植株 — Whole plant
摄影：丁涛 — Photo by: Ding Tao

叶 — Leaves
摄影：丁涛 — Photo by: Ding Tao

31 抱茎菝葜 | bào jīng bá qiā

Smilax ocreata A. DC.

菝葜科 Smilacaceae

菝葜属

攀援灌木，茎长可达7 m。茎杆木质，通常疏生刺。叶革质，卵形或椭圆形，长9～20 cm，宽4.5～15 cm；叶柄长2～3.5 cm，基部两侧具耳状的鞘，长约为叶柄的1/3～1/2，宽5～20 mm（一侧），作穿茎状抱茎。圆锥花序长4～10 cm，具2～4（～7）个伞形花序。浆果熟时暗红色，球状。花期3～6月，果期7～10月。

Vines climbing, up to 7 m. Stem and branches usually woody, sparsely prickly. Petiole 2–3.5 cm, broadly winged for 1/3–1/2 its length; wings 0.5–2 cm wide. Leaf blade ovate to elliptic, 9–20 × 4.5–15 cm, leathery. Inflorescence a raceme of 2–4 (–7) umbels, 4–10 cm, basally prophyllate. Berries dark red, globose. Fl. Mar.–Jun., fr. Jul.–Oct..

叶 Leaves
摄影：丁涛　Photo by: Ding Tao

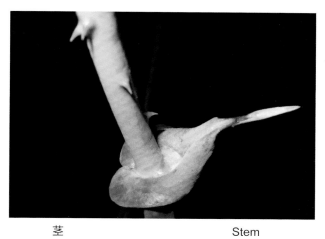

茎 Stem
摄影：丁涛　Photo by: Ding Tao

32 竹叶兰 | zhú yè lán

Arundina graminifolia (D. Don) Hochr.

兰科 Orchidaceae

竹叶兰属

植株高40～100（～150）cm，有时可达1.5 m以上。茎直立，通常为叶鞘所包。叶多数，薄革质或坚纸质，通常长8～20 cm，宽1～2 cm，先端渐尖；鞘抱茎，长2～4 cm；花粉红色或略带紫色或白色；蒴果近长圆形，长28～35 mm，宽8～15 mm。花果期主要为6～11月，但1～4月也有。

Plants 40–100 (–150) cm tall. Stem rigid, enclosed by leaf sheaths. Leaves numerous, 8–20 × 1–2 cm, leathery or papery, apex acuminate; sheaths 2–4 cm. Flowers white or pink, sometimes slightly tinged with purple. Capsule 28–35×8–15 mm. Fl. and fr. Jun.–Nov., sometimes Jan.–Apr..

植株 Whole plant
摄影：丁涛　Photo by: Ding Tao

花 Flower
摄影：丁涛　Photo by: Ding Tao

33 半柱毛兰 | bàn zhù máo lán

Eria corneri Rchb. f.   兰科 Orchidaceae

假鳞茎密集着生，卵状长圆形或椭圆状，长 2~5 cm，粗 1~2.5 cm，顶端具 2~3 枚叶。叶椭圆状披针形至倒卵状披针形，长（15~）20~45 cm，宽 1.5~6 cm，先端渐尖或长渐尖，基部收狭成长 2~3 cm 的柄。花序具 10 余朵花；花白色或略带黄色。蒴果倒卵状圆柱状。花期 8~9 月，果期 10~12 月。

Pseudobulbs ± clustered, ovoid-oblong or ellipsoid, 2–5 × 1–2.5 cm, apex 2- or 3-leaved. Leaf blade elliptic-lanceolate or obovate-lanceolate, (15–) 20–45 × 1.5–6 cm, apex acuminate or long acuminate; petiole 2–3 cm. Inflorescence more than 10-flowered; Flowers white or slightly tinged with yellow. Capsule obovoid-cylindric. Fl. Aug.–Sep., fr. Oct.–Dec..

植株 Whole plant 摄影：丁涛 Photo by: Ding Tao

果序 Infructescence 摄影：丁涛 Photo by: Ding Tao

34 橙黄玉凤花 | chéng huáng yù fèng huā

Habenaria rhodocheila Hance   兰科 Orchidaceae

植株高 8~35 cm。块茎长圆形，肉质，长 2~4 cm，直径 1~2 cm。茎粗壮，直立，圆柱形，下部具 4~6 枚叶，向上具 1~3 枚苞片状小叶。叶片线状披针形至近长圆形，长 10~15 cm，宽 1.5~2 cm，先端渐尖，基部抱茎。总状花序具 2~10 余朵疏生的花。蒴果纺锤形，先端具喙。花期 7~8 月，果期 10~11 月。

Plants 8–35 cm tall. Tubers oblong-cylindric, 2–4 × 1–2 cm, fleshy. Stem erect, terete, stout, with 4–6 leaves below middle and 1–3 bractlike leaflets above leaves. Leaf blade linear-lanceolate to suboblong, 10–15 × 1.5–2 cm, base amplexicaul, apex acuminate. Raceme loosely 2–10-flowered. Capsule fusiform, apex beaked. Fl. Jul.–Aug., fr. Oct.–Nov..

植株 Whole plants 摄影：丁涛 Photo by: Ding Tao

花序 Inflorescences 摄影：丁涛 Photo by: Ding Tao

35 宽叶羊耳蒜 | kuān yè yáng ěr suàn

羊耳蒜属

Liparis latifolia (Bl.) Lindl.

兰科 Orchidaceae

附生草本。肉质茎卵球形，长6~8 cm，宽4~5 cm，包藏于叶鞘之内。叶通常4~5枚，斜卵状椭圆形、卵形，长2.5~7.5（~12）cm，宽1~3（~6.5）cm，先端渐尖或长渐尖，基部收狭成柄；叶柄鞘状，长3~6.5（~8）cm，抱茎。花紫红色至绿黄色，较小。蒴果倒卵状椭圆形。花期5~8月，果期8~12月。

Herbs epiphytic. Pseudobulbs ovoid, relatively small, usually 6–8 × 4–5 mm, ± enclosed in white membranous sheaths. Leaf usually 4–5, ovate, oblong, or subelliptic, 2.5–7.5 (–12) × 1–3 (–6.5) cm, base contracted into ± amplexicaul petiole 3–6.5 (–8) cm, apex obtuse or subacute. Flowers pale yellowish green to pale green, small. Capsule obovoid or obovoid-ellipsoid. Fl. May–Aug., fr. Aug.–Dec..

植株　Whole plants
摄影：丁涛　Photo by: Ding Tao

叶　Leaves
摄影：丁涛　Photo by: Ding Tao

36 长茎羊耳蒜 | cháng jīng yáng ěr suàn

羊耳蒜属

Liparis viridiflora (Blume) Lindl.

兰科 Orchidaceae

附生草本。假鳞茎稍密集，通常为圆柱形，基部常多少平卧，自下向上渐狭，长（3~）7~18 cm，直径3~8（~12）mm，顶端具2叶。叶线状倒披针形或线状匙形，纸质，长8~25 cm，宽1.2~3 cm，先端渐尖并有细尖，基部收狭成柄，有关节。花绿白色或淡绿黄色。蒴果倒卵状椭圆形。花期9~12月，果期次年1~4月。

Herbs epiphytic. Pseudobulbs somewhat densely arranged, base often ± prostrate, attenuate from base to apex, usually cylindric, (3–) 7–18 cm × 3–8 (–12) mm. Leaves 2; articulate; blade linear-oblanceolate or linear-spatulate, 8–25 × 1.2–3 cm, papery, apex acuminate and apiculate. Flowers greenish white or pale greenish yellow. Capsule obovoid-ellipsoid. Fl. Sep.–Dec., fr. Jan.–Apr. of next year.

植株　Whole plants
摄影：丁涛　Photo by: Ding Tao

假鳞茎　Pseudobulbs
摄影：丁涛　Photo by: Ding Tao

37 宽叶线柱兰 | kuān yè xiàn zhù lán （亲种线柱兰） 线柱兰属

Zeuxine affinis (Lindl.) Benth. ex Hook. f. 兰科 Orchidaceae

地生草本。植株高 13~30cm。根状茎伸长，匍匐，肉质，具节。茎直立，具 4~6 枚叶。叶片卵形、卵状披针形或椭圆形，长 2.5~4cm，宽 1.2~2.5cm，先端急尖或钝。花茎淡褐色，长 5~20cm，被柔毛，具 1~2 枚鞘状苞片。总状花序具几朵至 10 余朵花；花较小，黄白色。花期 2~4 月。

Herbs terrestrial. Plants 13–30 cm tall. Rhizomes elongate, prostrate, fleshy, knobby. Stem erect, with 4–6 leaves. Leaf blade ovate, ovate-lanceolate or elliptic, 2.5–4×1.2–2.5cm, apex acute or obtuse. Flowers stem light brown, 5–20 cm, pilose, with 1–2 sheath bracts. Racemes with several to more than 10 flowers，flowers small，yellow and white. Fl. Feb.–Apr..

植株　Whole plant
摄影：丁涛　Photo by: Ding Tao

花序　Inflorescence
摄影：丁涛　Photo by: Ding Tao

38 裸花水竹叶 | luǒ huā shuǐ zhú yè 水竹叶属

Murdannia nudiflora (L.) Brenan 鸭跖草科 Commelinaceae

一年生草本。根须状，纤细，无毛或被长绒毛。茎多条自基部发出，披散，下部节上生根，长 10~50 cm，分枝或否，无毛，主茎发育。叶几乎全部茎生，叶片禾叶状或披针形，顶端钝或渐尖，长 2.5~10 cm，宽 5~10 mm。蝎尾状聚伞花序数个，排成顶生圆锥花序。蒴果卵圆状三棱形。花果期（6~）8~9（~10）月。

Herbs annual. Roots fibrous, slender, glabrous or tomentose. Rhizomes absent. Stems numerous, diffuse, creeping proximally, simple or branched, 10–50 cm, glabrous. Leaves nearly all cauline. Leaf blade linear or lanceolate, 2.5–10 × 0.5–1 cm, apex obtuse or acuminate. Cincinni several, in terminal panicles. Capsule ovoid-globose, trigonous. Fl. and fr. (Jun.–)Aug.–Sep(–Oct).

叶　Leaves
摄影：丁涛　Photo by: Ding Tao

果序　Infructescence
摄影：丁涛　Photo by: Ding Tao

39 长花枝杜若 | cháng huā zhī dù ruò

杜若属

***Pollia secundiflora* (Blume) Bakh. f.**

鸭跖草科 **Commelinaceae**

多年生草本。茎直立，高 1~2 m，疏被白色柔毛。叶无柄，椭圆形，长约 20 cm，宽约 5 cm，顶端渐尖，基部楔状渐窄，上面具瘤状突起，下面密生细柔毛；叶鞘长约 2.5 cm，相当密地被柔毛。花序长长地超出叶子，下部的花序分枝具长达 20 cm 以上的总梗。果成熟时黑色，直径约 6 mm。花期 4 月，果期 7~11 月。

Herbs perennial. Stems erect, 1–2 m tall, sparsely white-pubescent. Leaves sessile. Leaf sheath ca. 2.5 cm, rather densely pubescent. Leaf blade elliptic, ca. 20 × 5 cm, verrucose adaxially, densely puberulent abaxially. Inflorescence longer than distal leaves; proximal inflorescence branches with peduncle to more than 20 cm. Fruit globose, ca. 6 mm in diam.. Fl. Apr., fr. Jul.–Nov..

果序 — Infructescence
摄影：丁涛 — Photo by: Ding Tao

花序 — Inflorescence
摄影：丁涛 — Photo by: Ding Tao

40 海南山姜 | hǎi nán shān jiāng （草豆蔻）

山姜属

***Alpinia hainanensis* K. Schum. in Engler**

姜科 **Zingiberaceae**

多年生草本。株高达 3 m。叶柄长 2 cm。叶片线状披针形，长 20~65 cm，宽 2~12 cm，顶端渐尖，并有一短尖头，基部渐狭，边缘被毛，两面均无毛或稀可于叶背被极疏的粗毛。总状花序顶生，直立，长 10~30 cm。果球形，熟时金黄色。花期 4~6 月，果期 5~8 月。

Herbs perennial. Pseudostems to 3 m. Ptiole absent to 2 cm. Leaf blade linear-lanceolate, 20–65 × 2–12 cm, glabrous or rarely abaxially sparsely hirsute, base obliquely attenuate, margin hairy, apex acuminate. Racemes erect, 10–30 cm. Apsule globose, yellow hirsute. Fl. Apr.–Jun., fr. May–Aug..

植株 — Whole plant
摄影：丁涛 — Photo by: Ding Tao

花序 — Inflorescence
摄影：丁涛 — Photo by: Ding Tao

41 华山姜 | huá shān jiāng （小良姜）　　　　　　　　山姜属

Alpinia oblongifolia Hayata　　　　　　　　姜科 Zingiberaceae

株高约 1 m。叶披针形或卵状披针形，长 20~30 cm，宽 3~10 cm，顶端渐尖或尾状渐尖，基部渐狭，两面均无毛；叶柄长约 5 mm；叶舌膜质，长 4~10 mm，2 裂，具缘毛。狭圆锥花序。果球形，红色，直径 5~8 mm。花期 5~7 月，果期 6~12 月。

Plants 1 m tall. Ligule 4–10 mm, membranous, margin ciliate; petiole ca. 5 mm. Leaf blade oblong, ovate-lanceolate, or lanceolate, 20–30 × 3–10 cm, glabrous, base rounded or attenuate, apex acuminate or caudate-acuminate. Panicles narrow. Capsule red, globose, 5–8 mm in diam.. Seeds 5–8. Fl. May–Jul., fr. Jun.–Dec..

叶　　Leaf
摄影：丁涛　　Photo by: Ding Tao

果序　　Infructescence
摄影：丁涛　　Photo by: Ding Tao

42 华南谷精草 | huá nán gǔ jīng cǎo　　　　　　　谷精草属

Eriocaulon sexangulare L.　　　　　　　　谷精草科 Eriocaulaceae

叶丛生，线形，长 10~35 cm，宽 4~13 mm，脉 15~37 条。花葶 5~20，长 20~60 cm，具 4~6 棱，鞘状苞片长 4~12 cm，口部斜裂，裂片禾叶状；花序熟时近球形，不压扁，灰白色，直径 6.5 mm，基部平截。种子卵形，长 0.6~0.7 mm，表面具横格及 T 字形毛。花果期 8 月至翌年 3 月。

Leaves linear, 10–35 cm × 0.4–1.3 cm, veins 15–37. Scapes 5–20, 20–60 cm, 4–6-ribbed; sheath 4–12 cm; receptacle glabrous; heads subglobose, ca. 6.5 mm in diam, glaucous, base truncate. Seeds ovoid, 0.6–0.7 mm; testa hexagonally reticulate, prickles 1 per cell, T-shaped. Fl. and fr. Aug.–Mar. of next year.

植株　　Whole plant
摄影：丁涛　　Photo by: Ding Tao

花序　　Inflorescences
摄影：丁涛　　Photo by: Ding Tao

43 苍白秤钩风 | cāng bái chèng gōu fēng

Diploclisia glaucescens (Blume) Diels in Engler   防己科 **Menispermaceae**

木质大藤本，长可达20余米或更长。直径达10余厘米。叶柄通常比叶片长很多，叶片厚革质，下面常有白霜。圆锥花序狭而长，常几个至多个簇生于老茎和老枝上；花淡黄色，微香。核果黄红色，长圆状狭倒卵圆形，下部微弯。花期4月，果期8月。

Large woody vines, to 20 m or longer. Stems up to 10 cm in diam.. Petiole usually much longer than lamina. Leaf blade not peltate to conspicuously peltate, glaucescent abaxially, leathery. Inflorescences cauliflorous, on old leafless stems, panicles, usually several to many fascicled; flowers light yellow, slightly fragrant. Drupes yellowish red, narrowly oblong-obovate, base curved. Fl. Apr., fr. Aug..

叶背　　　　Leaf backs
摄影：丁涛　　Photo by: Ding Tao

果序　　　　Infructescence
摄影：丁涛　　Photo by: Ding Tao

44 天仙藤 | tiān xiān téng （大黄藤）

Fibraurea recisa Pierre   防己科 **Menispermaceae**

木质大藤本，长可达10余米或更长。小枝和叶柄具直纹。叶革质，长圆状卵形，长约10～25 cm，宽约2.5～9（～13）cm，顶端近骤尖或短渐尖，基部圆或钝，有时近心形或楔形，两面无毛。圆锥花序生无叶老枝或老茎上。核果长圆状椭圆形，黄色。花期春夏季，果期秋季。

Large woody vines up to 10 m or longer. Branchlets and petioles longitudinally striate. Leaf blade oblong-ovate, 10–25 × 2.5–9 (–13) cm, leathery, glabrous, base rounded or obtuse, sometimes subcordate or cuneate, apex subcuspidate or acutely acuminate. Inflorescences arising from leafless old stems, paniculate. Drupes yellow, oblong-elliptic. Fl. spring and summer, fr. autumn.

叶背　　　　Leaf backs
摄影：丁涛　　Photo by: Ding Tao

枝叶　　　　Branch and leaves
摄影：丁涛　　Photo by: Ding Tao

45 锡叶藤 | xī yè téng （老糠藤）

Tetracera sarmentosa (L.) Vahl 五桠果科 Dilleniaceae

锡叶藤属

常绿木质藤本，长达20m或更长。多分枝，枝条粗糙，幼嫩时被毛，老枝秃净。叶革质，极粗糙，圆形，长4~12cm，宽2~5cm，先端钝或圆，有时略尖，基部阔楔形或近圆形，通常斜，初时有刚毛，全缘或上半部有小钝齿。果实长约1cm，成熟时黄红色。花期4~5月，果期7~12月。

Evergreen woody climbers to 20 m, ramose. Branchlets scabrous, hairy when young, later glabrous. Leaf blade orbicular, 4–12×2–5 cm, leathery, very scabrous, setose when young, base broadly cuneate or approximately rounded, usually oblique, margin entire or finely serrate distally, apex obtuse or rounded, sometimes slightly acute. Follicles ca. 1 cm, orange. Fl. Apr.–May, fr. Jul.–Dec..

叶 / Leaf 摄影：丁涛 / Photo by: Ding Tao

枝叶 / Branch and leaves 摄影：丁涛 / Photo by: Ding Tao

46 广东蛇葡萄 | guǎng dōng shé pú táo

Ampelopsis cantoniensis (Hook. Arn.) K. Koch 葡萄科 Vitaceae

蛇葡萄属

藤本。小枝圆柱形，有纵棱纹。卷须2叉分枝。叶为二回羽状复叶或小枝上部着生有一回羽状复叶，羽状复叶者基部一对小叶常为3小叶，通常卵形、卵椭圆形或长椭圆形，长3~11cm，宽1.5~6cm，顶端急尖、渐尖或骤尾尖，基部多为阔楔形。花序为伞房状多歧聚伞花序。果实近球形。花期4~7月，果期8~11月。

Lianas。Branchlets terete, with longitudinal ridges. Tendrils bifurcate. Leaves bipinnate, or pinnate on upper branches, basal pinnae of bipinnate leaves usually 3-foliolate. Leaflets ovate, ovate-elliptic, or oblong, 3–11 × 1.5–6 cm, base truncate, apex acute, acuminate, or cuspidate. Inflorescence corymbose. Berry globose. Fl. Apr.–Jul., fr. Aug.–Nov..

复叶 / Compound leaf 摄影：丁涛 / Photo by: Ding Tao

枝叶 / Branch and leaves 摄影：丁涛 / Photo by: Ding Tao

47 藤黄檀 | téng huáng tán （大香藤）　　　　黄檀属

***Dalbergia hancei* Benth.**　　　豆科 **Fabaceae**

木质藤本。枝纤细，幼枝略被柔毛，小枝有时变钩状或旋扭。羽状复叶长 5～8 cm。小叶 7～13 对，较小狭长圆或倒卵状长圆形，长 10～20 mm，宽 5～10 mm，先端钝或圆，微缺，基部圆或阔楔形。花小，在腋生，紧密，短圆锥花序。荚果扁平，长圆形或带状。花期 3～5 月，果期 6～11 月。

Woody climbers. Branches slender; young shoots slightly pubescent; branchlets sometimes hooked or twisted. Leaves 5–8 cm. Leaflets 7–13, narrowly oblong or obovate-oblong, 10–20 × 5–10 mm, base rounded or broadly cuneate, apex obtuse or rounded, emarginate. Flowers small, in axillary, compact, short panicles. Legume dis-tinctly stipitate, oblong or strap-shaped. Fl. Mar.–May, fr. Jun.–Nov..

复叶　Compound leaf　摄影：丁涛　Photo by: Ding Tao

枝叶　Branch and leaves　摄影：丁涛　Photo by: Ding Tao

48 粉叶鱼藤 | fěn yè yú téng （粉背鱼藤）　　　　鱼藤属

***Derris glauca* Merr. Chun**　　　豆科 **Fabaceae**

木质藤本。枝、芽有黄色柔毛，小枝变无毛或被极稀疏柔毛，有小瘤体。羽状复叶；小叶 9～13 对，膜质，倒卵状长圆形，长 5～7 cm，宽 2～3.5 cm，先端尾状渐尖，基部楔形或阔楔形。聚伞花序组成圆锥花序。荚果薄，长椭圆形或舌状，长 4～8 cm，宽 1.5～2.5 cm。花期 4～5 月，果期 7～8 月。

Woody lianas. Branches and young shoots yellowish pubescent; branchlets tuberculate, very sparse pilose or glabrescent. Leaves 9–13-foliolate. Leaflet blades obovate-oblong, 5–7 × 2–3.5 cm, membranous, base cuneate to broadly cuneate, apex caudate-acuminate. Cymose pseudopanicles. Legume oblong to ligulate, 4–8 × 1.5–2.5 cm, thin. Fl. Apr.–May, fr. Jul.–Aug..

枝叶　Branch and leaves　摄影：丁涛　Photo by: Ding Tao

花枝　Flowering branches　摄影：丁涛　Photo by: Ding Tao

49　千斤拔 | qiān jīn bá　（蔓性千斤拔）

千斤拔属

***Flemingia prostrata* Roxb. f. ex Roxb.**　豆科 **Fabaceae**

直立灌木。幼枝三棱柱状，密被灰褐色短柔毛。叶具指状 3 小叶；叶柄长 1.5～2.5 cm；小叶厚纸质，长椭圆形或卵状披针形，偏斜长 4～7 cm，宽 1.5～3 cm，先端钝，有时有小凸尖，基部圆形，上面被疏短柔毛，背面密被灰褐色柔毛。总状花序腋生，荚果椭圆状，被短柔毛。花期 3～6 月，果期 5～10 月。

Subshrubs, erect. Young branchlets trigonous-prismatic, densely pubescent. Leaves digitately 3-foliolate; petiloe 1.5-2.5 cm; petiolules extremely short, densely pubes-cent; terminal leaflet oblong or ovate-lanceolate, 4–7 × 1.5–3 cm, thickly papery, sparsely pubescent, base rounded, apex obtuse, sometimes with small mucro. Raceme axillary. Legume elliptic. Fl. Mar.–Jun., fr. May–Oct..

复叶　Compound leaf
摄影：丁涛　Photo by: Ding Tao

枝叶　Branch and leaves
摄影：丁涛　Photo by: Ding Tao

50　葛 | gé

葛属

***Pueraria montana* var. *lobata* (Willd.) Maesen S. M. Almeida ex Sanjappa Predeep**　豆科 **Fabaceae**

粗壮藤本，长可达 8 m，全体被黄色长硬毛，茎基部木质，有粗厚的块状根。羽状复叶具 3 小叶；小叶三裂，顶生小叶宽卵形或斜卵形，长 7～15（～19）cm，宽 5～12（～18）cm，先端长渐尖，侧生小叶斜卵形，稍小。总状花序长 15～30 cm。荚果长椭圆形，扁平，被褐色长硬毛。花期 7～10 月，果期 10～12 月。

Robust lianas. Stems to 8 m, woodyat base, hirsute with yellowish hairs in all parts. Leaflets 3-lobed, terminal onebroadly ovate, 7–15 (–19) × 5–12 (–18) cm, apex acuminate, lateral ones obliquely ovate, smaller. Racemes 15–30 cm. Legumes long elliptic , flattened, brown hirsute. Fl. Jul.–Oct., fr. Oct.–Dec..

复叶　Compound leaf
摄影：丁涛　Photo by: Ding Tao

枝叶　Branch and leaves
摄影：丁涛　Photo by: Ding Tao

51 葫芦茶 | hú lú chá

Tadehagi triquetrum (L.) H. Ohashi

葫芦茶属　豆科 Fabaceae

灌木或亚灌木。茎直立，高 1~2 m。叶仅具单小叶；叶柄长 1~3 cm，翅宽 4~8 mm；小叶纸质，狭披针形至卵状披针形，长 5.8~13 cm，宽 1.1~3.5 cm，先端急尖，基部圆形或浅心形。总状花序长 15~30 cm。荚果长 2~5 cm，宽 5 mm，无网脉，有荚节 5~8，荚节近方形。花期 6~10 月，果期 10~12 月。

Shrubs or subshrubs. Stem erect, 1–2 m tall. Leaves 1-foliolate; petiole 1–3 cm, wing 4–8 mm wide; blade narrowly lanceolate to ovate-lanceolate, 5.8–13 × 1.1–3.5 cm, base rounded or shallowly cordate, apex acute or acuminate. Inflorescences 15–30 cm. Legume 5–8-jointed, 2-5 cm × 0.5 cm; articles not reticulate veined. Fl. Jun.–Oct., fr. Oct.–Dec..

花　Flowers　摄影：丁涛　Photo by: Ding Tao

花枝　Flowering branch　摄影：丁涛　Photo by: Ding Tao

52 蛇藨筋 | shé biāo jīn （蛇泡筋）

Rubus cochinchinensis Tratt.

悬钩子属　蔷薇科 Rosaceae

攀援灌木。掌状复叶常具 5 小叶，上部有时具 3 小叶，小叶片椭圆形、倒卵状椭圆形或椭圆状披针形，长 5~10（~15）cm，宽 2~3.5（~5）cm，顶生小叶比侧生者稍宽大，顶端短渐尖，基部楔形，边缘有不整齐锐锯齿。花成顶生圆锥花序。果实球形，幼时红色，熟时变黑色。花期 3~5 月，果期 7~8 月。

Shrubs climbing. Leaves palmately compound, 5-foliolate, sometimes 3-foliolate; blade of leaflets elliptic, obovate-elliptic, or elliptic-lanceolate, 5–10 (–15) × 2–3.5 (–5) cm, central leaflet slightly larger than lateral leaflets, base cuneate, margin irregularly sharply serrate, apex shortly acuminate. Inflorescences terminal ones cymose paniculate. Aggregate fruit red when immature, black at maturity, globose. Fl. Mar.–May, fr. Jul.–Aug..

复叶　Compound leaf　摄影：丁涛　Photo by: Ding Tao

花序　Inflorescence　摄影：丁涛　Photo by: Ding Tao

53 藤构 | téng gòu

Broussonetia kaempteri auct. non Siebold: Merr. Chun

构属　桑科 Moraceae

蔓生藤状灌木；树皮黑褐色。叶互生，螺旋状排列，近对称的卵状椭圆形，长 3.5～8 cm，宽 2～3 cm，先端渐尖至尾尖，基部心形或截形，边缘锯齿细，齿尖具腺体，不裂，稀为 2～3 裂，表面无毛，稍粗糙。雄花序短穗状，雌花集生为球形头状花序。聚花果直径 1 cm。花期 4～6 月，果期 5～7 月。

Shrubs, scandent. Bark blackish brown. Leaves spirally arranged. Leaf blade ± ovate-elliptic, simple or occasionally 2- or 3-lobed, 3.5–8 × 2–3 cm, scabrous and glabrous, base cordate to cuneate, margin finely serrate with glandular serrations at apex, apex attenuate to shortly acuminate. Male inflorescences spicate, female inflorescences globose. Syncarp ca. 1 cm in diam.. Fl. Apr.–Jun., fr. May–Jul..

叶背　Leaf back　摄影：丁涛　Photo by: Ding Tao

枝叶　Branch and leaves　摄影：丁涛　Photo by: Ding Tao

54 滇南赤车 | diān nán chì chē （波缘赤车）

Pellionia paucidentata (H. Schroet.) S. S. Chien

赤车属　荨麻科 Urticaceae

多年生草本。叶互生。叶片纸质，斜长椭圆形或斜倒披针形，长 2～15.5 cm，宽 1.5～6.5 cm，顶端骤尖或渐尖，基部斜楔形，钟乳体明显，密，有半离基三出脉。花序雌雄同株或异株。雄花序花梗长 2～7 cm，雌花序花梗长 0.2～5.5 cm。瘦果椭圆球形，有小瘤状突起。花期 4～11 月，果期 10～11 月。

Herbs perennial. Leaves alternate. Leaf blade papery, obliquely elliptic or oblanceolate, 2–15.5 × 1.5–6.5 cm, major lateral veins asymmetric, one basal, the other arising above base, base obliquely cuneate, apex cuspidate or acuminate; cystoliths conspicuous, dense or sparse; nanophyll absent. Staminate inflorescences, peduncle 2–7 cm. Pistillate inflorescences, peduncle 0.2–5.5 cm. Achenes ellipsoidal, tuberculate. Fl. Apr.–Dec., fr. Oct.–Nov..

植株　Whole plant　摄影：丁涛　Photo by: Ding Tao

花序　Inflorescence　摄影：丁涛　Photo by: Ding Tao

55 小叶红叶藤 | xiǎo yè hóng yè téng （红叶藤） 红叶藤属

Rourea microphylla (Hook. Arn.) Planch. 牛栓藤科 Connaraceae

木质藤本或攀援灌木，多分枝，高 1~4 m。枝褐色。奇数羽状复叶，小叶通常 7~17 片，小叶片坚纸质至近革质，卵形、披针形或长圆披针形，长 1.5~4 cm，宽 0.5~2 cm，先端渐尖而钝，基部楔形至圆形，常偏斜，全缘。圆锥花序，丛生于叶腋内。蓇葖果椭圆形或斜卵形，成熟时红色。花期 3~9 月，果期 5 月至翌年 3 月。

Woody lianas or climbing shrubs, much branched, 1–4 m tall. Branchlets brown. Leaves odd-pinnate. Leaflets 7–17 paired. Leaflet blade ovate or lanceolate to oblong-lanceolate, 1.5–4 × 0.5–2 cm, papery to subleathery, base cuneate to rounded, often oblique, margin entire, apex obtuse and acuminate. Inflorescences axillary in distal leaf axils. Follicle red when mature, cylindric or obliquely obovoid-cylindric. Fl. Mar.–Sep., fr. May–Mar. of next year.

复叶 / Compound leaves
摄影：丁涛 / Photo by: Ding Tao

枝叶 / Branch and leaves
摄影：丁涛 / Photo by: Ding Tao

56 赤苍藤 | chì cāng téng （蚂蟥藤） 赤苍藤属

Erythropalum scandens Blume 赤苍藤科 Erythropalaceae

常绿藤本，长 5~10 m，无毛。叶纸质至厚纸质或近革质，卵形、长卵形或三角状卵形，长 8~20 cm，宽 4~15 cm，顶端渐尖，钝尖或突尖，稀为圆形，基部变化大，微心形、圆形、截平或宽楔形。花排成腋生的二歧聚伞花序。核果卵状椭圆形或椭圆状，全为增大成壶状的花萼筒所包围。花果期 3~9 月。

Evergreen lianas, 5–10 m tall, glabrous. Leaf blade ovate, oblong-ovate, or triangular-ovate, 8–20 × 4–15 cm, papery to ± leathery, base obtuse, truncate, or ± cordate, and usually peltate, apex acuminate. Cymes, many-flowered. Drupe ellipsoid to obovoid, crowned by persistent calyx. Fl. and fr. Mar.–Sep..

果 / Fruits
摄影：丁涛 / Photo by: Ding Tao

枝叶 / Branch and leaves
摄影：丁涛 / Photo by: Ding Tao

57 异叶蒴莲 | yì yè shuò lián （蒴莲）

蒴莲属

***Adenia heterophylla* (Blume) Koord.**

西番莲科 **Passifloraceae**

草质藤本，长可达 30 m。茎圆柱形。叶长约 11 cm，宽约 10 cm，基部宽截形或宽心形，掌状 3 裂，中间裂片宽披针形，两侧裂片镰刀形，略短于中间裂片；叶柄长顶端与叶背基部之间具 2 个圆而扁平的腺体。花序通常有卷须从中心出现。蒴果，倒卵球形，猩红色，外果皮革质。花果期全年。

Canopy lianas, to 30 m long. Stems terete. Leaf 11cm × 10cm, blade glands often present, dotlike, base of mature leaf oblong, unlobed or (2 or) 3-lobed with distal 1/3 of leaf often slightly constricted from slight lobation. Inflorescences often with tendril emerging from center. Capsules 1–3 per inflorescence, outside deep red at dehiscence, ellipsoid. Fruit wall leathery. Fl. and fr. throughout year.

果枝　Fruiting branch　摄影：丁涛　Photo by: Ding Tao

蒴果　Capsule　摄影：丁涛　Photo by: Ding Tao

58 野牡丹 | yě mǔ dān （爆牙狼）

野牡丹属

***Melastoma malabathricum* L.**

野牡丹科 **Melastomataceae**

灌木，高 0.5~1（~5）m，茎钝四棱形或近圆柱形，密被平展的长粗毛及短柔毛。叶片坚纸质，卵形至椭圆形或椭圆状披针形，顶端渐尖，基部圆形或近心形，长 4~14 cm，宽 1.7~3.5（~6）cm，5 基出脉。伞房花序生于分枝顶端，具花 3~7 朵。蒴果坛状球形，密被鳞片状糙伏毛。花期 2~8 月，果期 7~12 月。

Shrubs 0.5–1 (–5) m tall. Stems 4-sided to subterete, procumbent, densely covered with appressed scales. leaf blade ovate, elliptic, or elliptic-lanceolate, 4–14 × 1.7–3.5 (–6) cm, stiffly papery, base rounded to subcordate, margin entire, apex acuminate. Inflorescences subcapitate corymbose, terminal, 3–7-flowered. Fruit urceolate-globular, densely squamose strigose. Fl. Feb.–Aug., fr. Jul.–Dec..

果序　Infructescence　摄影：丁涛　Photo by: Ding Tao

花　Flower　摄影：丁涛　Photo by: Ding Tao

59 刺果藤 | cì guǒ téng （大滑藤）

刺果藤属

***Byttneria grandifolia* DC.**

锦葵科 **Malvaceae**

木质大藤本，小枝幼嫩略被短柔毛。叶柄长 2~8 cm，被毛。叶广卵形、心形或近圆形，长 7~23 cm，宽 5.5~16 cm，顶端钝或急尖，基部心形，上面几无毛，下面被白色星状短柔毛，基生脉 5 条。花小，淡黄白色，内面略带紫红色。蒴果圆球形或卵状圆球形，直径 3~4 cm，具短而粗的刺。花期春夏季，果期 7 月至翌年 4 月。

Woody, big lianas. Branchlets sparsely puberulent when young. Petiole 2–8 cm, hairy. Leaf blade broadly ovate, cordate, or nearly orbicular, 7–23 × 5.5–16 cm, abaxially white stellate puberulent, adaxially glabrous, basal veins 5, base cordate, margin entire, apex obtuse or acute. Petals yellowish white, and purple-red adaxially. Capsule globose or ovoid-globose, 3–4 cm in diam., spiny, spines short and robust. Fl. spring and summer, fr. Jul.-Apr. of next year.

植株 Whole plant
摄影：丁涛　Photo by: Ding Tao

叶背 Leaf backs
摄影：丁涛　Photo by: Ding Tao

60 黄花稔 | huáng huā rěn （黄花稔）

黄花稔属

***Sida acuta* Burm. f.**

锦葵科 **Malvaceae**

直立亚灌木状草本，高 1~2 m。分枝多，小枝被柔毛至近无毛。叶披针形，长 2~5 cm，宽 4~10 mm，先端短尖或渐尖，基部圆或钝，具锯齿，两面均无毛或疏被星状柔毛，上面偶被单毛。花单朵或成对生于叶腋。蒴果近圆球形，分果爿（4~）6（~9），顶端具 2 短芒。花期冬春季，果期 4~11 月。

Subshrubs or herbs erect, 1–2 m tall. Branchlets pilose or subglabrous. Leaves lanceolate, or linear-lanceolate, 2–5 × 0.4–1 cm, both surfaces glabrous or sparsely stellate pilose, rarely with simple hairs adaxially, base obtuse, margin dentate, apex acute or acuminate. Flowers solitary or paired, axillary. Schizocarp nearly globose; mericarps (4–) 6 (–9),± extending into 2 awns. Fl. winter-spring, Fr. Apr.-Nov..

花 Flower
摄影：丁涛　Photo by: Ding Tao

花枝 Flowering branch
摄影：丁涛　Photo by: Ding Tao

61 寄生藤 | jì shēng téng （叉脉寄生藤） 寄生藤属

Dendrotrophe varians (Blume) Miq. 檀香科 Santalaceae

木质藤本，常呈灌木状。叶厚，倒卵形至阔椭圆形，长 3~7 cm，宽（1.4~）2~4.5 cm，顶端圆钝，有短尖，基部收狭而下延成叶柄，基出脉 3~5 条，侧脉大致沿边缘内侧分出；叶柄长 0.5~1 cm，扁平。雄花 5~6 朵集成聚伞状花序；雌花序通常单生。核果卵状或卵圆形，带红色。花期 1~3 月，果期 6~8 月。

Woody vines, usually shrubby. Petiole flat, 5–10 mm. Leaf blade obovate to broadly elliptic, 3–7 × (1.4-) 2–4.5 cm, thick, ± leathery, basal veins 3–5, prominent when dry, base narrowed and decurrent to form petiole, less often rounded, apex obtuse. Male inflorescences umbellate or cymose, 5- or 6-flowered. Female inflorescences usually of solitary flowers. Drupe reddish, ovoid. Fl. Jan.–Mar., fr. Jun.–Aug..

果 Fruit
摄影：丁涛　Photo by: Ding Tao

果枝 Fruiting branch
摄影：丁涛　Photo by: Ding Tao

62 火炭母 | huǒ tàn mǔ （硬毛火炭母） 蓼属

Persicaria chinensis (L.) H. Gross 蓼科 Polygonaceae

多年生草本。叶卵形或长卵形，长 4~16 cm，宽 1.5~8 cm，顶端短渐尖，基部截形或宽心形，边缘全缘，下部叶具叶柄，叶柄长 1~2 cm，通常基部具叶耳，上部叶近无柄或抱茎；托叶鞘膜质，无毛，具脉纹，顶端偏斜，无缘毛。头状花序，通常数个排成圆锥状，顶生或腋生。花期 7~11 月，果期 7~12 月。

Herbs perennial. Petiole 1–2 cm, usually auriculate at base, upper leaves subsessile. Leaf blade ovate, elliptic, or lanceolate, 4–16 × 1.5–8 cm cm, base truncate or broadly cordate, margin entire, apex shortly acuminate; ocrea tubular, membranous, glabrous, much veined, apex oblique, not ciliate. Inflorescence terminal or axillary, capitate. Achenes black, broadly ovoid, trigonous. Fl. Jul.–Nov., fr. Jul.–Dec..

植株 Whole plant
摄影：丁涛　Photo by: Ding Tao

叶 Leaf
摄影：丁涛　Photo by: Ding Tao

63 朱砂根 | zhū shā gēn （硃砂根） 紫金牛属

Ardisia crenata Sims 报春花科 Primulaceae

灌木，高 1~1.5（~3）m。叶柄长 6~10 mm，无毛。叶片革质或坚纸质，椭圆形、椭圆状披针形至倒披针形，顶端急尖或渐尖，基部楔形，长 7~15 cm，宽 2~4 cm，边缘具皱波状或波状齿，具明显的边缘腺点，两面无毛。伞形花序或聚伞花序。果球形，鲜红色，具腺点。花期 5~6 月，果期 10~12 月。

Shrubs 1–1.5 (–3) m tall. Petiole narrowly marginate, 6–10 mm, glabrous. Leaf blade elliptic, narrowly lanceolate, or oblanceolate, 7–15 × 2–4 cm, leathery or papery, prominently punctate, base cuneate, margin subrevolute, crenate, or undulate, apex acute or acuminate. Inflorescences terminal, umbellate or cymose. Fruit red, globose, punctate. Fl. May–Jun., fr. Oct.–Dec..

植株　Whole plant
摄影：丁涛　Photo by: Ding Tao

叶　Leaves
摄影：丁涛　Photo by: Ding Tao

64 密齿酸藤子 | mì chǐ suān téng zǐ （长圆叶酸藤子） 酸藤子属

Embelia vestita Roxb. 报春花科 Primulaceae

攀援灌木。小枝弯曲，圆柱状。叶片坚纸质，卵形至卵状长圆形，稀椭圆状披针形，顶端急尖、渐尖或钝，基部楔形或圆形，长（3.5~）7~11（~18）cm，宽（1.3~）2~4（~7.5）cm，边缘具细锯齿，稀成重锯齿；叶柄两侧微折皱。总状花序，腋生。果球形或略扁，红色，具腺点。花期 10~11 月，果期 10 月至翌年 7 月。

Shrubs scandent。Branchlets flexuous, terete. Petiole canaliculate and marginate, margin. Leaf blade oblong or lanceolate to ovate, (3.5–) 7–11 (–18) × (1.3–) 2–4 (–7.5) cm, papery to thin leathery, glossy, glabrous, red or black punctate, base obtuse, rounded or truncate, margin serrate or serrulate, apex acute to acuminate. Inflorescences racemose or rarely panicle of 2 racemes. Fruit red, globose, red punctate. Fl. Oct.–Dec., fr. Oct.–Jul. of next year.

叶背　Leaf backs
摄影：丁涛　Photo by: Ding Tao

枝叶　Branch and leaves
摄影：丁涛　Photo by: Ding Tao

65 中越猕猴桃 | zhōng yuè mí hóu táo

Actinidia indochinensis Merr. 猕猴桃科 Actinidiaceae

大型落叶木质藤本。叶幼时膜质，老时软革质，卵形至椭圆形、或矩状卵形至矩状椭圆形，长 4~10 cm，宽 3.5~5 cm，顶端钝至钝短尖或急尖至渐尖，基部圆形、钝形至阔楔形，两侧基本对称，边缘有小锯齿或略呈圆齿状小锯齿。聚伞花序。果近球形，皮孔黄棕色。花期 3~4 月，果期 9~10 月。

Woody vines, large, deciduous. Leaf blade ovate to elliptic to oblong-ovate or oblong-elliptic, 4–10 × 3.5–5 cm, membranous when young, leathery when old, base broadly cuneate to rounded, margin subentire to inconspicuously and remotely crenate toward apex, apex obtuse to acute to acuminate. Inflorescences cymose. Fruit subglobose, lenticels yellowish brown. Fl. Mar.–Apr., fr. Sep.–Oct..

果 / Fruit — 摄影：丁涛 / Photo by: Ding Tao
枝叶 / Branch and leaves — 摄影：丁涛 / Photo by: Ding Tao

66 白花苦灯笼 | bái huā kǔ dēng lóng （乌口树）

Tarenna mollissima (Hook. Arn.) B. L. Rob. 茜草科 Rubiaceae

灌木或小乔木，高 1~6 m。全株密被灰色或褐色柔毛或短绒毛，但老枝毛渐脱落。叶纸质，披针形、长圆状披针形或卵状椭圆形，长 4.5~22.5 cm，宽 1~10 cm，顶端渐尖或长渐尖，基部楔尖、短尖或钝圆，干后变黑褐色，顶端尖。伞房状的聚伞花序顶生。果近球形。花期 5~7 月，果期 5 月至翌年 2 月。

Shrubs or small trees, 1–6 m tall. Branches densely gray or brown pilosulous or tomentulose, becoming glabrescent when old. leaf blade drying papery and blackish brown, lanceolate, oblong-lanceolate, or ovate-elliptic, 4.5–22.5 × 1–10 cm, base cuneate, acute, or obtuse, apex acuminate or long acuminate; acute to cuspidate. Inflorescences corymbose. Berry subglobose. Fl. May–Jul., fr. May–Jan. of next year.

叶背 / Leaf backs — 摄影：丁涛 / Photo by: Ding Tao
果序 / Infructescence — 摄影：丁涛 / Photo by: Ding Tao

67 海南链珠藤 | hǎi nán liàn zhū téng
链珠藤属

Alyxia hainanensis Merr. Chun 夹竹桃科 **Apocynaceae**

木质藤本，长达 4 m。小枝压扁或方形后成圆形。叶对生或 3 叶轮生，坚纸质，椭圆形至长圆形，顶端渐尖，稀钝头，基部急尖或近圆形，边缘微向外卷，长 2～12 cm，宽 1～4.5 cm。花序腋生或近顶生的花束。核果近球形，通常长圆状椭圆形。花期 3～10 月，果期 6～12 月。

Woody vines, to 4 m. Branches slightly angled when young, later terete. Leaves opposite or in whorls of 3. Leaf blade elliptic, oblong, narrowly elliptic, or obovate, 2–12 × 1–4.5 cm, papery, glabrous, apex acute or short acuminate. Cymes fascicled, terminal and axillary. Fruit ellipsoid-globose. Fl. Mar.–Oct., fr. Jun.–Dec..

果枝　　　Fruiting branch
摄影：丁涛　Photo by：Ding Tao

枝叶　　　Branch and leaves
摄影：丁涛　Photo by：Ding Tao

68 头花银背藤 | tóu huā yín bèi téng
银背藤属

Argyreia capitiformis (Poir.) Ooststr. 旋花科 **Convolvulaceae**

木质藤本，长 10～15 m。茎被褐色或黄色开展的长硬毛。叶柄长 3～16 cm。叶卵形至圆形，稀长圆状披针形，长 8～18 cm，宽 4～13 cm，先端锐尖或渐尖，基部心形，两面被黄色长硬毛，侧脉 13～15 对，开展。聚伞花序密集成头状。果球形，橙红色。花期 9～12 月，果期 2 月。

Woody vines, 10–15 m tall. Stems spreading hirsute, with brown or dull yellow hairs. Petiole 3–16 cm. Leaf blade ovate to circular, rarely oblong-lanceolate, 8–18×4–13 cm, dull yellow hirsute, base cordate, apex acute or acuminate; lateral veins 13–15 pairs. Cymes capitate, dense. Berries orange-red, globose. Fl. Sep.–Dec., fr. Feb..

叶　　　Leaf
摄影：丁涛　Photo by：Ding Tao

枝　　　Branch
摄影：丁涛　Photo by：Ding Tao

69 飞蛾藤 | fēi é téng

飞蛾藤属

Dinetus racemosus (Wall.) Buch.-Ham. ex Sweet　　旋花科 **Convolvulaceae**

草质藤本，被淡黄到银色硬毛。叶柄 2.9～7.7 cm。叶卵形，长 6～16.7 cm，宽 3.3～9.4 cm，先端渐尖或尾状，具钝或锐尖的尖头，基部深心形；两面极疏被紧贴疏柔毛，背面稍密，稀被短柔毛至绒毛。圆锥花序腋生，长 13～45cm。蒴果卵形，具小短尖头。花期夏天至秋天，果期秋天至冬天。

Herbaceous vines; indumentum yellowish to silvery. Petiole 2.9–7.7 cm. Leaf blade deeply cordate, 6–16.7×3.3–9.4 cm, abaxially puberulent to tomentellous, adaxially strigose. Panicle axillary, 13–45 cm; Fruit slenderly ellipsoid-obovoid, glabrous, apex acute or apiculate. Fl. summer-autumn, fr. autumn-winter.

叶　Leaf　摄影：丁涛　Photo by: Ding Tao

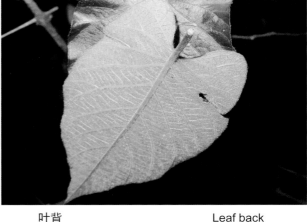

叶背　Leaf back　摄影：丁涛　Photo by: Ding Tao

70 匍茎短筒苣苔 | pú jīng duǎn tǒng jù tái

短筒苣苔属

Boeica stolonifera K. Y. Pan　　苦苣苔科 **Boeica**

草本。茎长 0.5～7cm。叶在茎顶端聚生，具柄；叶片宽椭圆形，长 3.5～10cm，宽 2.8～5.5cm，顶端锐尖，基部楔形，边缘具细钝齿，上面具较密的褐色长柔毛，下面被长柔毛和短柔毛。聚伞花序，具 5～15 花。蒴果长 1.8～2.2cm。花期 5～7 月，果期 7 月。

Herbs. Stems 0.5–7 cm. Leaf blade broadly elliptic or ovate to obovate, 3.5–10×2.8–5.5 cm, adaxially villous, abaxially villous and puberulent, densely villous along veins, base cuneate to rounded, margin serrulate to serrate, apex acute to obtuse. Cymes 5–15-flowered. Capsule 1.8–2.2 cm. Fl. May–Jul., fr. Jul.

植株　Whole plant　摄影：丁涛　Photo by: Ding Tao

花　Flower　摄影：丁涛　Photo by: Ding Tao

71 毛麝香 | máo shè xiāng （土茵陈）

***Adenosma glutinosum* (L.) Druce**　　车前科 **Plantaginaceae**

草本，高 30～100 cm。密被多细胞长柔毛和腺毛。叶对生，上部的多少互生，有长 3～10（～15）mm 的柄；叶片披针状卵形至宽卵形，长 2～10 cm，宽 1～5 cm，先端锐尖，基部楔形至截形或亚心形。花单生叶腋或在茎、枝顶端集成较密的总状花序。蒴果卵形。花果期 7～10 月。

Herbs, 30–100 cm tall. Stems erect, densely villous with eglandular and glandular hairs. Petiole 3–10 (–15) mm. Leaf blade lanceolate-ovate to broadly ovate, 2–10×1–5 cm; base cuneate, truncate, or subcordate, margin irregularly serrate and sometimes double serrate, apex acute. Flowers axillary and solitary or in dense racemes apically on stems and branches. Capsule ovoid. Fl. and fr. Jul.–Oct..

花　　Flower　　摄影：丁涛　　Photo by: Ding Tao

花枝　　Flowering branch　　摄影：丁涛　　Photo by: Ding Tao

72 曲枝假蓝 | qū zhī jiǎ lán

***Strobilanthes dalzielii* (W. W. Sm.) Benoist**　　爵床科 **Acanthaceae**

草本或灌木，茎直立，高 0.4～1 m，枝细瘦，"之"字形曲折。上部叶无柄或近无柄，近相等或极不等，大叶长 9～14 cm，宽 3～5 cm，小叶长 2～5 cm，宽 1～2 cm，卵形或卵状披针形，先端渐尖或急尖或稀钝，边缘疏锯齿。顶生花序和上部腋生穗状花序。蒴果线状长圆形，两侧压扁。花期 10 月至翌年 1 月，果期 7～10 月。

Subshrubs or perennial herbs, 40–100 cm tall, branched, strongly anisophyllous. Stems slender, erect, zigzag. Leaf blade ovate to ovate-lanceolate, smaller of pair 2–5 × 1–2 cm and larger one 9–14 × 3–5 cm, base rounded but cordate for apical leaves, margin serrate, apex acuminate to acute and sometimes falcate. Inflorescences axillary or terminal, spikes. Capsule linear-oblong, compressed, glabrous. Fl. Oct.–Jan. of next year, fr. Jul.–Oct..

枝叶　　Branch and leaves　　摄影：丁涛　　Photo by: Ding Tao

花　　Flowers　　摄影：丁涛　　Photo by: Ding Tao

附录 I 植物名录
Appendix I Plant Name Checklist

里白科 Gleicheniaceae
  芒萁属 Dicranopteris
  铁芒萁 Dicranopteris linearis
海金沙科 Lygodiaceae
  海金沙属 Lygodium
  曲轴海金沙 Lygodium flexuosum
鳞始蕨科 Lindsaeaceae
  鳞始蕨属 Lindsaea
  团叶鳞始蕨 Lindsaea orbiculata
碗蕨科 Dennstaedtiaceae
  鳞盖蕨属 Microlepia
  边缘鳞盖蕨 Microlepia marginata
凤尾蕨科 Pteridaceae
  铁线蕨属 Adiantum
  扇叶铁线蕨 Adiantum flabellulatum
  凤尾蕨属 Pteris
  剑叶凤尾蕨 Pteris ensiformis
  半边旗 Pteris semipinnata
  林下凤尾蕨 Pteris grevilleana
  百越凤尾蕨 Pteris fauriei var. chinensis
铁角蕨科 Aspleniaceae
  铁角蕨属 Asplenium
  长叶铁角蕨 Asplenium prolongatum
金星蕨科 Thelypteridaceae
  毛蕨属 Cyclosorus
  华南毛蕨 Cyclosorus parasiticus
乌毛蕨科 Blechnaceae
  乌毛蕨属 Blechnum
  乌毛蕨 Blechnum orientale
鳞毛蕨科 Dryopteridaceae
  复叶耳蕨属 Arachniodes
  中华复叶耳蕨 Arachniodes chinensis
肾蕨科 Nephrolepidaceae
  肾蕨属 Nephrolepis
  肾蕨 Nephrolepis cordifolia
三叉蕨科 Tectariaceae
  三叉蕨属 Tectaria
  三叉蕨 Tectaria subtriphylla
骨碎补科 Davalliaceae
  骨碎补属 Davallia
  大叶骨碎补 Davallia divaricata
水龙骨科 Polypodiaceae
  槲蕨属 Drynaria
  团叶槲蕨 Drynaria bonii
  伏石蕨属 Lemmaphyllum
  抱针骨牌蕨 Lemmaphyllum diversum
  薄唇蕨属 Leptochilus
  宽羽线蕨 Leptochilus ellipticus var. pothifolius
  绿叶线蕨 Leptochilus leveillei
买麻藤科 Gnetaceae
  买麻藤属 Gnetum
  买麻藤 Gnetum montanum
松科 Pinaceae
  松属 Pinus
  马尾松 Pinus massoniana
柏科 Cupressaceae
  杉木属 Cunninghamia
  杉木 Cunninghamia lanceolata
五味子科 Schisandraceae
  八角属 Illicium
  八角 Illicium verum
金粟兰科 Chloranthaceae
  草珊瑚属 Sarcandra
  草珊瑚 Sarcandra glabra
胡椒科 Piperaceae
  胡椒属 Piper
  石南藤 Piper wallichii
木兰科 Magnoliaceae
  长喙木兰属 Lirianthe
  香港木兰 Lirianthe championii
番荔枝科 Annonaceae
  假鹰爪属 Desmos
  假鹰爪 Desmos chinensis
  瓜馥木属 Fissistigma
  白叶瓜馥木 Fissistigma glaucescens
  紫玉盘属 Uvaria
  紫玉盘 Uvaria macrophylla
莲叶桐科 Hernandiaceae
  青藤属 Illigera
  心叶青藤 Illigera cordata
樟科 Lauraceae
  樟属 Cinnamomum
  黄樟 Cinnamomum parthenoxylon
  山胡椒属 Lindera
  滇粤山胡椒 Lindera metcalfiana
  香叶树 Lindera communis
  鼎湖钓樟 Lindera chunii
  木姜子属 Litsea
  假柿木姜子 Litsea monopetala
  尖脉木姜子 Litsea acutivena
  豺皮樟 Litsea rotundifolia var. oblongifolia
  轮叶木姜子 Litsea verticillata
  黄丹木姜子 Litsea elongata
  黄椿木姜子 Litsea variabilis
  润楠属 Machilus
  华润楠 Machilus chinensis
  狭叶润楠 Machilus rehderi
  短序润楠 Machilus breviflora
  绒毛润楠 Machilus velutina
  赛短花润楠 Machilus parabreviflora
  新木姜子属 Neolitsea
  长圆叶新木姜子 Neolitsea oblongifolia
天南星科 Araceae
  魔芋属 Amorphophallus
  魔芋 Amorphophallus konjac
  石柑属 Pothos
  石柑子 Pothos chinensis
阿福花科 Asphodelaceae
  山菅兰属 Dianella
  山菅兰 Dianella ensifolia
百部科 Stemonaceae
  百部属 Stemona
  大百部 Stemona tuberosa
露兜树科 Pandanaceae
  露兜树属 Pandanus
  露兜树 Pandanus tectorius
菝葜科 Smilacaceae
  菝葜属 Smilax
  抱茎菝葜 Smilax ocreata
兰科 Orchidaceae
  竹叶兰属 Arundina
  竹叶兰 Arundina graminifolia
  毛兰属 Eria
  半柱毛兰 Eria corneri
  玉凤花属 Habenaria

橙黄玉凤花 *Habenaria rhodocheila*
羊耳蒜属 *Liparis*
宽叶羊耳蒜 *Liparis latifolia*
长茎羊耳蒜 *Liparis viridiflora*
线柱兰属 *Zeuxine*
宽叶线柱兰 *Zeuxine affinis*
天门冬科 Asparagaceae
　龙血树属 *Dracaena*
　　细枝龙血树 *Dracaena elliptica*
鸭跖草科 Commelinaceae
　水竹叶属 *Murdannia*
　　裸花水竹叶 *Murdannia nudiflora*
　杜若属 *Pollia*
　　长花枝杜若 *Pollia secundiflora*
姜科 Zingiberaceae
　山姜属 *Alpinia*
　　华山姜 *Alpinia oblongifolia*
　　海南山姜 *Alpinia hainanensis*
谷精草科 Eriocaulaceae
　谷精草属 *Eriocaulon*
　　华南谷精草 *Eriocaulon sexangulare*
防己科 Menispermaceae
　秤钩风属 *Diploclisia*
　　苍白秤钩风 *Diploclisia glaucescens*
　天仙藤属 *Fibraurea*
　　天仙藤 *Fibraurea recisa*
清风藤科 Sabiaceae
　泡花树属 *Meliosma*
　　狭叶泡花树 *Meliosma angustifolia*
山龙眼科 Proteaceae
　山龙眼属 *Helicia*
　　小果山龙眼 *Helicia cochinchinensis*
　　枇杷叶山龙眼 *Helicia obovatifolia* var. *mixta*
　假山龙眼属 *Heliciopsis*
　　假山龙眼 *Heliciopsis henryi*
　　调羹树 *Heliciopsis lobata*
五桠果科 Dilleniaceae
　锡叶藤属 *Tetracera*
　　锡叶藤 *Tetracera sarmentosa*
蕈树科 Altingiaceae
　枫香树属 *Liquidambar*
　　枫香树 *Liquidambar formosana*
虎皮楠科 Daphniphyllaceae
　虎皮楠属 *Daphniphyllum*
　　牛耳枫 *Daphniphyllum calycinum*
鼠刺科 Iteaceae

鼠刺属 *Itea*
　鼠刺 *Itea chinensis*
葡萄科 Vitaceae
　蛇葡萄属 *Ampelopsis*
　　广东蛇葡萄 *Ampelopsis cantoniensis*
豆科 Fabaceae
　海红豆属 *Adenanthera*
　　海红豆 *Adenanthera pavonina* var. *microsperma*
　合欢属 *Albizia*
　　楹树 *Albizia chinensis*
　猴耳环属 *Archidendron*
　　亮叶猴耳环 *Archidendron lucidum*
　　碟腺棋子豆 *Archidendron kerrii*
　　薄叶猴耳环 *Archidendron utile*
　黄檀属 *Dalbergia*
　　藤黄檀 *Dalbergia hancei*
　鱼藤属 *Derris*
　　粉叶鱼藤 *Derris glauca*
　千斤拔属 *Flemingia*
　　千斤拔 *Flemingia prostrata*
　老虎刺属 *Pterolobium*
　　老虎刺 *Pterolobium punctatum*
　葛属 *Pueraria*
　　葛 *Pueraria montana* var. *lobata*
　葫芦茶属 *Tadehagi*
　　葫芦茶 *Tadehagi triquetrum*
远志科 Polygalaceae
　黄叶树属 *Xanthophyllum*
　　黄叶树 *Xanthophyllum hainanense*
蔷薇科 Rosaceae
　枇杷属 *Eriobotrya*
　　香花枇杷 *Eriobotrya fragrans*
　臀果木属 *Pygeum*
　　臀果木 *Pygeum topengii*
　石斑木属 *Rhaphiolepis*
　　石斑木 *Rhaphiolepis indica*
　悬钩子属 *Rubus*
　　蛇莓筋 *Rubus cochinchinensis*
大麻科 Cannabaceae
　朴属 *Celtis*
　　假玉桂 *Celtis timorensis*
　　朴树 *Celtis sinensis*
　白颜树属 *Gironniera*
　　白颜树 *Gironniera subaequalis*
桑科 Moraceae

波罗蜜属 *Artocarpus*
　二色波罗蜜 *Artocarpus styracifolius*
构属 *Broussonetia*
　藤构 *Broussonetia kaempteri* auct. non
榕属 *Ficus*
　变叶榕 *Ficus variolosa*
　垂叶榕 *Ficus benjamina*
　大果榕 *Ficus auriculata*
　尖叶榕 *Ficus henryi*
　斜叶榕 *Ficus tinctoria* subsp. *gibbosa*
　杂色榕 *Ficus variegata*
　水同木 *Ficus fistulosa*
　白肉榕 *Ficus vasculosa*
　粗叶榕 *Ficus hirta*
　褐叶榕 *Ficus pubigera*
　青藤公 *Ficus langkokensis*
　黄毛榕 *Ficus esquiroliana*
　黄葛树 *Ficus virens*
橙桑属 *Maclura*
　构棘 *Maclura cochinchinensis*
鹊肾树属 *Streblus*
　刺桑 *Streblus ilicifolius*
荨麻科 Urticaceae
　赤车属 *Pellionia*
　　滇南赤车 *Pellionia paucidentata*
壳斗科 Fagaceae
　栎属 *Quercus*
　　上思青冈 *Quercus delicatula*
卫矛科 Celastraceae
　卫矛属 *Euonymus*
　　疏花卫矛 *Euonymus laxiflorus*
牛栓藤科 Connaraceae
　红叶藤属 *Rourea*
　　小叶红叶藤 *Rourea microphylla*
杜英科 Elaeocarpaceae
　杜英属 *Elaeocarpus*
　　中华杜英 *Elaeocarpus chinensis*
　　山杜英 *Elaeocarpus sylvestris*
　　绢毛杜英 *Elaeocarpus nitentifolius*
小盘木科 Pandaceae
　小盘木属 *Microdesmis*
　　小盘木 *Microdesmis caseariifolia*
红树科 Rhizophoraceae
　竹节树属 *Carallia*
　　旁杞木 *Carallia pectinifolia*
赤苍藤科 Erythropalaceae

赤苍藤属 Erythropalum
　　赤苍藤 Erythropalum scandens
大戟科 Euphorbiaceae
　　巴豆属 Croton
　　　巴豆 Croton tiglium
　　粗毛野桐属 Hancea
　　　粗毛野桐 Hancea hookeriana
　　血桐属 Macaranga
　　　印度血桐 Macaranga indica
　　　轮苞血桐 Macaranga andamanica
　　三宝木属 Trigonostemon
　　　勐仑三宝木 Trigonostemon bonianus
　　油桐属 Vernicia
　　　木油桐 Vernicia montana
叶下珠科 Phyllanthaceae
　　五月茶属 Antidesma
　　　黄毛五月茶 Antidesma fordii
　　银柴属 Aporosa
　　　银柴 Aporosa dioica
　　秋枫属 Bischofia
　　　重阳木 Bischofia polycarpa
　　黑面神属 Breynia
　　　黑面神 Breynia fruticosa
　　土蜜树属 Bridelia
　　　禾串树 Bridelia balansae
　　　膜叶土蜜树 Bridelia glauca
　　算盘子属 Glochidion
　　　毛果算盘子 Glochidion eriocarpum
　　叶下珠属 Phyllanthus
　　　余甘子 Phyllanthus emblica
核果木科 Putranjivaceae
　　核果木属 Drypetes
　　　青枣核果木 Drypetes cumingii
西番莲科 Passifloraceae
　　蒴莲属 Adenia
　　　异叶蒴莲 Adenia heterophylla
杨柳科 Salicaceae
　　山桂花属 Bennettiodendron
　　　山桂花 Bennettiodendron leprosipes
　　脚骨脆属 Casearia
　　　爪哇脚骨脆 Casearia velutina
　　　膜叶脚骨脆 Casearia membranacea
　　箣柊属 Scolopia
　　　箣柊 Scolopia chinensis
　　柞木属 Xylosma
　　　南岭柞木 Xylosma controversa

藤黄科 Clusiaceae
　　藤黄属 Garcinia
　　　岭南山竹子 Garcinia oblongifolia
金丝桃科 Hypericaceae
　　黄牛木属 Cratoxylum
　　　黄牛木 Cratoxylum cochinchinense
桃金娘科 Myrtaceae
　　子楝树属 Decaspermum
　　　子楝树 Decaspermum gracilentum
　　桃金娘属 Rhodomyrtus
　　　桃金娘 Rhodomyrtus tomentosa
　　蒲桃属 Syzygium
　　　子凌蒲桃 Syzygium championii
　　　狭叶蒲桃 Syzygium levinei
　　　红鳞蒲桃 Syzygium hancei
　　　黑嘴蒲桃 Syzygium bullockii
野牡丹科 Melastomataceae
　　柏拉木属 Blastus
　　　柏拉木 Blastus cochinchinensis
　　野牡丹属 Melastoma
　　　毛棯 Melastoma sanguineum
　　　野牡丹 Melastoma malabathricum
　　谷木属 Memecylon
　　　谷木 Memecylon ligustrifolium
橄榄科 Burseraceae
　　橄榄属 Canarium
　　　毛叶榄 Canarium subulatum
漆树科 Anacardiaceae
　　漆树属 Toxicodendron
　　　野漆 Toxicodendron succedaneum
无患子科 Sapindaceae
　　鳞花木属 Lepisanthes
　　　赤才 Lepisanthes rubiginosa
　　韶子属 Nephelium
　　　韶子 Nephelium chryseum
芸香科 Rutaceae
　　山油柑属 Acronychia
　　　山油柑 Acronychia pedunculata
　　黄皮属 Clausena
　　　云南黄皮 Clausena yunnanensis
　　　假黄皮 Clausena excavata
　　山小橘属 Glycosmis
　　　少花山小橘 Glycosmis oligantha
　　蜜茱萸属 Melicope
　　　三桠苦 Melicope pteleifolia
　　九里香属 Murraya

　　　九里香 Murraya exotica
　　吴茱萸属 Tetradium
　　　楝叶吴萸 Tetradium glabrifolium
　　花椒属 Zanthoxylum
　　　两面针 Zanthoxylum nitidum
　　　勒𣗪花椒 Zanthoxylum avicennae
楝科 Meliaceae
　　米仔兰属 Aglaia
　　　米仔兰 Aglaia odorata
锦葵科 Malvaceae
　　刺果藤属 Byttneria
　　　刺果藤 Byttneria grandifolia
　　破布叶属 Microcos
　　　破布叶 Microcos paniculata
　　翅子树属 Pterospermum
　　　翻白叶树 Pterospermum heterophyllum
　　梭罗树属 Reevesia
　　　梭罗树 Reevesia pubescens
　　黄花棯属 Sida
　　　黄花棯 Sida acuta
　　苹婆属 Sterculia
　　　假苹婆 Sterculia lanceolata
檀香科 Santalaceae
　　寄生藤属 Dendrotrophe
　　　寄生藤 Dendrotrophe varians
蓼科 Polygonaceae
　　蓼属 Persicaria
　　　火炭母 Persicaria chinensis
山茱萸科 Cornaceae
　　八角枫属 Alangium
　　　八角枫 Alangium chinense
五列木科 Pentaphylacaceae
　　柃属 Eurya
　　　大叶五室柃 Eurya quinquelocularis
　　　大果毛柃 Eurya megatrichocarpa
　　　细齿叶柃 Eurya nitida
　　　黑柃 Eurya macartneyi
山榄科 Sapotaceae
　　梭子果属 Eberhardtia
　　　锈毛梭子果 Eberhardtia aurata
　　紫荆木属 Madhuca
　　　紫荆木 Madhuca pasquieri
　　肉实树属 Sarcosperma
　　　肉实树 Sarcosperma laurinum
柿科 Ebenaceae
　　柿属 Diospyros

岭南柿 *Diospyros tutcheri*
罗浮柿 *Diospyros morrisiana*

报春花科 Primulaceae
紫金牛属 Ardisia
大罗伞树 *Ardisia hanceana*
朱砂根 *Ardisia crenata*
罗伞树 *Ardisia quinquegona*
酸藤子属 Embelia
密齿酸藤子 *Embelia vestita*
杜茎山属 Maesa
米珍果 *Maesa acuminatissima*
铁仔属 Myrsine
密花树 *Myrsine seguinii*

山茶科 Theaceae
山茶属 Camellia
东兴金花茶 *Camellia indochinensis* var. *tunghinensis*
显脉金花茶 *Camellia euphlebia*
硬叶糙果茶 *Camellia gaudichaudii*
长尾毛蕊茶 *Camellia caudata*
紫茎属 Stewartia
柔毛紫茎 *Stewartia villosa*

山矾科 Symplocaceae
山矾属 Symplocos
光叶山矾 *Symplocos lancifolia*
老鼠屎 *Symplocos stellaris*
薄叶山矾 *Symplocos anomala*

安息香科 Styracaceae
赤杨叶属 Alniphyllum
赤杨叶 *Alniphyllum fortunei*
山茉莉属 Huodendron
西藏山茉莉 *Huodendron tibeticum*

猕猴桃科 Actinidiaceae
猕猴桃属 Actinidia
中越猕猴桃 *Actinidia indochinensis*
水东哥属 Saurauia
水东哥 *Saurauia tristyla*

茜草科 Rubiaceae
水团花属 Adina
水团花 *Adina pilulifera*
茜树属 Aidia
茜树 *Aidia cochinchinensis*
猪肚木属 Canthium
大叶猪肚木 *Canthium simile*
猪肚木 *Canthium horridum*
山石榴属 Catunaregam
山石榴 *Catunaregam spinosa*
龙船花属 Ixora
白花龙船花 *Ixora henryi*
粗叶木属 Lasianthus
斜基粗叶木 *Lasianthus attenuatus*
粗叶木 *Lasianthus chinensis*
西南粗叶木 *Lasianthus henryi*
玉叶金花属 Mussaenda
玉叶金花 *Mussaenda pubescens*
大沙叶属 Pavetta
香港大沙叶 *Pavetta hongkongensis*
南山花属 Prismatomeris
南山花 *Prismatomeris tetrandra*
九节属 Psychotria
九节 *Psychotria asiatica*
黄脉九节 *Psychotria straminea*
鱼骨木属 Psydrax
鱼骨木 *Psydrax dicocca*
裂果金花属 Schizomussaenda
裂果金花 *Schizomussaenda henryi*
乌口树属 Tarenna
白皮乌口树 *Tarenna depauperata*
白花苦灯笼 *Tarenna mollissima*
水锦树属 Wendlandia
水锦树 *Wendlandia uvariifolia*
短花水金京 *Wendlandia formosana* subsp. *breviflora*

夹竹桃科 Apocynaceae
链珠藤属 Alyxia
海南链珠藤 *Alyxia hainanensis*
山橙属 Melodinus
思茅山橙 *Melodinus cochinchinensis*
倒吊笔属 Wrightia
蓝树 *Wrightia laevis*

紫草科 Boraginaceae
厚壳树属 Ehretia
厚壳树 *Ehretia acuminata*
长花厚壳树 *Ehretia longiflora*

旋花科 Convolvulaceae
银背藤属 Argyreia
头花银背藤 *Argyreia capitiformis*
飞蛾藤属 Dinetus
飞蛾藤 *Dinetus racemosus*

木樨科 Oleaceae
木樨属 Osmanthus
牛屎果 *Osmanthus matsumuranus*

苦苣苔科 Boeica
短筒苣苔属 Boeica
匍茎短筒苣苔 *Boeica stolonifera*

车前科 Plantaginaceae
毛麝香属 Adenosma
毛麝香 *Adenosma glutinosum*

唇形科 Lamiaceae
大青属 Clerodendrum
垂茉莉 *Clerodendrum wallichii*
大青 *Clerodendrum cyrtophyllum*
牡荆属 Vitex
山牡荆 *Vitex quinata*
微毛布惊 *Vitex quinata* var. *puberula*

爵床科 Acanthaceae
马蓝属 Strobilanthes
曲枝假蓝 *Strobilanthes dalzielii*

粗丝木科 Stemonuraceae
粗丝木属 Gomphandra
粗丝木 *Gomphandra tetrandra*

冬青科 Aquifoliaceae
冬青属 Ilex
三花冬青 *Ilex triflora*
棱枝冬青 *Ilex angulata*
榕叶冬青 *Ilex ficoidea*
铁冬青 *Ilex rotunda*

五福花科 Adoxaceae
荚蒾属 Viburnum
南方荚蒾 *Viburnum fordiae*
常绿荚蒾 *Viburnum sempervirens*
海南荚蒾 *Viburnum hainanense*
淡黄荚蒾 *Viburnum lutescens*

海桐科 Pittosporaceae
海桐属 Pittosporum
广西海桐 *Pittosporum kwangsiense*
海桐 *Pittosporum tobira*

五加科 Araliaceae
罗伞属 Brassaiopsis
罗伞 *Brassaiopsis glomerulata*
鹅掌柴属 Schefflera
鹅掌柴 *Schefflera heptaphylla*

附录 II 植物中文名索引
Appendix II Plant Chinese Name Index

B

巴豆 57, 213
八角 15
八角枫 97, 229
白花苦灯笼 280
白花龙船花 123
白皮乌口树 133
白肉榕 50, 209
白颜树 43
白叶瓜馥木 259
柏拉木 81
百越凤尾蕨 251
半边旗 252
半柱毛兰 264
薄叶猴耳环 164
薄叶山矾 115
抱茎菝葜 263
边缘鳞盖蕨 249
变叶榕 49, 209

C

苍白秤钩风 269
草珊瑚 258
蒴柊 73, 219
豹皮樟 25, 200
长花厚壳树 137
长花枝杜若 267
长茎羊耳蒜 265
常绿菝葜 149, 244
长尾毛蕊茶 110
长叶铁角蕨 252
长圆叶新木姜子 158
橙黄玉凤花 264
赤才 85, 223
赤苍藤 275
赤杨叶 118, 236
重阳木 65, 215
垂茉莉 140, 241
垂叶榕 169
刺果藤 277
刺桑 175
粗毛野桐 58
粗丝木 143, 243
粗叶木 125
粗叶榕 46, 207

D

大百部 262
大果毛柃 99, 230
大果榕 44, 207
大罗伞树 106, 233
大青 139, 241
大叶骨碎补 255
大叶五室柃 101, 231
大叶猪肚木 121
淡黄荚蒾 195
滇南赤车 274
滇粤山胡椒 22
调羹树 160
碟腺棋子豆 37, 204
鼎湖钓樟 20, 199
东兴金花茶 113
短花水金京 134, 240
短序润楠 156

E

鹅掌柴 152, 246
二色波罗蜜 168

F

翻白叶树 94, 227
飞蛾藤 282
粉叶鱼藤 271
枫香树 35, 204

G

葛 272
构棘 174
谷木 182
光叶山矾 116
广东蛇葡萄 270
广西海桐 150, 245

H

海红豆 162
海南荚蒾 194
海南链珠藤 281
海南山姜 267
海桐 196
禾串树 67, 216
褐叶榕 172
黑柃 98, 229
黑面神 66, 216
黑嘴蒲桃 78, 221
红鳞蒲桃 79, 221

厚壳树 193
葫芦茶 273
华南谷精草 268
华南毛蕨 253
华润楠 28, 201
华山姜 268
黄椿木姜子 26, 200
黄丹木姜子 155
黄葛树 51, 210
黄花稔 277
黄脉九节 131
黄毛榕 170
黄毛五月茶 63, 214
黄牛木 76, 220
黄叶树 166
黄樟 19, 198
火炭母 278

J

寄生藤 278
假黄皮 87, 224
假苹婆 96, 228
假山龙眼 34, 203
假柿木姜子 24
假鹰爪 17, 197
假玉桂 42, 206
尖脉木姜子 23, 199
尖叶榕 45
剑叶凤尾蕨 250
九节 130, 239
九里香 90, 225
绢毛杜英 176

K

宽叶线柱兰 266
宽叶羊耳蒜 265
宽羽线蕨 257

L

蓝树 192
老虎刺 165
老鼠屎 117, 235
箭檬花椒 91, 226
棱枝冬青 144, 243
楝叶吴萸 185
两面针 186
亮叶猴耳环 38, 205
裂果金花 191

林下凤尾蕨 ...251	绒毛润楠 ...30	西南粗叶木 ...126
岭南山竹子 ...75, 220	榕叶冬青 ...145	锡叶藤 ...270
岭南柿 ...187	柔毛紫茎 ...114	西藏山茉莉 ...188
露兜树 ...262	肉实树 ...104, 232	细齿叶柃 ...100, 230
轮苞血桐 ...59	**S**	细枝龙血树 ...31, 202
轮叶木姜子 ...27	赛短花润楠 ...29, 201	狭叶泡花树 ...32, 202
罗浮柿 ...105, 232	三叉蕨 ...255	狭叶润楠 ...157
罗伞 ...151, 245	三花冬青 ...147	狭叶蒲桃 ...80, 222
罗伞树 ...107, 233	三桠苦 ...89, 225	显脉金花茶 ...111, 235
裸花水竹叶 ...266	山杜英 ...55, 212	香港大沙叶 ...128
绿叶线蕨 ...257	山桂花 ...70, 218	香港木兰 ...16
M	山菅兰 ...261	香花枇杷 ...39
马尾松 ...14, 197	山牡荆 ...141, 242	香叶树 ...21
买麻藤 ...258	杉木 ...154	小果山龙眼 ...33, 203
毛果算盘子 ...69, 217	山石榴 ...122, 237	小盘木 ...177
毛棯 ...82	山油柑 ...183	小叶红叶藤 ...275
毛麝香 ...283	扇叶铁线蕨 ...250	斜基粗叶木 ...124, 238
毛叶榄 ...83	上思青冈 ...52, 210	斜叶榕 ...173
勐仑三宝木 ...61	韶子 ...86, 223	心叶青藤 ...260
米仔兰 ...92, 226	少花山小橘 ...88, 224	锈毛梭子果 ...102
米珍果 ...108, 234	蛇藨筋 ...273	**Y**
密齿酸藤子 ...279	肾蕨 ...254	野牡丹 ...276
密花树 ...109, 234	石斑木 ...41, 206	野漆 ...84, 222
膜叶脚骨脆 ...71	石柑子 ...261	异叶蒴莲 ...276
膜叶土蜜树 ...68, 217	石南藤 ...259	银柴 ...64, 215
魔芋 ...260	疏花卫矛 ...53, 211	印度血桐 ...60, 213
木油桐 ...62, 214	鼠刺 ...36	楹树 ...163
N	水东哥 ...189	硬叶糙果茶 ...112
南方荚蒾 ...148	水锦树 ...135, 240	余甘子 ...178
南岭柞木 ...74, 219	水同木 ...171	鱼骨木 ...132, 239
南山花 ...129, 238	水团花 ...190	玉叶金花 ...127
牛耳枫 ...161	思茅山橙 ...136	云南黄皮 ...184
牛屎果 ...138	梭罗树 ...95, 228	**Z**
P	**T**	杂色榕 ...48, 208
旁杞木 ...56, 212	桃金娘 ...77	中华杜英 ...54, 211
披针骨牌蕨 ...256	藤构 ...274	中华复叶耳蕨 ...254
枇杷叶山龙眼 ...159	藤黄檀 ...271	中越猕猴桃 ...280
破布叶 ...93, 227	天仙藤 ...269	猪肚木 ...120, 237
朴树 ...167	铁冬青 ...146, 244	朱砂根 ...279
匍茎短筒苣苔 ...282	铁芒萁 ...248	竹叶兰 ...263
Q	头花银背藤 ...281	爪哇脚骨脆 ...72, 218
千斤拔 ...272	团叶槲蕨 ...256	紫荆木 ...103, 231
茜树 ...119, 236	团叶鳞始蕨 ...249	子楝树 ...180
青藤公 ...47, 208	臀果木 ...40, 205	子凌蒲桃 ...181
青枣核果木 ...179	**W**	紫玉盘 ...18, 198
曲枝假蓝 ...283	微毛布惊 ...142, 242	
曲轴海金沙 ...248	乌毛蕨 ...253	
R	**X**	

附录 III 植物学名索引
Appendix III   Plant Scientific Name Index

A

Acronychia pedunculata 183
Actinidia indochinensis 280
Adenanthera pavonina var. *microsperma* 162
Adenia heterophylla 276
Adenosma glutinosum 283
Adiantum flabellulatum 250
Adina pilulifera 190
Aglaia odorata 92, 226
Aidia cochinchinensis 119, 236
Alangium chinense 97, 229
Albizia chinensis 163
Alniphyllum fortunei 118, 236
Alpinia hainanensis 267
Alpinia oblongifolia 268
Alyxia hainanensis 281
Amorphophallus konjac 260
Ampelopsis cantoniensis 270
Antidesma fordii 63, 214
Aporosa dioica 64, 215
Arachniodes chinensis 254
Archidendron kerrii 37, 204
Archidendron lucidum 38, 205
Archidendron utile 164
Ardisia crenata 279
Ardisia hanceana 106, 233
Ardisia quinquegona 107, 233
Argyreia capitiformis 281
Artocarpus styracifolius 168
Arundina graminifolia 263
Asplenium prolongatum 252

B

Bennettiodendron leprosipes 70, 218
Bischofia polycarpa 65, 215
Blastus cochinchinensis 81
Blechnum orientale 253
Boeica stolonifera 282
Brassaiopsis glomerulata 151, 245
Breynia fruticosa 66, 216
Bridelia balansae 67, 216
Bridelia glauca 68, 217
Broussonetia kaempteri auct. *non* 274
Byttneria grandifolia 277

C

Camellia caudata 110
Camellia euphlebia 111, 235
Camellia gaudichaudii 112
Camellia indochinensis var. *tunghinensis* 113
Canarium subulatum 83
Canthium horridum 120, 237
Canthium simile 121
Carallia pectinifolia 56, 212
Casearia membranacea 71
Casearia velutina 72, 218
Catunaregam spinosa 122, 237
Celtis sinensis 167
Celtis timorensis 42, 206
Cinnamomum parthenoxylon 19, 198
Clausena excavata 87, 224
Clausena yunnanensis 184
Clerodendrum cyrtophyllum 139, 241
Clerodendrum wallichii 140, 241
Cratoxylum cochinchinense 76, 220
Croton tiglium 57, 213
Cunninghamia lanceolata 154
Cyclosorus parasiticus 253

D

Dalbergia hancei 271
Daphniphyllum calycinum 161
Davallia divaricata 255
Decaspermum gracilentum 180
Dendrotrophe varians 278
Derris glauca 271
Desmos chinensis 17, 197
Dianella ensifolia 261
Dicranopteris linearis 248
Dinetus racemosus 282
Diospyros morrisiana 105, 232
Diospyros tutcheri 187
Diploclisia glaucescens 269
Dracaena elliptica 31, 202
Drynaria bonii 256
Drypetes cumingii 179

E

Eberhardtia aurata 102
Ehretia acuminata 193
Ehretia longiflora 137
Elaeocarpus chinensis 54, 211

Elaeocarpus nitentifolius 176
Elaeocarpus sylvestris 55, 212
Embelia vestita 279
Eria corneri 264
Eriobotrya fragrans 39
Eriocaulon sexangulare 268
Erythropalum scandens 275
Euonymus laxiflorus 53, 211
Eurya macartneyi 98, 229
Eurya megatrichocarpa 99, 230
Eurya nitida 100, 230
Eurya quinquelocularis 101, 231

F
Fibraurea recisa 269
Ficus auriculata 44, 207
Ficus benjamina 169
Ficus esquiroliana 170
Ficus fistulosa 171
Ficus henryi 45
Ficus hirta 46, 207
Ficus langkokensis 47, 208
Ficus pubigera 172
Ficus tinctoria subsp. *gibbosa* 173
Ficus variegata 48, 208
Ficus variolosa 49, 209
Ficus vasculosa 50, 209
Ficus virens 51, 210
Fissistigma glaucescens 259
Flemingia prostrata 272

G
Garcinia oblongifolia 75, 220
Gironniera subaequalis 43
Glochidion eriocarpum 69, 217
Glycosmis oligantha 88, 224
Gnetum montanum 258
Gomphandra tetrandra 143, 243

H
Habenaria rhodocheila 264
Hancea hookeriana 58
Helicia cochinchinensis 33, 203
Helicia obovatifolia var. *mixta* 159
Heliciopsis henryi 34, 203
Heliciopsis lobata 160
Huodendron tibeticum 188

I
Ilex angulata 144, 243
Ilex ficoidea 145
Ilex rotunda 146, 244

Ilex triflora 147
Illicium verum 15
Illigera cordata 260
Itea chinensis 36
Ixora henryi 123

L
Lasianthus attenuatus 124, 238
Lasianthus chinensis 125
Lasianthus henryi 126
Lemmaphyllum diversum 256
Lepisanthes rubiginosa 85, 223
Leptochilus ellipticus var. *pothifolius* 257
Leptochilus leveillei 257
Lindera chunii 20, 199
Lindera communis 21
Lindera metcalfiana 22
Lindsaea orbiculata 249
Liparis latifolia 265
Liparis viridiflora 265
Liquidambar formosana 35, 204
Lirianthe championii 16
Litsea acutivena 23, 199
Litsea elongata 155
Litsea monopetala 24
Litsea rotundifolia var. *oblongifolia* 25, 200
Litsea variabilis 26, 200
Litsea verticillata 27
Lygodium flexuosum 248

M
Macaranga andamanica 59
Macaranga indica 60, 213
Machilus breviflora 156
Machilus chinensis 28, 201
Machilus parabreviflora 29, 201
Machilus rehderi 157
Machilus velutina 30
Maclura cochinchinensis 174
Madhuca pasquieri 103, 231
Maesa acuminatissima 108, 234
Melastoma malabathricum 276
Melastoma sanguineum 82
Melicope pteleifolia 89, 225
Meliosma angustifolia 32, 202
Melodinus cochinchinensis 136
Memecylon ligustrifolium 182
Microcos paniculata 93, 227
Microdesmis caseariifolia 177
Microlepia marginata 249
Murdannia nudiflora 266

Murraya exotica 90, 225
Mussaenda pubescens 127
Myrsine seguinii 109, 234

N

Neolitsea oblongifolia 158
Nephelium chryseum 86, 223
Nephrolepis cordifolia 254

O

Osmanthus matsumuranus 138

P

Pandanus tectorius 262
Pavetta hongkongensis 128
Pellionia paucidentata 274
Persicaria chinensis 278
Phyllanthus emblica 178
Pinus massoniana 14, 197
Piper wallichii 259
Pittosporum kwangsiense 150, 245
Pittosporum tobira 196
Pollia secundiflora 267
Pothos chinensis 261
Prismatomeris tetrandra 129, 238
Psychotria asiatica 130, 239
Psychotria straminea 131
Psydrax dicocca 132, 239
Pteris ensiformis 250
Pteris fauriei var. *chinensis* 251
Pteris grevilleana 251
Pteris semipinnata 252
Pterolobium punctatum 165
Pterospermum heterophyllum 94, 227
Pueraria montana var. *lobata* 272
Pygeum topengii 40, 205

Q

Quercus delicatula 52, 210

R

Reevesia pubescens 95, 228
Rhaphiolepis indica 41, 206
Rhodomyrtus tomentosa 77
Rourea microphylla 275
Rubus cochinchinensis 273

S

Sarcandra glabra 258
Sarcosperma laurinum 104, 232
Saurauia tristyla 189
Schefflera heptaphylla 152, 246
Schizomussaenda henryi 191

Scolopia chinensis 73, 219
Sida acuta 277
Smilax ocreata 263
Stemona tuberosa 262
Sterculia lanceolata 96, 228
Stewartia villosa 114
Streblus ilicifolius 175
Strobilanthes dalzielii 283
Symplocos anomala 115
Symplocos lancifolia 116
Symplocos stellaris 117, 235
Syzygium bullockii 78, 221
Syzygium championii 181
Syzygium hancei 79, 221
Syzygium levinei 80, 222

T

Tadehagi triquetrum 273
Tarenna depauperata 133
Tarenna mollissima 280
Tectaria subtriphylla 255
Tetracera sarmentosa 270
Tetradium glabrifolium 185
Toxicodendron succedaneum 84, 222
Trigonostemon bonianus 61

U

Uvaria macrophylla 18, 198

V

Vernicia montana 62, 214
Viburnum fordiae 148
Viburnum hainanense 194
Viburnum lutescens 195
Viburnum sempervirens 149, 244
Vitex quinata 141, 242
Vitex quinata var. *puberula* 142, 242

W

Wendlandia formosana subsp. *breviflora* 134, 240
Wendlandia uvariifolia 135, 240
Wrightia laevis 192

X

Xanthophyllum hainanense 166
Xylosma controversa 74, 219

Z

Zanthoxylum avicennae 91, 226
Zanthoxylum nitidum 186
Zeuxine affinis 266

致　谢

在南方地形崎岖的热带森林中建立大面积森林监测样地殊为不易。广西防城金花茶保护区季节性雨林2个1hm^{2}样地的建设从2015年7月选址，2015年9月完成样地节点的测量标定，2016年4月完成两个样地的首次植物调查，累计投入的人力达到500多个工作日。

样地的顺利建成，离不开各方面的大力支持和帮助，值此样地手册出版之际，诚挚感谢相关单位的领导、专家和同行为样地建设做出的重要贡献。

（1）生态环境部南京环境科学研究所副所长徐海根研究员对监测示范基地建设的顶层设计，是样地顺利建成的重要保证；南京环境科学研究所王智研究员和丁晖研究员给予人力、经费和技术规范方面的大力支持，徐网谷参加了样地技术规范的制定及样地选点等工作。

（2）广西防城金花茶国家级自然保护区管理中心陈勇棠、潘子来、杨海娟、黄瑞斌等在样地基础资料和后勤保障方面给与诸多帮助，上岳保护站陈广棠、陈拾棠、李勇儒、潘锦福、潘子强、覃毅等参加了野外调查工作，保证了样地野外工作顺利完成。李文儒师傅提供了周到的后勤保障服务。

（3）广西植物研究所领导对样地监测网络的建设非常重视并提供大量支持，蒋裕良、张兴明、何运林等参加了野外调查工作。吴望辉、覃永华、刘静、黄俞淞等植物分类专家在百忙之中帮助鉴定部分标本；区内外同行和相关数据库提供大量精美的植物图片。

样地建设得到了中央级科学事业单位修缮购置项目"全国生物多样性野外监测示范基地修缮项目"、广西重点研发计划、广西喀斯特植物保育与恢复生态学重点实验室运行专项等项目的支持。专著出版得到国家科技基础资源调查项目、国家重点研发计划、国家自然科学基金、桂林市创新平台和人才计划、广西植物研究所学科发展基金等多个项目的经费资助。

在此一并致谢。

ACKNOWLEDGEMENTS

It is not easy to establish a large-scale forest monitoring sample plot in tropical forests with rugged terrain in the south of China. Two 1hm^2 sample plots of seasonal rain forest in the Fangcheng Golden Camellia Nature Reserve in Guangxi were selected since July, 2015, the survey and calibration of the plots were completed in September, 2015. The first plant survey of the two sample plots was completed in April, 2016, with a total investment of more than 500 working days. The successful completion of the plots received the strong support and help from many people and organizations. On the occasion of the publication of this book, we sincerely thank all the leaders, experts and relevant organizations for their important contributions to the plot construction.

(1) The support and the top-level design of Xu Haigen, deputy director of Nanjing Institute of Environmental Sciences, MEE are important guarantees for the construction of the plots; Professor Wang Zhi and Ding Hui of Nanjing Institute of Environmental Sciences, MEE gave strong support in terms of manpower, funds and technical specifications; Associate Research Fellow Xu Wanggu of Nanjing Institute of Environmental Sciences, MEE participated in the formulation of technical guidance and site selection.

(2) Chen Yongtang, Pan Zilai, Yang Haijuan and Huang Ruibin of the management centers of Fangcheng Golden Camellia National Nature Reserve in Guangxi Zhuang Autonomous Region, gave a lot of help in the basic data and logistics support of the plots. Chen Guangtang, Chen Shigtang, Li Yongru, Pan Jinfu, Pan Ziqiang and Qin Yi of the Shangyue protection station participated in the field investigation, which ensured the completion of the field work. Master Li Wenru provided thoughtful logistics support services.

(3) The leaders of Guangxi Institute of Botany attached great importance to the construction of monitoring plot network and provided a lot of support. Jiang Yuliang, Zhang Xingming and He Yunlin participated in the field investigation. Phytotaxonomists Wu Wanghui, Qin Yonghua, Liu Jing and Huang Yusong helped identify some specimens in their busy schedules; Colleagues and some websites of plant photo bank provided a large number of beautiful plant pictures.

The construction of the plots was supported by the project of Chinese Biodiversity Field Monitoring Demonstration Base Repair Project III, Guangxi Key research and development Plan, and the independent research project of Guangxi Key Laboratory of Karst Plant Conservation and Restoration Ecology. The Special Foundation for National Science and Technology Basic Research Program of China, The National Science and Technology Basic Resources Investigation Project, The National Natural Science Foundation of China, The Guilin Innovation Platform and Talent Plan Project, and the discipline development fund project of Guangxi Institute of Botany gave financial support for this book publication.

We would like to extend our sincere gratitude to all these people and projects.